D0861748

1001 INVENTIONS

THE ENDURING LEGACY *of* MUSLIM CIVILIZATION

SALIM T. S. AL-HASSANI, CHIEF EDITOR

NATIONAL GEOGRAPHIC

Washington, D.C.

CONTENTS

Foreword 6 • Introduction 8 • Map of Major Contributions in Muslim Civilization 14

THE GOLDEN AGE • TIME LINE OF DEVELOPMENT IN MUSLIM CIVILIZATION	CHAPTER ONE 16 THE STORY BEGINS	
ON THE COFFEE TRAIL • FINE DINING THREE-COURSE MENU • CLOCKS • CHESS • MUSIC CLEANLINESS • TRICK DEVICES • VISION AND CAMERAS FASHION AND STYLE • CARPETS	CHAPTER TWO 34 HOME	
SCHOOLS • UNIVERSITIES • HOUSE OF WISDOM • LIBRARIES AND BOOKSHOPS • TRANSLATING KNOWLEDGE • MATHEMATICS TRIGONOMETRY • CHEMISTRY • COMMERCIAL CHEMISTRY GEOMETRY • ART AND THE ARABESQUE • THE SCRIBE • WORD POWER	CHAPTER THREE 62 SCHOOL	
AGRICULTURAL REVOLUTION • FARMING MANUALS WATER MANAGEMENT • WATER SUPPLY • DAMS WINDMILLS • TRADE • TEXTILES • PAPER • POTTERY GLASS INDUSTRY • JEWELS • CURRENCY	CHAPTER FOUR 108 MARKET	
HOSPITAL DEVELOPMENT • INSTRUMENTS OF PERFECTION SURGERY • BLOOD CIRCULATION • IBN SINA'S BONE FRACTURES • NOTEBOOK OF THE OCULIST • INOCULATION HERBAL MEDICINE • PHARMACY • MEDICAL KNOWLEDGE	CHAPTER FIVE 152 HOSPITAL	
TOWN PLANNING • ARCHITECTURE • ARCHES VAULTS • THE DOME • THE SPIRE • INFLUENTIAL IDEAS CASTLES AND KEEPS • PUBLIC BATHS • THE TENT FROM KIOSK TO CONSERVATORY • GARDENS • FOUNTAINS	CHAPTER SIX 186 TOWN	
PLANET EARTH • EARTH SCIENCE • NATURAL PHENOMENA GEOGRAPHY • MAPS • TRAVELERS AND EXPLORERS • NAVIGATION NAVAL EXPLORATION • GLOBAL COMMUNICATION WAR AND WEAPONRY • SOCIAL SCIENCE AND ECONOMICS	CHAPTER SEVEN 226 WORLD	
ASTRONOMY • OBSERVATORIES • ASTRONOMICAL INSTRUMENTS ASTROLABE • ARMILLARY SPHERE • SIGNS FOR WISE PEOPLE THE MOON • LUNAR FORMATIONS • CONSTELLATIONS • FLIGHT	CHAPTER EIGHT 264 UNIVERSE	
PERSONALITIES FROM THE PAST • EUROPE'S LEADING MINDS A THOUSAND YEARS OF SCHOLARSHIP • AUTHORS AND TREATISES FURTHER READING • GLOSSARY • ILLUSTRATIONS CREDITS INDEX • ACKNOWLEDGMENTS	REFERENCE 302 WEALTH OF KNOWLEDGE	

OPPOSITE: *A Persian manuscript shows a scholar teaching male and female students in a classroom.*

FOREWORD

CLARENCE HOUSE

In 1993, I gave a lecture at the Oxford Centre for Islamic Studies called "Islam and the West", in which I addressed the dangerous level of misunderstanding between these two worlds and drew attention to the joint scientific and cultural heritage that in fact links us together.

Today, in times when these links matter more than ever, I am delighted to see the success of the initiative called *1001 Inventions*, which presents and celebrates the many scientific, technological and humanitarian developments shared by the Islamic world and the West. For while the West struggled in a period often called the "Dark Ages", enormous intellectual and cultural developments were taking place in Muslim civilization from the seventh century onward. In science, astronomy, mathematics, algebra (itself an Arabic word), law, history, medicine, pharmacology, optics, agriculture, architecture, theology, music — a "Golden Age" of discovery was flourishing in the Islamic world, which would contribute enormously to the European Renaissance.

It is a matter of great pride that, although global in its outreach and operation, *1001 Inventions* is in fact an initiative of a British-based team supported by a network of academics around the world. The *1001 Inventions* exhibitions, books and films have achieved significant success in popularizing public understanding of the cultural roots of science and, thereby, encouraging intercultural respect and appreciation.

This book, now in its third edition and published by the National Geographic Society, introduces the fascinating legacy of the Golden Age to a new audience. It is part of a global effort to increase understanding of how men and women of many different faiths and ethnicities, living under the umbrella of Muslim Civilization, made extraordinary advances during this period and how these advances touch every part of society today.

I can only pray that the whole concept of *1001 Inventions*, and especially this book, will inspire future generations of scientists and scholars, men and women, Muslims and non-Muslims alike, to build a better world for their fellow man, whatever their creed or colour. I look forward to the further success and expansion of this important and much-needed initiative.

Charles

His Royal Highness The Prince of Wales

An illustration from Al-Jazari's The Book of Knowledge of Ingenious Mechanical Devices *depicts a robot that serves drinks. The writing at right describes how the device works.*

INTRODUCTION

The development of this book tells an interesting story. In 1975, Lord B. V. Bowden, the principal at the time of the University of Manchester Institute of Science and Technology, or UMIST, became fascinated by the manner in which the Muslims managed a domain that stretched from China to Spain for so many centuries. Of particular interest was how they introduced the concept of "indexation" in combating inflation, which was rampant in the Roman Empire. He announced in the House of Lords that in order to guide the United Kingdom's economy, then riddled with inflation, we should learn from the Muslims' experience and consider the economic principles laid down some 1,400 years ago in the Quran as revealed to the Prophet Muhammad (pbuh).*

Lord Bowden set up the Institute for the History of Muslim Science, Technology, and Commerce, recruiting me and a few professors from UMIST and the Victoria University of Manchester, and we were augmented by a few dignitaries. Although this initiative did not thrive for long, it gave me the opportunity to encounter historians and scholars outside my engineering discipline and, more significantly, it revealed to me the frightening level of their ignorance of the traditions and beliefs of other cultures. Lord Bowden passed away in 1989, and with him went that institute.

Then in 1993, Professor Donald Cardwell, head of the Department of History of Science and Technology, and the founder of the Museum of Science and Industry in Manchester, presented me with a challenge. Much in the spirit of Lord Bowden he said to me: "Salim, you should by now know there are a thousand years missing from the history of engineering, a period we call the Dark Ages. Most of the missing knowledge is contained in Arabic manuscripts filling the cellars of many famous libraries. You are a distinguished professor of engineering at a prestigious university and you know the Arabic language. Therefore, you are best suited to do something about filling this gap."

That wake-up call propelled me to follow a line of inquiry that eventually changed my life. That was when the story of this book began.

Before taking this challenge, however, I looked up various books and journal papers and consulted numerous friends. Book after book, journal after journal, all pointed to this incredible gap. Take, for instance, a typical popular book at the time: *Scientists and Inventors: The People Who Made Technology from Earliest Times to Present Day* by Anthony Feldman and Peter Ford, published in 1979. The authors explain that the book gives in chronological order humanity's scientific and technological progress from the invention of movable type to the discovery of penicillin. The names of the great inventors, to whom they devote short chapters, follow in chronological order: Empedocles (circa 490-430 B.C.E.), Democritus (460-370 B.C.E.), Hippocrates (460-377 B.C.E.), Aristotle (383-322 B.C.E.), Archimedes (287-212 B.C.E.),

* I use the term "pbuh," meaning "Peace Be Upon Him/Her" to indicate the respect paid by Muslims to the Prophet Muhammad and other prophets including Jesus, Mary, Moses, and Isaac.

A 13th-century illustration from Animals and Their Uses *by Ibn Bakhtishu shows Aristotle and Alexander the Great.*

Johannes Gutenberg (1400-1468 C.E.), followed by others such as Da Vinci.

The remarkable jump of 1,600 years from the time of Archimedes to Gutenberg was amazing and troubling. Further reading of other books revealed that the whole period, 450-1492 C.E., is in fact passed over as the Dark Ages. The period is altogether ignored as far as science and civilization are concerned, termed variously as "a middle age," an intermediary period, a uniform bloc, "vulgar centuries," and, most disconcerting of all, an "obscure time." Some books include a bit more on the Romans, but still leap over one thousand years. More disquieting were the gaps in school textbooks and other sources of learning, which form the views and perceptions of pupils on other cultures aside from their own.

Later that same year, on October 27, 1993, I attended an inspiring lecture by HRH Prince Charles at the Sheldonian Theatre, Oxford, titled "Islam and the West." Addressing a galaxy of eminent scholars in one of the strongholds of orientalism, his speech was received like fire in dry woods. The eye-opening extract below reinforced my findings:

If there is much misunderstanding in the West about the nature of Islam, there is also much ignorance about the debt our own culture and civilization owe to the Islamic world. It is a failure, which stems, I think, from the strait-jacket of history, which we have inherited. The medieval Islamic world, from central Asia to the shores of the Atlantic, was a world where scholars and men of learning flourished. But because we have tended to see Islam as the enemy of the West, as an alien culture, society, and system of belief, we have tended to ignore or erase its great relevance to our own history.

All students are trained to think critically; yet when faced with the darkness of ten centuries in Europe, they are told things appeared, as if by miracle, all at once during the Renaissance. This defies logic. Discoveries, inventions, and developments that alter the course of humanity, as any scientist knows, do not appear by chance. Continuity is fundamental, especially in the birth and rise of the sciences; it is almost so in every other field of study.

A couple of years later and just before passing away, Professor Cardwell arranged for me to give a presentation at the esteemed Literary and Philosophical Society, titled the "Muslim Contribution to Science and Technology." The amount of amazement and surprise expressed by the audience on the little I had to say reinforced the assertion of Prince Charles. From then on, whenever I lectured on the topic I felt like a one-eyed man among the blind. Of special excitement was the fascination of young people in the subject of knowing where our present civilization came from.

The ambition to write a book on the subject was pushed aside by the demands of being a professor of mechanical engineering, in a university world invaded by market forces with all the pressures of lecturing, researching, publishing, fund-raising, administration, and running two consulting companies. The practical solution was to hire historians and initiate undergraduate projects on the virtual reconstruction of ancient machines. This, together with the support of like-minded academics and professionals, saw the emergence of the Foundation for Science, Technology and Civilisation, or FSTC. The would-be book instead began to take shape in the form of a website, www.MuslimHeritage.com, which attracted excellent peer-reviewed papers from renowned writers and researchers.

Very quickly, the website became the first destination and source of information for many institutions of learning, schools, media groups, and young people from all over the English-speaking world. It now attracts more than 50,000 daily page views.

The spotlight fell on the relationship between the Muslim world and the West immediately after the 9/11 terrorist attacks on September 11, 2001, on New York's World Trade Center and the Pentagon. Quite amazing was a daring speech given just two weeks later by businesswomen and historian Carly Fiorina, chief executive officer at the time of Hewlett-Packard Corporation. At a meeting of all the corporation's worldwide managers, on September 26, 2001, Carly Fiorina announced:

> *There was once a civilization that was the greatest in the world. It was able to create a continental super-state that stretched from ocean to ocean and from northern climes to tropics and deserts. Within its dominion lived hundreds of millions of people, of different creeds and ethnic origins.*
>
> *One of its languages became the universal language of much of the world, the bridge between the peoples of a hundred lands. Its armies were made up of people of many nationalities, and its military protection allowed a degree of peace and prosperity that had never been known. The reach of this civilization's commerce extended from Latin America to China, and everywhere in between.*
>
> *And this civilization was driven more than anything by invention. Its architects designed buildings that defied gravity. Its mathematicians created the algebra and algorithms that would enable the building of computers, and the creation of encryption. Its doctors examined the human body, and found new cures for disease. Its astronomers looked into the heavens, named the stars, and paved the way for space travel and exploration. Its writers created thousands of stories. Stories of courage, romance, and magic. Its poets wrote of love, when others before them were too steeped in fear to think of such things.*
>
> *When other nations were afraid of ideas, this civilization thrived on them, and kept them alive. When censors threatened to wipe out knowledge from past civilizations, this civilization kept the knowledge alive, and passed it on to others.*
>
> *While modern Western civilization shares many of these traits, the civilization I'm talking about was the Islamic world from the year 800 to 1600, which included the Ottoman Empire and the courts of Baghdad, Damascus, and Cairo, and enlightened rulers like Suleyman the Magnificent.*
>
> *Although we are often unaware of our indebtedness to this other civilization, its gifts are very much a part of our heritage. The technology industry would not exist without the contributions of Arab mathematicians.*

A number of colleagues, well established in the subject, began a lecturing campaign in Britain, Europe, and abroad. A large number of people from all walks of life derived pleasure and inspiration from this knowledge. Presentations to the younger generation, especially the ones I gave to the Youth NGOs at the European Parliament in Brussels, sparked enormous interest in science and technology, and especially in the lives of Muslim pioneers in chemistry, physics, medicine, biology, algebra, engineering, architecture, art, agriculture, and in numerous manufacturing industries who have impacted so positively on our modern civilization. It was clear this underappreciated subject was finally coming of age.

In 2006, FSTC launched the 1001 Inventions initiative, and since then, public interest in the scientific achievements of Muslim civilization has increased exponentially. Our first exhibition was sponsored by numerous United Kingdom government, scientific, and academic establishments and charitable organizations. It toured British science museums for two years and subsequently visited the British Houses of Parliament and the United Nations. The first two editions of the *1001 Inventions* books

Professor Salim T. S. al-Hassani demonstrates a mechanical interactive at the 1001 Inventions exhibition during its run in Istanbul, Turkey. The professor is the chief editor of 1001 Inventions: The Enduring Legacy of Muslim Civilization—*the companion volume to the exhibit.*

sold more than 100,000 copies. However, this was just the start of what would be a much greater flowering of international interest in our work, alongside increased dialogue about the cultural roots of science and new opportunities to promote social cohesion and intercultural respect and appreciation.

In 2010, thanks to the generous sponsorship of the Jameel Foundation (later ALJCI), FSTC launched a much larger, state-of-the-art, exhibition, which embarked upon a global tour, starting at the world-renowned Science Museum in London. As part of the exhibition's production process, the Science Museum retained an independent panel of expert historians to conduct a complete review of the content to ensure the highest standards of historical accuracy were maintained.

The public demand for the exhibition in London far exceeded expectations, attracting more than 400,000 visitors—four times the expected number—in five months, many of whom had never visited the Science Museum before. A few months after launch, Turkish prime minister Recep Tayyip Erdogan took time out from his state visit to

Britain to see the exhibition for himself. He insisted that the next venue on our global tour be Istanbul, before it began its North America leg. Thus, the exhibition also had a very well-received seven-week residency in Turkey.

At the Istanbul launch event the prime minister expressed his deep appreciation of our initiative:

> *1001 Inventions relives the 1,000-year-long adventure of science and technology in Muslim civilization [and] provides a positive message for our youth, for the Muslim world, and for humanity. Exhibits displayed here about the history of medicine, astronomy, mathematics, geometry, chemistry, and so on, still manage to amaze us even today.*
>
> *It is our common responsibility to make sure we do not forget, or let be forgotten, these important underappreciated scientists from Muslim civilization. In that sense, the 1001 Inventions initiative has a very important and propitious task. I congratulate each and every individual involved in the creation of this exhibition and I am optimistic this exhibition will provide a brand-new perspective on our modern scientific world.*

The people of Istanbul responded with similar enthusiasm, with more than 450,000 walking through the doors to experience our bilingual Turkish-English exhibition situated in the city's historic Sultanahmet Square, next to the Hagia Sofia and the Blue Mosque. The Turkish media was unanimous in its praise, and the scientific legacy from the Ottoman period resonated strongly with our Turkish audience. This served as a magnificent European send-off for the North American tour that would follow.

At the time of this writing, the exhibition has so far enjoyed a warm reception in New York City where it was displayed at the Hall of Science and is currently attracting more than 50,000 visitors a week at the California Science Center in Los Angeles.

We are truly grateful for the support we have received from world leaders, diplomats, and education establishments. More satisfying yet is the enthusiasm of the millions of people, many of them teenagers and young people, who have engaged with our educational books and exhibitions, as well as online and through social media channels.

An integral part of the exhibition is a short educational film, starring Oscar-winning actor Sir Ben Kingsley, titled *1001 Inventions and the Library of Secrets*, which is available for free via our website. The movie was a revelation. Downloaded more than ten million times, it went on to win more than 20 international film awards, including "Best Film" recognition at Cannes and the New York Film Festival.

The momentum we've created continues to grow and we have ambitious plans for the future. An Arabic-language version of the exhibition began touring the Middle East in autumn 2011, bringing 1001 Inventions to a new audience that is hungry for a greater understanding of their own scientific history. The original exhibition arrives in Washington, D.C., the summer of 2012. Furthermore, we intend to produce more educational films and documentaries, and new educational material and translations of the 1001 Inventions exhibition, in other Asian and European languages.

This book is just one of the much labored over fruits of the 1001 Inventions initiative. Its painstaking completion is an achievement of no single individual, but of all those mentioned on the Acknowledgments page. The book identifies in an enjoyable, easy-to-read format aspects of our modern lives that are linked with inventions from Muslim civilization. It is our hope that through these pages we can enhance intercultural respect while at the same time inspire young people from both Muslim and non-Muslim backgrounds to find career role models in science, technology, engineering, and mathematics.

Professor Salim T. S. al-Hassani
Chief Editor and Chairman, FSTC

www.1001inventions.com
www.MuslimHeritage.com

Map of Major Contributions in Muslim Civilization

Dar al-Islam, or the Muslim world, stretched over three vast continents, from Toledo in Spain, through Arabia and Indonesia to China, and as far south as East Africa. It reached its peak in the 12th century under the Abbasids. Cities in the Middle East and Spain became global centers of culture, trade, and learning. Their atmosphere of tolerance and creativity stimulated groundbreaking advances in medicine, engineering, philosophy, mathematics, astronomy, and architecture. Explore the map below to see what happened, where—and when.

World Map
Al-Idrisi (1099–1166)
Al-Idrisi was commissioned by the Norman king of Sicily, Roger II, to make a map. He produced an atlas of 70 maps called the *Book of Roger*, showing that the Earth was round, which was a common notion held by Muslim scholars.
[page 236]

Gothic Rib Vaulting
(1000)
The gothic ribs of the Toledo and Córdoba Mosque vaults inspired European architects and their patrons to adopt them in the Romanesque and Gothic movements.
[page 199]

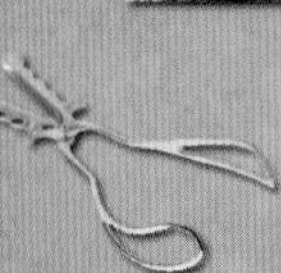

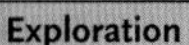

Surgical Instruments
Al-Zahrawi (936–1013)
Cutting-edge surgeon Al-Zahrawi introduced more than 200 surgical tools that revolutionized medical science. These instruments would not look out of place in today's 21st-century hospitals.
[page 158]

Exploration
Ibn Battuta (1304–1368/70)
Ibn Battuta traveled more than 75,000 miles in 29 years through more than 40 modern countries, compiling one of the best eyewitness accounts of the customs and practices of the medieval world.
[page 250]

Foundation of Sociology and Economics
Ibn Khaldun (1332–1406)
This man traced the rise and fall of human societies in a science of civilization, recording it all in his famous *Al-Muqaddimah*, or *Introduction to a History of the World*, which forms the basis of sociology and economic theory.
[page 262]

Horseshoe Arch
(715)
Resembling a horseshoe, this arch was first used in the Umayyad Great Mosque of Damascus. In Britain, it is known as the Moorish arch and was popular in Victorian times; it was often used in railway station entrances.
[page 195]

Al-Nuri Hospital
(1156)
Hospitals provided free health care to all. Al-Nuri was an immense and sophisticated hospital where druggists, barbers, orthopedists, oculists, and physicians were all examined by "market inspectors" to make sure they met the highest standards.
[page 154]

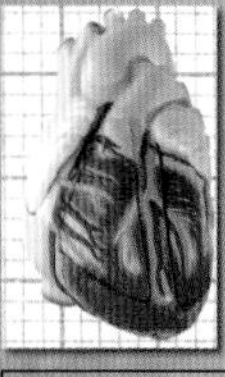

Blood Circulation
Ibn al-Nafis (1210–1288)
Ibn al-Nafis of Egypt first described pulmonary circulation of venous blood passing into the heart and lungs via the ventricles, thus becoming oxygenated and arterial blood. He was finally accredited with this discovery in 1957.
[page 166]

Pointed Arch
(ninth century)
The pointed arch concept, on which Gothic architecture is based, came to Europe from Egypt's beautiful Ibn Tulun Mosque of Cairo via Sicily with Amalfitan merchants. It enabled European architects to overcome problems in Romanesque vaulting.
[page 196]

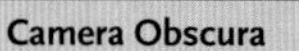

Camera Obscura
Ibn a-Haytham (965–1039)
In a darkened room (qamara in Arabic), Ibn al-Haytham observed light coming through a small hole in the window shutters producing an upside-down image on the opposite wall. This early pinhole camera has led to the cameras of today.
[page 56]

Castles
(12th century)
The invincible design of the castles of Syri and Jerusalem were imitated in Western lands with key features like round towers, arrow slits, barbicans, machicolations, parapets, and battlements.
[page 210]

London • Paris • Moscow • Toledo • Córdoba • Granada • Tangier • Fez • Tunis • Sicily • Istanbul • Diyarba • Damascus • Jerusalem (Al Quds) • Cairo • Timbuktu • Mecc • Mombasa

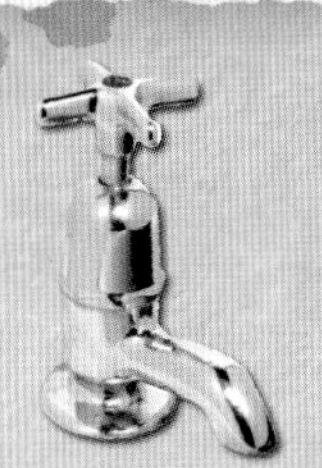

Water-Raising Machine

Al-Jazari (early 13th century)

Al-Jazari's greatest legacy is the application of the crank and connecting-rod system, which transmits rotary motion into linear motion. His machines were able to raise huge amounts of water without anyone lifting a finger.
[page 121]

Chemistry

(722–815)

This was a period when chemical instruments and processes that form the basis of today's chemistry were created and developed. Jabir ibn Hayyan discovered vitally important acids like sulfuric, nitric, and nitromuriatic acid, while Al-Razi set up a modern laboratory, designing and using more than 20 instruments like the crucible and still.
[page 90]

Trick Devices

(ninth century)

Three brothers, the Banu Musa brothers, were great mathematicians who funded the translation of Greek scientific treatises; they also invented fabulous trick devices that, some say, are precursors to executive toys.
[page 52]

House of Wisdom

(eighth-fourteenth century)

This immense scientific academy was the brainchild of four generations of caliphs who drew together the cream of Muslim scholars. It was an unrivaled center for the study of humanities and for sciences, where the greatest collection of worldly knowledge was accumulated and developed.
[page 72]

Cryptology

Al-Kindi (801–873)

Second World War problem solvers carried on the code-breaking tradition first written about by polymath Al-Kindi from Baghdad when he described frequency analysis and laid the foundation of cryptography.
[page 258]

Distillation

Jabir ibn Hayyan (722–815)

Jabir ibn Hayyan perfected the distillation process using the alembic still, which is still used today. The Muslim world produced rose water, essential oils, and pure alcohol for medical use. Today, distillation has given us products ranging from plastics to gasoline.
[page 92]

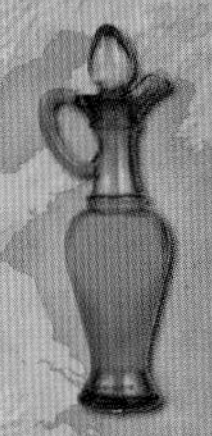

Khwarizm

Kabul

Baghdad

Kufa

Delhi

Patna

Canton

Mocha

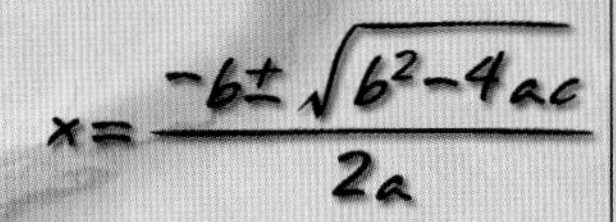

Algebra

Al-Khwarizmi (780–850)

Al-Khwarizmi introduced the beginnings of algebra; it then developed into a form still used today by many who lived after him.
[page 84]

Shampooing

Sake Dean Mohamed (18th century)

Shampooing was introduced to Britain at Brighton by Sake Dean Mohamed, from Patna, India, who became the "Shampooing Surgeon" to King George IV and King William IV.
[page 51]

Coffee

(eighth century)

Khalid the goat herder noticed his excitable animals had eaten red berries, which led to the early Arabic drink *al-qahwa*. Coffee drinking flourished across the Muslim world in the 1500s and spread to Europe through trade in 1637.
[page 36]

Lands encompassed by Muslim civilization at various times from the seventh century onward.

"We ought not to be ashamed of appreciating the truth and of obtaining it wherever it comes from, even if it comes from races distant and nations different from us."

AL-KINDI, NINTH-CENTURY MUSLIM SCHOLAR WHO STUDIED IN BAGHDAD, IRAQ

THE STORY BEGINS

THE GOLDEN AGE

TIME LINE OF DEVELOPMENT IN MUSLIM CIVILIZATION

FOR CENTURIES AFTER THE FALL OF ANCIENT ROME, SCIENTIFIC PROGRESS IN western Europe slowed almost to a standstill. In the developing Muslim world, however, a golden age of discovery flourished from the seventh century until the sixteenth century. During this period in the Muslim world, scholars of various faiths and cultures built and improved upon the knowledge of ancient Egypt, ancient Mesopotamia, Persia, China, India, and of the Greeks and Romans, making breakthroughs that helped pave the way for the European Renaissance.

Great men and women of the past—mathematicians, astronomers, chemists, physicians, architects, engineers, economists, sociologists, artists, artisans, historians, geographers, and educators—expressed their faith by making beneficial contributions to society and humanity. They did so with open-mindedness and, in many instances, in collaboration with people of other faiths, cultures, and backgrounds.

Now, you can uncover the stories of the talented men and women who lived in the golden age of Muslim civilization, and discover the ways in which their inventions shaped the way we live today.

OPPOSITE: *Al-Jazari designed a water-powered Elephant Clock in the 13th century, reconstructed for the 1001 Inventions exhibition.*

01 The Golden Age

This volume looks at the scientific legacy of Muslim civilization—from the theories, inventions, devices, processes, and ideas first conceived, to the discoveries adopted, developed, and spread during the age of Muslim scholarship. In seven chapters, Home, School, Hospital, Market, Town, World, and Universe, the book aims to uncover the cultural roots of science to enhance intercultural respect and appreciation in our world today.

Home

In the Home chapter, you will encounter the thousand-year-old inventions that still shape daily life. From chess to cameras, today's home life is packed with objects influenced by early Muslim civilization.

Chess came to the Persian court via India in the ninth century, and spread across Muslim civilization. Today, we retain the word that ends a chess match, "checkmate," which comes from the Persian phrase *Shahmat,* meaning "the king is defeated."

Muslim Spain was one of the sources of a new trend to eat three-course meals. Other new ideas from Muslim civilization included new fashions, highly prized carpets, and ingenious devices like the fountain pen. Incredible mechanical clocks were just one type of device invented by 13th-century engineer Al-Jazari.

As early as the 14th century, large quantities of spices came to Europe from Egypt and Syria. And drinking coffee flourished across the Muslim world for years before it spread through trade into Europe in the 17th century.

School

The School chapter tells the story of the considerable influence of Muslim civilization on the development and spread of knowledge. Mathematics, science, arts, languages ... whatever subject interests you most, discover its links with the distant past, and the people, like ninth-century university patron Fatima al-Fihri and late eighth-century chemist Jabir ibn Hayyan, whose innovations live on today.

Early chemists distilled fragrances from plants and flowers in the eighth and ninth centuries, a practice that spread to Europe, along with other developments in chemistry. Mathematicians developed the concept of zero and the system of decimal math we use today; they also spotted intricate geometrical designs in the patterns of flowers and shells, which in turn influenced characteristic designs in architecture and decorative arts.

The House of Wisdom, a prestigious academy and library, was founded a thousand years ago in Baghdad. There, Muslim, Christian, and Jewish scholars cooperated in translating knowledge, fueling scientific debate and discovery.

Schools and universities sprang up in towns and cities from Cairo to Timbuktu, and one generous-spirited woman called Fatima al-Fihri founded a university for her community in Fez, Morocco, using her own fortune. Al-Qarawiyin today is known as the world's oldest university, continuing to grant degrees to students.

Market

The Market chapter explores the ways in which influential ideas from Muslim civilization spread around the world. Across three continents, a buzzing network of trade and travel developed

and grew from the eighth century, encouraging a creative exchange of ideas for supplying energy, growing food, and producing goods, many of which are familiar today.

As trade expanded, so did knowledge and prosperity. Merchants, rulers, and pilgrims traveled between cities in Africa, Asia, and Europe, taking with them wealth and new ideas. The Silk Road stretched thousands of miles, linking China with the Middle East and Europe.

Sea trade flowed through thriving ports in Málaga and Alexandria. In busy markets, traders dealt in brocade from Herat, carpets from Damascus, and fruit from Spain. Merchants, rulers, and pilgrims traveled between cities in Africa, Asia, and Europe, resting at caravansaries along popular routes. By offering free shelter, food, and sometimes entertainment, these Muslim charitable foundations helped promote trade.

Al-Masudi, a tenth-century Muslim geographer, traveler, and historian, recorded a boom in food production in early Muslim civilization, as crops and the knowledge to grow them spread far and wide—including growing peaches, aubergines, and oranges in Spain. Grafting and crop rotation added to farms' productivity and diversity, while irrigation became easier with the water pumps later developed by 16th-century Ottoman engineer Taqi al-Din ibn Ma'rouf.

Actor Ben Kingsley plays a mysterious librarian in an award-winning short film titled 1001 Inventions and the Library of Secrets. *In the movie Kingsley takes schoolchildren on a journey to meet pioneering scientists and engineers from Muslim civilization. The librarian is revealed to be 13th-century polymath Al-Jazari.*

Hospital

In the Hospital chapter, you will see the many ways in which medical knowledge and treatment from Muslim civilization influenced the medicine we experience today. The health care system that developed in early Muslim societies offered pioneering surgery, hospital care, and an increasing variety of drugs and medicines developed from ancient knowledge and new research.

Patients in early Muslim societies might have taken pills, pastilles, syrups, and powders; undergone surgery; or even have had a cataract removed. In the ninth century, detailed diagrams of the eye were produced by scholars such as Hunayn ibn Ishaq, a Nestorian Christian highly respected by his Muslim peers. This knowledge was developed to a tremendous extent by Ibn al-Haytham in the 11th century, giving a solid foundation for mathematical and physiological optics.

Al-Zahrawi, a tenth-century surgeon, developed many surgical instruments we still use today,

Trained guides play characters from the history of science and invention during the 1001 Inventions global exhibition tour.

"It is admitted with difficulty that a nation in majority of nomads could have had known any form of agricultural techniques other than sowing wheat and barley . . . If we took the bother to open up and consult the old manuscripts so many views will be changed."

A. CHERBONNEAU, TRANSLATOR

while 11th-century doctor and philosopher Ibn Sina taught widely on medicine, philosophy, and natural sciences. He also developed methods of treating fractured bones that doctors still adhere to today. Medical books written by these men and

The 1001 Inventions exhibition brings to life a forgotten period in history using models, interactive activities, video, text, and imagery.

other scholars in Muslim civilization influenced European medicine for centuries.

Town

The Town chapter explores the shared heritage of architecture between the modern world and Muslim civilization. Domes, vaults, arches, and towers: The architecture of Muslim civilization demonstrated a huge variety of new ideas, many of which were reused and adapted all over the world. The lasting legacy of the interchange of architectural and decorative ideas between East and West over hundreds of years is clear.

Master architect Sinan built the earthquake-resistant domed Suleymaniye Mosque in 16th-century Istanbul, designing the interior with a filter room to cleanse smoke from the expelled air and collect the soot for making ink.

Towns of a thousand years ago centered around public life with the mosque, market, and public baths centrally placed, surrounded by residential areas. Discover how towns of Muslim civilization were surprisingly advanced, with paved roads, litter collection, covered sewers, and sometimes street lighting.

World

In the World chapter, you will discover how geographers, navigators, explorers, and scholars in Muslim civilization influenced mapmaking and the way we see the world today. In their efforts to explain rainbows, measure the Earth's circumference, and determine how mountains form, scholars from a thousand years ago made huge leaps of intuition and insight in their search to understand our planet.

Extraordinary travelers give a flavor of this age of discovery, like Ibn Battuta, who left home in 1325 to perform a pilgrimage to Mecca, and returned three decades later having explored the limits of Muslim lands. Or Zheng He, a Chinese Muslim navy admiral who led seven voyages of discovery in the 15th century aboard giant wooden ships.

You can also explore the maps left to us by pioneers of this period, like the global map drawn by Moroccan scholar Al-Idrisi, who created it centuries before Marco Polo or Columbus, and the oldest-surviving detailed map showing the Americas, drawn by 16th-century Turkish naval captain Piri Reis.

Universe

Finally, the Universe chapter examines how astronomers, natural philosophers, and instrument makers in Muslim civilization expanded our knowledge of the universe. Astronomical instruments were refined and made by craftspeople like Merriam al-Ijliyah, who constructed astrolabes for the ruler of Aleppo in northern Syria in the tenth century. By the 16th century, astronomer Taqi al-Din was using huge versions of stargazing tools like quadrants and sextants to increase the accuracy of measurements.

The first large-scale observatory in the Muslim world was that built by Sultan Malikshah in Isfahan in the late 11th century. The influential 13th-century Maragha Observatory in Iran was a scientific institution where astronomers challenged received wisdom about the universe and developed new mathematical models on which Renaissance scholars relied.

By building on knowledge that had come before, and adding new observations and insight, Muslim scholars left us with a rich shared heritage of astronomy from East and West, commemorated today in the Greek and Arabic names of many stars and constellations.

Muslim Civilization: Where and When?

After Prophet Muhammad died in the year 632, the caliphs who came after him built an empire that stretched from southern Spain, through North Africa and the Middle East, to India and China, and by the 15th century as far as Indonesia and eastern Europe. Generally tolerant to the faiths they found in the lands they ruled, the caliphs oversaw an incredible expansion of knowledge and prosperity.

As trade flourished, so did knowledge and new ideas. Scholars worked to translate the writings of ancient thinkers like Brahmagupta, Aristotle, Euclid, Ptolemy, and Hippocrates into Arabic to allow debate and development of mathematics, astronomy, chemistry, medicine, and engineering. It was a golden age of thought, development, and wealth creation.

How did such an enlightened era come to an end? This is a question many have tried to answer and is beyond the remit of this book. However, by the early 15th century, Muslim civilization had experienced attacks both from Crusaders' campaigns in Spain, Turkey, and Palestine, and the Mongol invasion of Persia, Iraq, and Syria. The famous libraries and learning of the Muslim world came under catastrophic threat during these times of conflict. When Baghdad was invaded in 1258, the attacking Mongol armies destroyed countless manuscripts, while in Córdoba, the vast majority of the city's 600,000 Islamic books were destroyed by crusading invaders.

Having lost Spain and Sicily, the Muslim world then suffered the onslaught of Timur the Lame, known as Tamerlane. These devastations together started the decline and eventual fall of Islamic civilization, and the end of this bright period of classical Muslim scholarship. This was accompanied by the rise of the West after the discovery of the New World, with all the wealth it brought to Europe, and the eventual demise of the Silk Road along which knowledge had flowed through various cultures.

At the same time, the Muslim world was weakened by the rise of new inward-looking ideas that deflected interest away from philosophy, logic, and the translation of faith into deeds to benefit society. And so the Muslim

At the California Science Center in Los Angeles, children enjoy a computer-based game that explores inventions in the home. The game is part of the award-winning 1001 Inventions exhibition.

world disintegrated into numerous independent nation-states, many of which subsequently experienced colonization or went to war with each other, even after independence. Neither the peoples of these states nor their colonial rulers had paid much attention to education or socioeconomic reform, leaving these countries in a long struggle for identity and survival in a fast-moving modern world.

But as this book shows, the knowledge of the scholars of Muslim civilization was far from lost. Thousands of those precious documents of the past, written mainly in Arabic, fill the archives of the British Library, Berlin's Staatsbibliothek, Paris's Bibliothèque Nationale, and many others. In Toledo's cathedral archive today you can still see about 2,500 of the surviving manuscripts that scholars translated from Arabic to Latin. And it was these Latin translations that fed the scientific and philosophical revolution of the 1600s and kept the flame of knowledge alive.

Great scientific minds like Robert Boyle, Edmund Halley, John Wallis, and Johannes Hevelius, who were the early pioneers of the Royal Society, took great interest in translating and learning from Arabic manuscripts. A recent exhibition, "Arabick Roots," organized to celebrate the 350th anniversary of the Royal Society, noted that King Charles I asked the Levant Company to send home a manuscript on every ship returning to England from the East.

Today, the influence of the thousand-year Muslim civilization is clear to see—in medicine and astronomy, architecture and engineering, mathematics and chemistry, history and geography, as well as in today's customs, fashions, and tastes. Through *1001 Inventions* you can now explore this legacy for yourself.

632–1796

For more than a thousand years from the seventh century onward, the Muslim world stretched from southern Spain as far as China. During this period scholars, male and female and of many beliefs, worked collaboratively to build and improve upon ancient knowledge. They made breakthroughs that led to an incredible expansion of knowledge and prosperity—a golden age of civilization.

Follow the time line below to trace the progress of mathematics, science, architecture, exploration, education, and medicine during Muslim civilization. And see how ideas and knowledge migrated from the East, paving the way for the Renaissance—another great age of development.

637
Islam spreads to Persia, Palestine, Syria, Lebanon, and Iraq, and later to Egypt.

ca 635
Al-Shifa is appointed the first female health and safety minister by Umar, second caliph, in the city of Medinah and then in Basrah.

654
Islam has spread to all of North Africa.

Dome of the Rock Mosque

691
Building begins of the Dome of the Rock Mosque in Jerusalem.

625 650 675 700

632
Prophet Muhammad dies and Abu Bakr becomes first caliph, or Muslim ruler.

644
A windmill powering a millstone is built in Persia.

A medieval windmill

661
The Umayyad dynasty rules the caliphate from Damascus.

"There can be no education without books."

ARABIC PROVERB

711
Islam reaches Spain.

Old street in Córdoba, Spain

■ MUSLIM EVENTS
■ EUROPEAN EVENTS

Distillation

ca 722
Jabir ibn Hayyan is born. He is considered the "father of chemistry."

Astrolabe

ca 777
Astrolabe maker and astronomer Al-Fazari dies.

787
Building begins of the Great Mosque of Córdoba.

786
Caliph Harun al-Rashid establishes the House of Wisdom in Baghdad.

785
King Offa mints a Gold Mancus coin, imitating the gold dinar of Caliph al-Mansur.

Gold Mancus

813
Caliph Al-Ma'mun expands the House of Wisdom; the translation movement intensifies.

750 — 775 — 800 — 825

750
Abbasids overthrow the Umayyads and in 762 build a new capital in Baghdad. Spain is ruled by an Umayyad family descendant.

780
Mathematician Al-Khwarizmi is born. His book *Algebr wal Muqabala* developed modern algebra.

795
First mention of a paper mill in Baghdad.

800
Caliph Harun al-Rashid presents Charlemagne with a clock that strikes the hour.

801
Al-Kindi is born. He was a mathematician, philosopher, physicist, chemist, and musician.

828
Abu Mansur opens Al-Shammasiyah Observatory, near Baghdad.

Al-Kindi

> *"If he [the teacher] is indeed wise he does not bid you enter the house of his wisdom, but rather leads you to the threshold of your own mind."*
>
> KHALIL GIBRAN IN HIS BOOK *THE PROPHET*

Al-Razi, known as "Hippocrates of the Arabs"

Abbas ibn Firnas's flight

850
Banu Musa brothers publish their *Book of Ingenious Devices.*

864
Al-Razi (Rhazes) is born. A physician, chemist, and medical teacher, he is considered the "father of clinical and experimental medicine." His writings were later translated into Latin.

887
Abbas ibn Firnas, pioneer of unpowered flight, dies in Córdoba.

880
Physician and inspector of Baghdad hospitals Sinan ibn Thabit ibn Qurra is born. He started mobile hospital services for rural and Bedouin areas.

913
Abbasid Caliph Al-Muqtader issues the first licensing regulation for medical practice. He established the hospital of Al-Muqtadiri in Bab al-Sham in Baghdad; his mother established Al-Sayyida Hospital in Souq Yehia in Baghdad.

850 875 900 925

858
Astronomer Al-Battani is born. He determined astronomical measurements with accuracy.

859
Al-Qarawiyin University in Fez is completed by Fatima al-Fihri.

872
Ahmad ibn Toloun, Abbasid governor of Egypt, establishes a hospital in Cairo known to be the first to include a department for mental diseases.

895
Ibn al-Jazzar al-Qayrawani (Aljizar) is born. He wrote the first independent book on pediatrics and social pediatrics: *Risalah fi Siyasat as-Sibyan wa Tadbirihim (A Treatise on Infant and Child Care and Treatment).*

900
Fatimids rule Egypt and North Africa, then, nine years later, Sicily.

936
Surgeon Al-Zahrawi (Albucasis) is born in Córdoba. He refined the science of surgery, invented dozens of surgical instruments, and wrote the first illustrated surgical book.

Al-Qarawiyin University

Al-Zahrawi

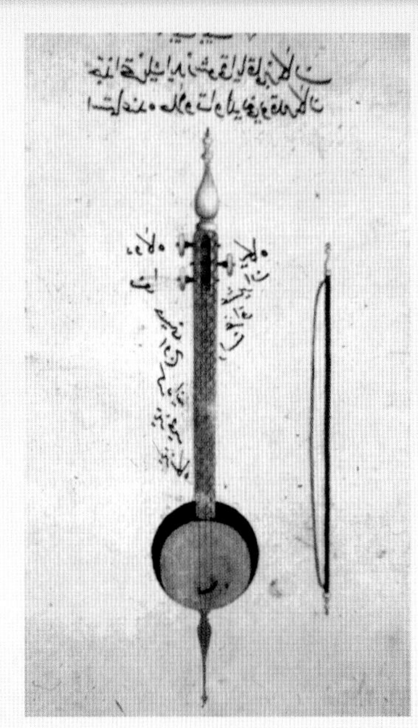

A rababah, ancestor of the violin

Al-Azhar University

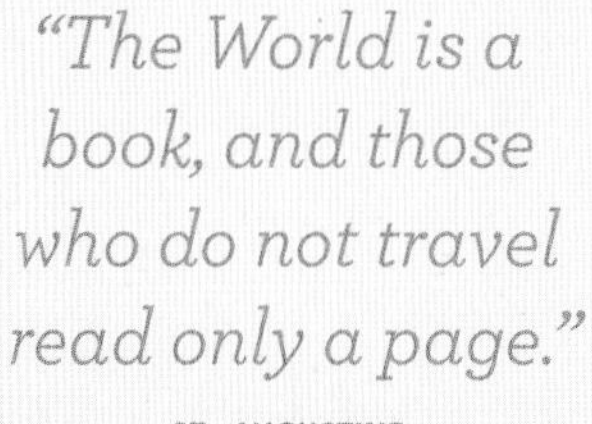

"The World is a book, and those who do not travel read only a page."

ST. AUGUSTINE, CHRISTIAN PHILOSOPHER

950
Al-Farabi from Baghdad dies. A philosopher and musician, he invented the ancestor of the violin.

970
Labna, a mathematician and scientist, is appointed private secretary to the Umayyad caliph Al-Hakim in Córdoba.

972
Al-Azhar University is founded by the Fatimids in Cairo.

973
Al-Biruni is born. The polymath, astronomer, mathematician, and geographer measured the circumference of Earth

1009
Astronomer Ibn Yunus dies in Cairo, leaving thousands of accurate records, including 40 planetary conjunctions and 30 lunar eclipses.

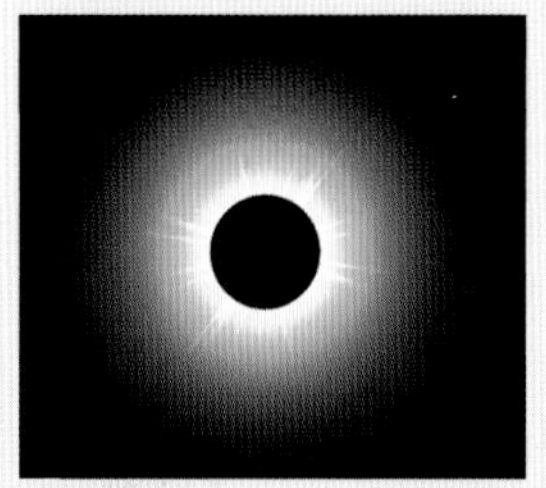

Mapping the sun's position

975 | 1000 | 1025 | 1050

957
Cartographer and writer Al-Masudi describes his visit to the oil fields of Baku.

965
Physicist Ibn al-Haytham is born. His discoveries and theories revolutionized optics.

Ibn Sina's *Canon*

980
Prince of physicians Ibn Sina (Avicenna) is born. He wrote the influential *Canon of Medicine*.

987
Sutaita al-Mahamli, a mathematician and expert witness in courts, dies in Baghdad.

999
Building begins of Bab Mardum Mosque in Toledo, which uses a unique form of rib vaulting.

Bab Mardum Mosque

"Seek knowledge from the cradle to the grave."

MUSLIM PROVERB

1050
Constantine the African moves from Tunisia to Salerno, initiating the transfer of Islamic medicine to Europe.

DE CONSERVANDA BONA VALETVDINE, Liber Scholæ Salernitanæ.

DE ANIMI PATHEMATIS, ET remedijs quibusdam generalibus.
CAPVT I.
1 ANglorum Regi scribit schola tota Salerni.
2 Si vis incolumem, si vis te reddere sanum,
3 Curas tolle graueis, irasci crede profanum.

Constantine the African

The Story of Hayy ibn Yaqzan

Al-Khwarizmi

1066
The Norman Conquest of England begins a flow of Muslim motifs and ideas into that country.

1085
Christians take Toledo. A center at Toledo is established, translating Arabic books into Latin.

Ibn Bassal's *Book of Agriculture* revolutionizes farming. He is from Toledo, Spain.

1110
Ibn Tufal, author of *Hayy ibn Yaqzan,* is born.

1143
Robert of Chester translates the Quran and works of Al-Khwarizmi.

1140
Daniel of Morley travels to Córdoba to learn mathematics and astronomy, returning to lecture at Oxford.

1075 1100 1125 1150

1065
The Nizamiyya madrasa, the first school in Baghdad, is established by Nizam al-Mulk, the Seljuk minister who appointed distinguished philosopher and theologian Al-Ghazali as a professor.

1091
Abu Marwan ibn Zuhr (Avenzoar) is born. He is a pioneer of experimental surgery and co-author, with Ibn Rushd (Averroes), of an original encyclopedic medical text. His two daughters became doctors.

1096
The first Crusades begin.

1099
Al-Idrisi is born. He produced a world map for Norman king Roger II of Sicily.

1126
Ibn Rushd (Averroes) is born. He wrote an extensive corpus of philosophy in which he stated significant theories in epistemology, natural philosophy, and metaphysics. An accomplished physician, he wrote the famous treatise *Al-Kulliyat fi al-tib*, known in Latin as the *Colliget*.

1145
Jabir ibn Aflah invents an observational instrument known as the *torquetum*, a mechanical device to convert between spherical coordinate systems.

1154
Nur al-Din Zangi establishes Al-Nuri Hospital in Damascus, a large teaching hospital.

> *"Who so ever treats people without knowledge of medicine, becomes liable."*
>
> PROPHET MUHAMMAD, NARRATED BY AL-BUKHARI AND MUSLIM

Al-Idrisi's world map

Botanical species by Ibn al-Baytar

Traditional carpets

1197
Botanist Ibn al-Baytar is born in Málaga. He wrote a famous pharmacopeia.

1187
Salah al-Din al-Ayyubi (Saladin) regains Jerusalem. He established Al-Nasiri Hospital in Cairo.

1186
Queen Dhaifa Khatoon is born in Aleppo, Syria. She was the daughter-in-law to Saladin, and a supporter of science and learning.

1185
Temple Church is built in London by the Templars, imitating the Dome of the Rock in Jerusalem.

1233
Ibn al-Quff is born. A Christian surgeon and author, he continued Al-Zahrawi's efforts to develop surgery as a science and independent medical specialty. He wrote *Kitab al-Umda fi al-Jirahah (The Main Pillars of Surgery).*

1255
Queen Eleanor, Castilian bride of King Edward I, brings Andalusian carpets to England in her dowry.

1254
King Alfonso el Sabio establishes Latin and Arabic colleges in Seville and commissions the translation of Arabic texts.

1260
Roger Bacon publishes *Secrets of Art and Nature* praising influences of Muslim scholars.

1200 1225 1250 1275

1202
Leonardo of Pisa, known as Fibonacci, introduces Arabic numerals and mathematics to Europe in his book *Liber Abaci.*

1206
Al-Jazari completes his *Book of Ingenious Mechanical Devices.*

1210
Ibn al-Nafis is born. He was a scholar of jurisprudence and doctor who was first to discover pulmonary circulation. He wrote *Al-Seerah al-Kamiliyah* refuting the ideas of Ibn Tufayl's novel *Hayy ibn Yaqzan* on the oneness of existence.

1229
Robert Grosseteste, who studied in Córdoba, becomes first chancellor of Oxford University. He was elected bishop of Lincoln in 1253.

Mamluk lusterware

1250
Mamluk dynasty rules Egypt after the Ayyubids and later defeats the Mongols.

1256
Ibn abi al-Mahasin al-Halabi writes his comprehensive scholarly and illustrated work on eye diseases, *Al-Kafi fi al-Kuhl (The Book of Sufficient Knowledge in Ophthalmology).*

1258
Mongols devastate and rule Baghdad and conquer Syria.

> *"And the leaves of the tree were for the healing and the restoration of the nations."*
>
> THE BIBLE, REVELATION 22:2

Ibn Battuta

"Beauty of style and harmony and grace and good rhythm depend on simplicity."

PLATO, GREEK PHILOSOPHER

The Muqaddimah
by Ibn Khaldun

1293
The first paper mill outside Islamic Spain in Europe is established in Bologna.

1325
Ibn Battuta leaves Tangier on his 29-year journey.

1332
Ibn Khaldun, the "father of sociology," is born.

1385
Serafeddin Sabuncuoglu is born. An Ottoman surgeon, he continued the work of Al-Zahrawi and Ibn al-Quff by writing an independent surgical textbook.

1383
Chemist Maryam al-Zanatiyeh dies in Qarawiyin, Tunisia.

1300 1325 1350 1375

1267
Marco Polo starts his 24-year journey

1284
Al-Mansuri Hospital in Cairo is completed after 11 months of construction.

1311
The Ecumenical Council of Vienne decides to establish schools of Arabic and Islamic studies at universities in Paris, Oxford, Bologna, and Salamanca.

1330
Giotto's painting "Madonna and Child" uses *tiraz*, bands of Arabic inscriptions, which mark royal garments and other textiles from the Muslim world.

1347
The Black Death reaches Alexandria and Cairo from Europe.

1354
Emir Mohammed V builds the Lion Fountain, a water-powered clock, in the Alhambra.

1405
Zheng He starts his seven epic sea voyages from China. In the largest wooden boats the world had seen, he established China as a leading power in the Indian Ocean, brought back exotic species like the giraffe, and drew tribute from many nations.

Lion Fountain

Statue of Zheng He

Ulugh Beg Observatory

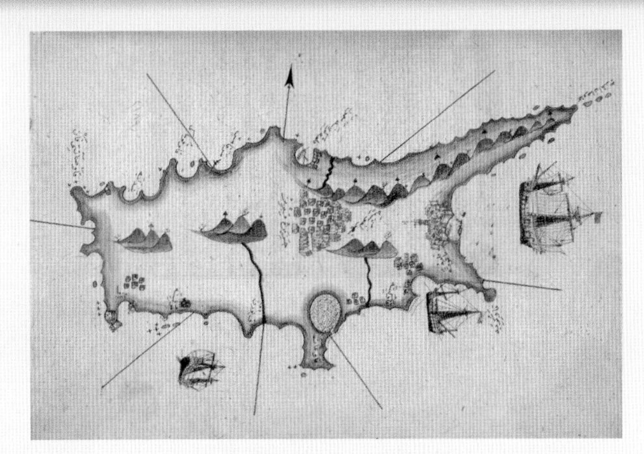
Detail of Piri Reis's map

1437
Ulugh Beg publishes his star catalog.

1432
Ibn Majid is born in Arabia. He was a master navigator and is said to have guided Vasco da Gama from South Africa to India.

1492
Christopher Columbus lands in the New World.

1489
Koca Mimar Sinan is born. A renowned architect, he built Turkey's Selimiye and Suleymaniye mosques, as well as many others.

1513
Piri Reis constructs the earliest known map showing America.

1450 1475 1500 1525

1452
Leonardo da Vinci is born. He was a major contributor to the foundation of the Renaissance.

1497
Venice publishes a translation of *Al-Tasrif* by Al-Zahrawi. Basel and Oxford follow suit.

1543
Nicolaus Copernicus publishes *De Revolutionibus*, drawing on the work of Nasir al-Din al-Tusi and Ibn al-Shatir.

"The earth is like a beautiful bride who needs no man-made jewels to heighten her loveliness."

KHALIL GIBRAN, LEBANESE WRITER

Leonardo da Vinci

Nicolaus Copernicus

"If anyone travels on a road in search of knowledge, Allah will cause him to travel on one of the roads of Paradise."

PROPHET MUHAMMAD, NARRATED BY ABU AL-DARDAH

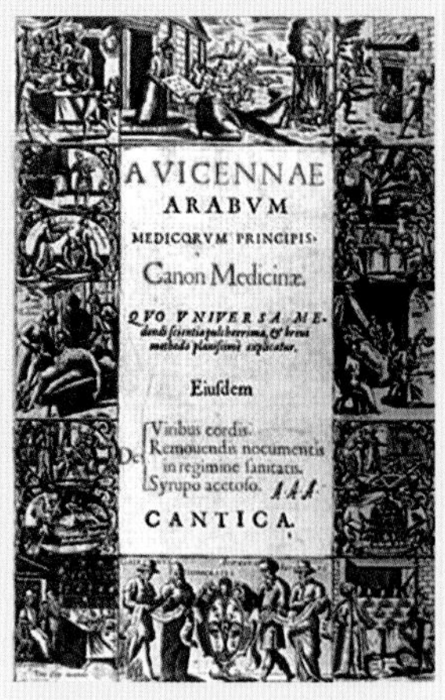
A VICENNAE
ARABVM
MEDICORVM PRINCIPIS.
Canon Medicinæ.
QVO VNIVERSA MEdendi scientia pulcherrima, & breui methodo planissime explicatur.
Eiusdem
De { Viribus cordis. Remouendis nocumentis in regimine sanitatis. Syrupo acetoso.
CANTICA.

Latin translation of Canon

1564
Galileo Galilei is born.

1571
Johannes Kepler is born. He drew on the work of Ibn al-Haytham in his work on optics.

1593
The Canon of Ibn Sina is printed in Rome and, along with *Al-Hawi* by Al-Razi, becomes a standard text in the European medical curriculum.

1611
Polish astronomer Johannes Hevelius is born. In the frontispiece of his *Selenographia* (Gdansk, 1647), he depicted Ibn al-Haytham as symbolizing knowledge through reason, and Galileo Galilei as symbolizing knowledge through the senses.

1550 | 1575 | 1600 | 1625

1558
The first German, and probably European, observatory is built in Kassel.

1577
Istanbul observatory of Taqi al-Din is founded. It will close a few years later, in 1580.

Taqi al-Din's observatory

1604
Edward Pococke is born. He spent five years in Aleppo learning Arabic; he also translated *Hayy ibn Yaqzan*, a precursor to Robinson Crusoe.

1606
Edmund Castell is born. He lectured on the use of Avicenna's medical work. For more than 18 years, he compiled a dictionary of seven Asian languages.

1616
John Wallis is born. A renowned mathematician and member of the Royal Society in London, he translated and lectured on the work of Arabic mathematicians. He included Al-Tusi's work in his *Opera Mathematica*.

1627
Robert Boyle, England's most famous chemist, is born. He sought Arabic manuscripts and had them translated.

Lagari Hasan Celebi's rocket flight

"The Cow-Pock" by James Gillray

1634
King Charles I requests that the Levant Company send home Arabic manuscripts on every ship returning to England.

1633
Lagari Hasan Celebi flies the first manned rocket over the Bosporus.

1664
At the request of Hevelius, the Royal Society agrees to translate the astronomical manuscript of Ulugh Beg from Persian to Latin in its entirety.

1656
Scientist and astronomer Edmund Halley is born. He translated Arabic editions of Greek mathematics and researched observations of Al-Battani.

1682
Moroccan ambassador to London Muhammed Ibn Haddu is elected a fellow of the Royal Society in London.

1729
Tripoli ambassador in England Cassem Aga writes about the widespread practice of smallpox inoculation in North Africa and is elected a fellow of Royal Society in London.

1796
Edward Jenner tests inoculation with cowpox.

1675 | 1700 | 1725 | 1750

1642
Isaac Newton is born. He kept a copy of the Latin translation of Ibn al-Haytham's *Book of Optics* in his library

1650
Turkish merchants bring coffee to the United Kingdom.

1678
John Greeves publishes a paper in the Royal Society Philosophical *Transactions* on Egyptians' use of large ovens to hatch thousands of chicken eggs at a time.

1721
Lady Mary Montagu tests smallpox inoculation in Britain, having witnessed the practice in Turkey.

1725
Moroccan ambassador to London Mohammed Ben Ali Abgali is elected a fellow of the Royal Society in London.

Edward Lloyd's Coffee House, London

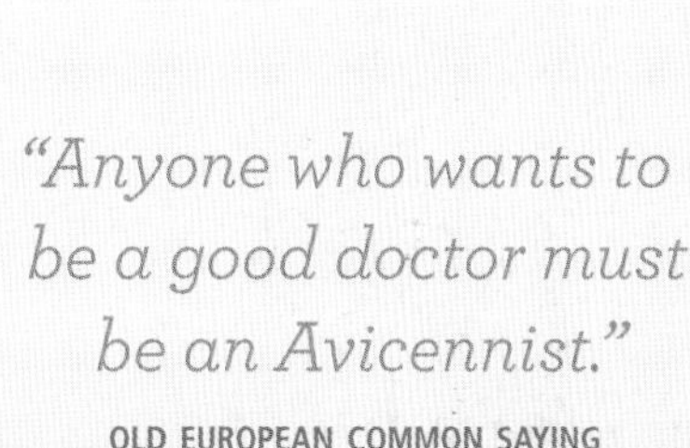

"Anyone who wants to be a good doctor must be an Avicennist."

OLD EUROPEAN COMMON SAYING

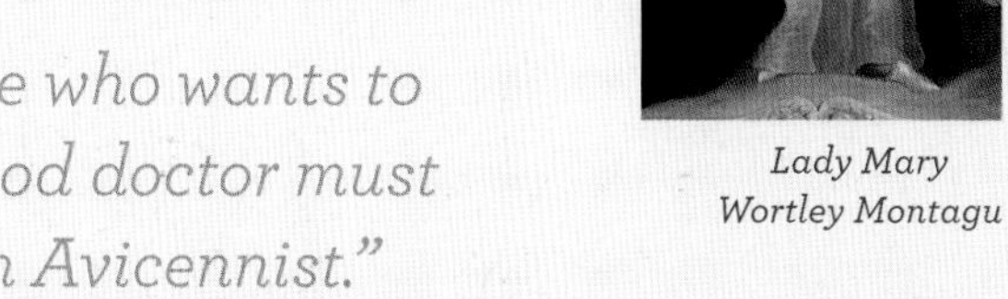

Lady Mary Wortley Montagu

CHAPTER TWO

"He is happiest, be he King or peasant, who finds peace in his home."

JOHANN WOLFGANG VON GOETHE, GERMAN WRITER

HOME

ON THE COFFEE TRAIL • FINE DINING • THREE-COURSE MENU
CLOCKS • CHESS • MUSIC • CLEANLINESS • TRICK DEVICES
VISION AND CAMERAS • FASHION AND STYLE • CARPETS

YOUR HOME IS YOUR PRIVATE DOMAIN, WHERE YOU CAN BE WHO YOU WANT TO BE, WHERE the big world stops at the front door. Your home represents who you are; it speaks your language. In the kitchen, maybe a favorite coffeepot sits by the kettle, under a clock that is beside a holiday photo taken from last year's vacation. Fragrant smells of fine soap and perfume waft out of the bathroom while music floats up the luxuriously carpeted stairs.

Read on and you will be intrigued to discover that the items mentioned above originated or were developed more than a millennium ago by industrious Muslims who sought to provide greater comforts in their world.

The humble roots of your trusted camera were in a dark room in 11th-century Egypt, and if you are late, hurriedly looking at your watch, think about the 7-meter-high (23 feet) clocks designed in 13th-century Turkey with state-of-the-art engineering technology. A man with the nickname of "the Blackbird" came from eighth-century Baghdad to Muslim Spain, bringing the etiquette of three-course meals and seasonal clothes, while chemists developed perfumes, and chess went from being a war game to household entertainment.

OPPOSITE: *A 16th-century manuscript shows a traditional Muslim coffeehouse.*

01 ON THE COFFEE TRAIL

More than 1.5 billion cups of coffee are drunk worldwide every day—enough to fill nearly 300 Olympic-size swimming pools. If you do not have a jar of coffee in your kitchen, you are in a minority. Coffee is a global industry and the second largest commodity-based product; only oil beats it.

More than 1,200 years ago hardworking people fought to stay awake without this stimulant until, as the story goes, a herd of curious goats and their watchful master, an Arab named Khalid, discovered this simple, life-changing substance. As his goats grazed on the Ethiopian slopes, he noticed they became lively and excited after eating a particular berry. Instead of just eating the berries, people boiled them to create *al-qahwa*.

Sufis in Yemen drank *al-qahwa* for the same reasons we do today, to stay awake. It helped them to concentrate during late night *Thikr* (prayers in remembrance of Allah). Coffee was spread to the rest of the Muslim world by travelers, pilgrims, and traders, reaching Mecca and Turkey in the late 15th century and Cairo in the 16th century.

It was a Turkish merchant named Pasqua Rosee who first brought coffee to England in 1650, selling it in a coffeehouse in George-yard, Lombard Street, London. Eight years later, another coffeehouse called Sultaness Head was opened in Cornhill. Lloyd's of London, today a famous insurance company, was originally a coffee shop called Edward Lloyd's Coffee House. By 1700, there were about 500 coffeehouses in London, and nearly 3,000 in the whole of England. They were known as "penny universities" because you could listen and talk with the great minds of the day for the price of a coffee.

The consumption of coffee in Europe was largely based on the traditional Muslim preparation of the drink. This entailed boiling the mixture of coffee powder, sugar, and water together, which

A goatherd herds his animals in Ethiopia. More than 1,200 years ago, an Ethiopian goatherd noticed his goats became lively after eating some red berries. People later boiled the same berries to make coffee.

left a coffee residue in the cup because it was not filtered. However, in 1683, a new way of preparing and drinking coffee was discovered, and it became a coffeehouse favorite.

Cappuccino coffee was inspired by Marco d'Aviano, a priest from the Capuchin monastic order, who was fighting against the Turks besieging Vienna in 1683. Following the retreat of the Turks, the Viennese made coffee from abandoned sacks of Turkish coffee. Finding it too strong for their taste, they mixed it with cream and honey. This made the color of coffee turn brown, resembling the color of the Capuchins' robes. Thus, the Viennese named it cappuccino in honor of Marco D'Aviano's order. Since then, cappuccino has been drunk for its enjoyable, smooth taste.

"Coffee is the common man's gold, and like gold it brings to every person the feeling of luxury and nobility."

SHEIKH 'ABD-AL-KADIR, WHO WROTE THE EARLIEST KNOWN MANUSCRIPT ON THE HISTORY OF COFFEE IN 1588

ORIGIN OF THE COFFEEHOUSE

An illustration depicts Edward Lloyd's Coffee House, established in the 17th century.

The first coffeehouse in Europe appeared in Venice in 1645, after coffee came to Europe through trade with North Africa and Egypt. Edward Lloyd's Coffee House in London, established in the late 17th century, was a meeting place for merchants and shipowners. Coffeehouses became forerunners of today's pubs. They were the places where the public discussed political affairs and also gave rise to the liberal movement.

02 Fine Dining

We can thank a ninth-century man with the nickname "the Blackbird" for introducing the concept of three-course meals into Europe. Eating habits were totally transformed when Ziryab landed in Andalusia in the ninth century and said meals should start with soup, followed by a main course of fish, meat, or fowl, and finish off with fruits and nuts.

Medieval Muslims, like many others, ate according to seasonal influences. Typical winter meals used vegetables such as sea kale, beets, cauliflower, turnips, parsnips, carrots, celery, peas, broad beans, lentils, chickpeas, olives, hard wheat, and nuts. These were usually eaten with meat dishes. Desserts usually consisted of dried fruits such as figs, dates, raisins, and prunes. The fruits were accompanied by drinks made from syrups of violet, jasmine, aloe, medicinal spices, fruit pastilles, and gums.

By contrast, their summer diet consisted of 11 types of green beans, radishes, lettuces, chicory, aubergines, carrots, cucumbers, gherkins, watercress, marrow, courgettes, and rice. The meat accompanying these was mainly poultry, ostrich, and beef. Desserts included fruits such as lemon, lime, quinces, nectarines, mulberries, cherries, plums, apricots, grapes, pomegranates, watermelon, pears, apples, and honeydew melon. Drinks were made from syrups and preserves of fruit pastilles, lemon, rose, jasmine, ginger, and fennel.

This banquet of food was presented on a tablecloth, the concept of which was spread in Andalusia by Ziryab. He also changed the heavy metal drinking goblets and gold cups used on the dinner tables of the Cordoban court to delicate crystal.

In most European aristocratic circles, the demand for Asian recipes and spices increased rapidly. Sources from the chronicles of the pope in Avignon in the 14th century tell us that ships from Beirut brought jams, preserves, rice, and special flour for making cakes, plus compensatory laxatives. Queen Christina of Denmark took care to follow the Muslim diet and imported their products and fruits. Since Denmark could mostly supply apples and rye, it is perhaps "food for thought" to consider the origin of Danish pastries.

Crystal was developed in Andalusia due to the ingenuity of another Muslim, 'Abbas ibn Firnas, who died in 887. In his experiments, he manufactured glass from sand and stone, establishing a crystal industry based on rocks mined north of Badajos. Most of the Andalusian rock crystal pieces that have survived are found in European churches and monasteries, the most famous among them being a

TOP: *A rock crystal ewer from the Fatimid period in Cairo, Egypt, dates from the tenth or eleventh century. Ziryab brought crystal to the dinner table in the ninth century, after 'Abbas ibn Firnas introduced it to Al-Andalus, the Arabic name given to the Iberian Peninsula during Muslim rule.*

*"Coffee makes us severe,
and grave,
and philosophical."*

JONATHAN SWIFT, IRISH WRITER

spherical bottle currently in the Astorga Cathedral, Spain. It bears vegetal patterns and a Kufic inscription, the common decorative elements on rock crystal pieces.

As well as introducing crystal that was used for drinking glasses, 'Abbas ibn Firnas was the same man who used glass in a most ingenious way to construct a planetarium, supplying it with artificial clouds, thunder, and lightning. Naturally this astounded the ninth-century public.

Muslim potters then introduced the art of stylish dining with a variety of ceramics and glazes. Málaga and Valencia were major centers of the industry, and Muslims revolutionized the production and decoration of pottery through their use of luster glaze, which you can read more about in the Pottery section of the Market chapter.

Both Valencian and Malagan potters exported their wares to Christian-populated regions like northern Spain and southern France and as far east as Italy. Here, Malagan potters were thought to have laid the foundations of the famous Majolica ware, which went on to dominate the Italian ceramic industry.

Next time you have a meal, look at the ceramics and glasses. Are the plates made of fine earthenware with designs that look like precious metals? Are the glasses delicate, chiming if you gently tap them?

A 16th-century manuscript from Gelibolulu Mustafa Ali's book, Nusratname, *shows a banquet given by commander-in-chief Lala Mustafa Pasha, seated at the head of the table, to leading dignitaries of the army in Izmit. On either side of the commander are officers who participated in the campaign. The soldiers are eating a variety of dishes served by servants carrying pitchers of rose water. Cutlery and napkins cover the diners' laps.*

"THE BLACKBIRD"

Abul-Hasan 'Ali ibn Nafi' was nicknamed Ziryab, "the Blackbird," because of his melodious voice and dark complexion. A musician and fashion designer, he came from Iraq in the ninth century to Córdoba, Andalusia, one of the leading cultural centers of Muslim civilization. Here he set fashions in eating, etiquette, clothes, and music that have lasted until today. Because of his impact, you can read more about him in many sections of this chapter.

The Blackbird became the foremost trendsetter of this time. His talent generated an invitation to Moorish Spain, where he received a salary of 200 golden dinars in addition to many privileges. With him, he brought fine etiquette, cooking, fashion, and even toothpaste.

03 Three-Course Menu

From an anonymous Andalusian cookbook of the 13th century
Translated by Charles Perry

Starters: Meat Soup with Cabbage
Take meat and cut it up as finely as possible. Take ripe cheese, the best you can obtain, cut it up, and throw on it an onion pounded with coriander. Take tender "eyes" of cabbage, boil, then pound them with all of that in a wooden mortar, and throw them in the pot, after boiling once or twice. Add some *murri,* a little vinegar and some pepper and caraway. Cover the contents of the pot with dough [or sourdough] and cover with eggs.

Main meal: *Mirkas* with Fresh Cheese
Take some meat, carefully pounded as described earlier. Add fresh cheese that is not too soft lest it should fall apart, and half a piece of diced meat and some eggs, for it is what holds it together, along with pepper, cloves, and dry coriander. Squeeze on it some mint juice and coriander juice. Beat it all and use it to stuff the innards, which are tied with threads in the usual way. Next, fry it in fresh oil, and eat it in nibbles, without sauce, or however you like.

Main meal: Roast in a *Tajine*
Take an entire side of a young, plump kid and place it in a large *tajine* [earthenware cooking dish with a lid still used in North Africa today]; put it in the oven and leave it there until the top is browned. Take it out, turn it and put it in the oven a second time until it is browned on both sides. Take it out and sprinkle it with salt, ground pepper and cinnamon. This is extremely good and is the most notable roast that exists, because the fat and moisture stay in the bottom of the pan and nothing is lost in the fire, as in the roast on a spit or the roast in a *tannur* [clay oven].

Main meal: Fish *Tharid*
Pound pieces of a big fish well and add egg white, pepper, cinnamon, enough of all the spices, and a little leavening yeast. Beat them until all is well mixed. Take a pot and put in it a spoonful of vinegar, two of cilantro juice, one and a half of onion juice, one of *murri naqi'* [pure type of barley flour], spices, flavourings, pine nuts, six spoonfuls of oil, and enough salt and water, and put it over a moderate fire. When it has boiled several times, make the pounded [fish] meat into the form of a fish and insert into its interior one or two boiled eggs, and put it carefully into the sauce while it is boiling. Cut the remainder into good meatballs; take boiled egg yolks and cloak them with that meat also. Throw all that in the pot and when all is done, take the fish from the pot and the meat-cloaked yolks, and fry them in a frying pan until browned. Then, cover the contents

of the pot with six eggs, pounded almonds and breadcrumbs, and dot the pot [with yolks].

Main meal: Roast Chickens

Take young, fat chickens, clean and boil them in a pot with water, salt and spices. Take them out of the pot and pour the broth with the fat in a dish and add to it what has been said for the roast over coals. Rub that onto the boiled chicken, arrange it on a spit and turn it over a moderate fire with a continuous movement and baste it constantly, until it is ready and browned; then sprinkle it with what remains of the sauce and serve. It tastes nicer than livestock meat, and is more uniform. Other birds may be roasted the exact same way.

Sweet: Tharda of the Emir

Knead white flour well with water, a little oil and leavening yeast, making four thin *raghifs* [flatbread, rolled out decidedly thinner than a pita, like a thin pancake]. Fry them in a frying pan with much fresh oil, until they brown a little. Take them out of the oil and pound them well. From the rest of the dough make little hollow things on the pattern of *mujabbana* [cheese pie], and make top crusts for them. Fry them in fresh oil, making sure they stay white and do not turn brown, fry the top crusts, too. Then, take peeled pistachios, almonds, and pine-nuts, and sufficient sugar; pound them coarsely, spice them and knead them with sharp rosewater and mix with ground *raghifs* and stir until completely mixed. Fill the hollow dumplings prepared earlier with that mix, and put on their covers, and proceed confident that they will not be overdone. Arrange them on a dish and put between them the rest of the filling and sprinkle them with sharp rosewater until the dish is full. Sprinkle with plenty of ground sugar and present it. And if some syrup of thickened, honeyed rosewater syrup is dripped on it, it will be good, God willing.

Drink: Syrup of Pomegranates

Take a *ratl* [500 grams approximately] of sour pomegranates and another of sweet pomegranates and add their juice to two *ratls* of sugar: cook all this until it reaches the consistency of syrup, and set it aside until needed. Its benefits: it is useful in cases of fevers, cuts the thirst, alleviates bilious fever and lightens the body gently.

04 CLOCKS

Whatever we do, wish, hope, dream, or fear, time will always go on, with or without us. Whether it is an examination we dread taking, an important interview, or a birthday, there will be a time when the activity begins and ends.

From the first sundial, people have wanted to record time. Now we have silent, digital timepieces as well as the ticktock of modern clocks. Their ancestors were the drip-drop of the clepsydra and of water clocks. The clepsydra, a simple vase marked with divisions that measured water flowing out of a small spout near the base, was used in Egypt before 1500 B.C.E.

Another ancient water timing device is from India and is called *ghatika-yantra*. It consists of a small, hemispherical bowl (made of copper or a coconut) with a small hole in its base. Floated in a larger pot of water, the bowl would gradually fill and sink. When it reached the bottom, an audible thud alerted the timekeeper, who would raise it up to start the process again. This became very popular in Buddhist and Hindu temples, and later was widely used in Indian Muslim mosques.

Our story begins with 13th-century water clocks and an ingenious man called Al-Jazari from Diyarbakir in southeast Turkey. He was a pious Muslim and a highly skilled engineer who gave birth to the concept of automatic machines. He was inspired by the history of machines and the technology of his predecessors, particularly ancient Greek and Indian scientific inventions.

TOP: *The history of mechanical clocks includes many timepieces conceived in the Muslim world from the 13th century on.* BOTTOM, FROM LEFT: *Images depict the evolution of measuring time—from sundials, clepsydras, and water clocks, to weight-driven grandfather clocks and today's digital watches.*

"By time,
surely mankind is in loss,
except for those
who have believed
and done useful deeds
and advised each other
to truth
and advised each other
to patience."

QURAN (103)

By 1206, Al-Jazari had made numerous clocks of all shapes and sizes while he was working for the Artuq kings of Diyarbakir. Then-king Nasir al-Din said to him, "You have made peerless devices, and through strength have brought them forth as works; so do not lose what you have wearied yourself with and have plainly constructed. I wish you to compose for me a book which assembles what you have created separately, and brings together a selection of individual items and pictures."

The outcome of this royal urging was an outstanding book on engineering called *The Book of Knowledge of Ingenious Mechanical Devices.* This book became an invaluable resource for people of different engineering backgrounds, as it described 50 mechanical devices in six categories, including water clocks.

Just as we need to know the time today, so did Muslims more than 700 years ago, and Al-Jazari was keeping to a Muslim tradition of clock making. Muslims knew that time could not be stopped, that we are always losing it, and that it was important to know the time to use it well doing good deeds. Muslims also needed to know when to pray at the right times each day. Mosques had to know the time so they could announce the call to prayer. Important annual events, such as when to fast in Ramadan, celebrate *Eid,* or go on pilgrimage to Mecca also had to be anticipated.

This inspiration meant that the "peerless devices" to which King Nasir al-Din referred included the Elephant Clock. As well as telling the time, this grand clock was a symbol of status, grandeur, and wealth; it also incorporated the first robotics with moving, time-telling figures.

Controlled Sinking of Perforated Bowl

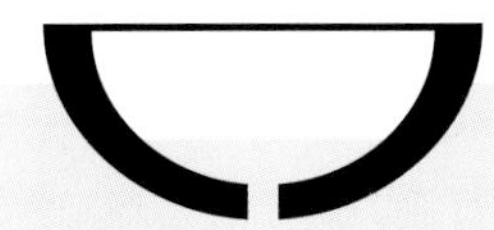

An Indian *ghatika-yantra*—as the bowl fills with water it sinks to the bottom of the tank after a preset time interval, depending on the weight and size of the bowl and size of the hole. As it hits the bottom, it makes a thud and alerts the timekeeper, who lifts it to start the process again.

AL-JAZARI'S MECHANICAL MARVEL

A Water-Powered Clock with a Multicultural Meaning

LEGACY: Al-Jazari's greatest legacy is the application of the "crank and connecting-rod" system crucial to pumps and engines.

LOCATION: Diyarbakir, modern-day Turkey

DATE: Early 13th century

INVENTOR: Al-Jazari, mechanical engineer

Without today's accurate digital and mechanical clocks, the pace of modern life would seem impossible. But more than 800 years ago, inventors were already developing sophisticated timekeeping devices to help keep track of when to pray, fast, celebrate Eid, or perform pilgrimage.

The Elephant Clock was a masterpiece that celebrated the diversity of humankind. Its moving parts were automated using an Indian-inspired water-powered timer. Combined with this were an Egyptian phoenix, Greek hydraulic technology, Chinese dragons, an Indian elephant, and mechanical figurines in Arabian dress. The clock cleverly reflected cultural and technological influences from across the world, from Spain to China.

The clock's inventor was the celebrated engineer Al-Jazari, who was from southern Turkey. He designed machines of all shapes and sizes, completing his *Book of Knowledge of Ingenious Mechanical Devices* in 1206. His crowning achievement was the combination of the crank wheel, connecting rod, and piston system, which converts rotational motion to linear motion—crucial to pumps, engines, and many other machines. But he was fascinated by every kind of mechanism, including in his book an automatic hand-washing machine and a mechanical musical band operated by a camshaft.

The Elephant Clock used as its timer a bowl that would slowly sink into a hidden water tank, a form of an Indian mechanism called *ghatika*. Many previous cultures had used water clocks—from the Egyptians to the Greeks, who marked passing time using the measurement of water flowing into or out of a bowl. Their devices were called clepsydras, meaning "water thieves."

Every half hour the timer would set off a series of dramatic sounds and movements. A ball would roll from the top of the clock and turn an hour dial, while the scribe and his pen would revolve automatically to show the minutes past the hour. As it dropped down, the ball triggered the elephant driver's mallet to strike a cymbal.

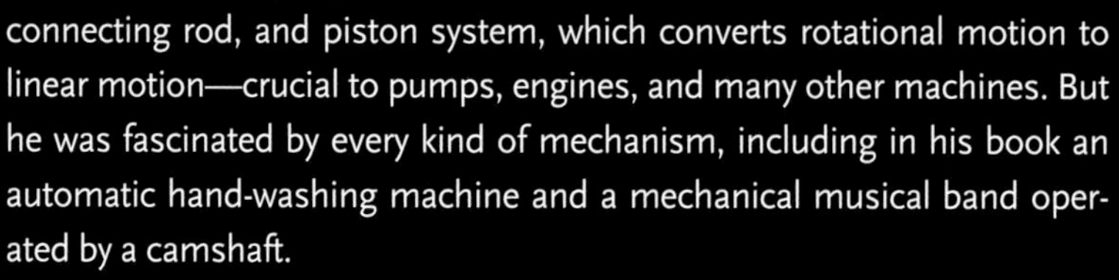

■ *Thirteenth-century engineer Al-Jazari designed his Elephant Clock to reflect influences from the nations and cultures across Muslim civilization.*

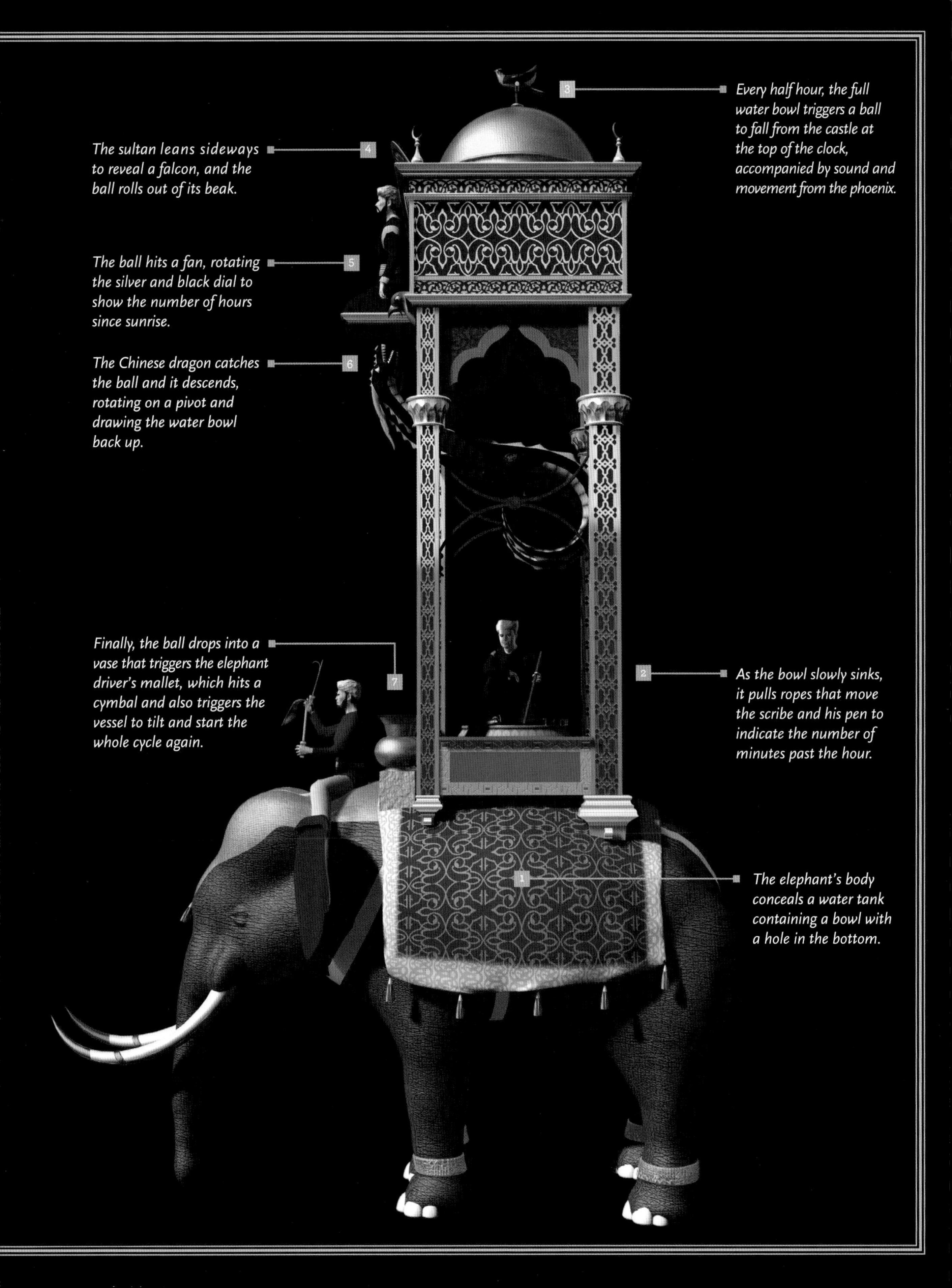
1
The elephant's body conceals a water tank containing a bowl with a hole in the bottom.
2
As the bowl slowly sinks, it pulls ropes that move the scribe and his pen to indicate the number of minutes past the hour.
3
Every half hour, the full water bowl triggers a ball to fall from the castle at the top of the clock, accompanied by sound and movement from the phoenix.
4
The sultan leans sideways to reveal a falcon, and the ball rolls out of its beak.
5
The ball hits a fan, rotating the silver and black dial to show the number of hours since sunrise.
6
The Chinese dragon catches the ball and it descends, rotating on a pivot and drawing the water bowl back up.
7
Finally, the ball drops into a vase that triggers the elephant driver's mallet, which hits a cymbal and also triggers the vessel to tilt and start the whole cycle again.

05 CHESS

Steam rises from the hot pools of Budapest's outdoor baths, hanging above gathering crowds as people crouch low over marble chessboards. In China, chessboards are laid out in the parks, as they are in Central Park, New York. Chess is a game of mental combat played by most nationalities on 64 squares with 32 pieces. Despite its size and unassuming appearance, the number of possible games that can be played is beyond counting.

The stories, figures, and individuals surrounding chess give it a mysterious dimension, and its definite origins remain unknown. It came from either India or Persia. In the 14th century, Ibn Khaldun connects chess to an Indian named Sassa ibn Dahir, an eminent man of wisdom.

There was an ancient Indian game called *Chaturanga,* which means "having four limbs," probably referring to the four branches of the Indian army of elephants, horsemen, chariots, and infantry. Chaturanga was not exactly chess but a precursor to the chess of today.

A 14th-century Persian manuscript describes how an Indian ambassador brought chess to the Persian court, from where it was taken to Europe by Arabs going to medieval Spain.

An illustration shows a Muslim and Christian playing chess in a tent, from King Alfonso X's 13th-century Libros del Ajedrez.

Before it reached Europe, the Persians modified the game into *Chatrang,* using it in their war games. Arabs came into contact with chess, or *Shatranj* as it was then called, in Persia and absorbed it into their culture.

At that time, the playing pieces were *Shah,* the king; *Firzan,* a general, who became the queen in modern times; *Fil* was an elephant that became the bishop; *Faras* was the horse; *Rukh* was a chariot that is now the castle or rook; and *Baidaq* was the foot soldier or pawn.

The game was very popular with the general public as well as the nobility, and the Abbasid caliphs particularly loved it. The great masters, though, were Al-Suli, Al-Razi, Al-Aadani, and Ibn al-Nadim. In the mid-20th century, Russian grand master Yuri Averbak played an astonishing move in one of his championship games, which he won. Many thought this to be an ingenious new idea, but it was actually devised more than a thousand years ago by Al-Suli.

Arab grand masters wrote copiously about chess, its laws, and its strategies, and these spread all over the Muslim world. There were

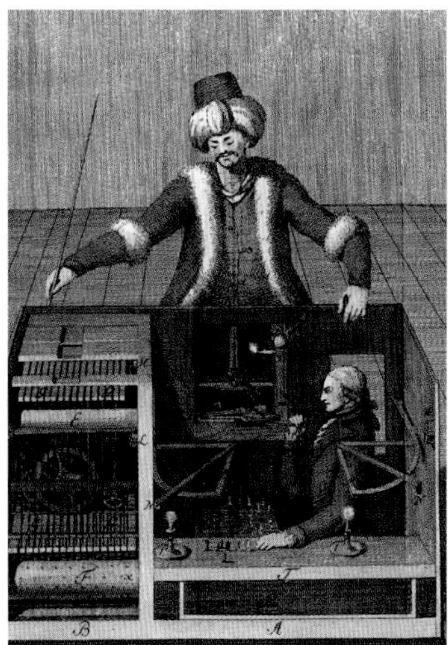

LEFT: *Abu Bakr al-Suli's* Muntahab Kitab al-Shatranj *depicts an early tenth-century chess table miniature. The Arabic reads, "The black is winning and it is his turn to play." Scholars are not sure whether this was a game through correspondence or an instruction manual of how to play.* CENTER AND RIGHT: *Kempelen's Iron Muslim "robot" of 1769 had a chess master inside the cabinet, who played skillfully and beat other master players of the day.*

books on chess history, openings, endings, and problems. *Book of the Examples of Warfare in the Game of Chess,* written around 1370, introduced the chess game "The Blind Abbess and Her Nuns" for the first time.

The whirlwind Ziryab, a great musician and trendsetter, brought chess to Andalusia in the early ninth century. The word "checkmate" is Persian in origin and a corruption of *Shahmat,* meaning "the king is defeated."

From Andalusia, the game spread among Christian Spaniards and the Mozarabs, and reached northern Spain as far as the Pyrenees, crossing the mountains into southern France. The first European records to mention chess go back to 1058, when the will of Countess Ermessind of Barcelona donated her crystal chess pieces to St. Giles monastery at Nimes. A couple of years later, Cardinal Damiani of Ostia wrote to Pope Gregory VII, urging him to ban the "game of the infidels" from spreading among the clergy.

Chess was also carried via the trade routes from central Asia to the southern steppes of early Russia: Seventh- and eighth-century Persian chess pieces have been found in Samarkand and Farghana. By 1000, chess had spread even farther on the Viking trade routes as the Vikings carried it back to Scandinavia. Those trade routes meant that by the 11th century, chess had made its way as far as Iceland, and an Icelandic saga written in 1155 talks of the Danish king, Knut the Great, playing the game in 1027.

By the 14th century, chess was widely known in Europe, and King Alfonso X, known as "the Wise," had produced the *Book of Chess and Other Games* in the 13th century. For the last eight centuries, chess has gone from strength to strength, producing a few funny stories along the way, such as the robotic chess master of 1769.

Hungarian Wolfgang de Kempelen decided to give a gift to his queen, Empress Maria Theresa, who was a chess fanatic. He gave her a robot machine called the Iron Muslim, later renamed Ottoman Turk, which played chess skillfully, beating high-ranked players of the day. Inside, all cramped up, was a chess master. People traveled miles to marvel at the turban-wearing robot. In fact, 15 separate chess players inhabited it for 85 years, in the guise of an Ottoman "robotic" Turk.

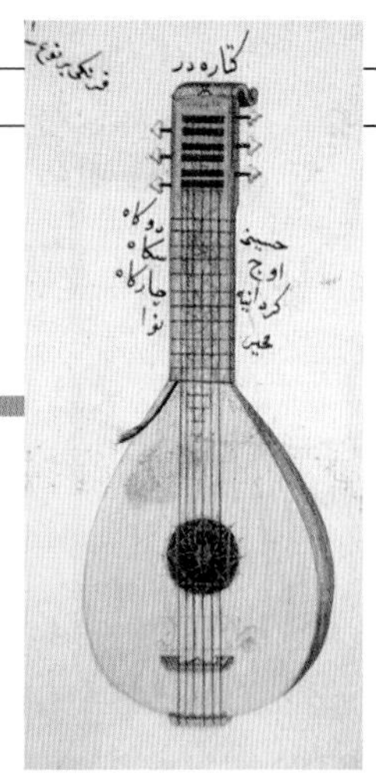

06 MUSIC

Music crosses continents, cultures, people, and nature. Like language, it enables us to communicate. But do 20th-century artists and singers know that much of their craft lies in the hands of ninth-century Muslims of the Middle East? These artists, particularly Al-Kindi, used musical notation, the system of writing music. They also named the notes of a musical scale with syllables instead of letters, called solmization. These syllables make up the basic scale in music today. We are all familiar with do, re, mi, fa, sol, la, ti. The Arabic alphabet for these notes is dal, ra, mim, fa, sad, lam, sin. The phonetic similarity between today's scale and the Arabic alphabet used in the ninth century is striking.

About 70 years after Al-Kindi, Al-Farabi developed the *rababah,* an ancestor of the violin family, and *qanun,* a table zither. He wrote five books on music, but *The Great Book of Music* on the theory of music was his masterpiece. In the 12th century, it was translated into Hebrew, and then into Latin. The influence of Al-Farabi and his book continued up to the 16th century.

Roving musicians, merchants, and travelers all helped Arabic music on its way into Europe, and this helped to shape the cultural and artistic life of Spain and Portugal under the 800 years of Muslim rule. One of the earliest examples of this is found in the collection of *Cantigas de Santa María.* Composed around 1252 upon the orders of Alfonso X el Sabio, king of Castilla and Aragón, this collection consists of 415 religious songs about the Virgin Mary.

"Arabs, when they came to Europe, in the beginning of the eighth century, were more advanced in the cultivation of music, . . . in the construction of musical instruments, than were European nations, thus only can their astounding musical influence be accounted for."

CARL ENGEL, A 20TH-CENTURY HISTORY OF MUSIC SCHOLAR

Many individuals also played a part in the spreading of new music into Europe. Appreciable influence lies with one man, Ziryab, known as "the Blackbird" because of his melodious voice

TOP: *An 18th-century manuscript on musical composition and rhythm shows a* qitara, *or guitar, from* Tafhim al-maqamat *by Kamani Khidir Aga.*
BOTTOM: *Traditional musicians perform in Morocco.*

Two musicians are depicted in Alfonso X's 13th-century Cantigas de Santa María.

and dark complexion. He was a gifted pupil of a renowned Baghdad musician, but his talent and excellence in music slowly overtook his teacher's, so the Umayyad caliph invited him to Andalusia.

Ziryab settled in the court of Córdoba in 822, which was under the Caliph 'Abd al-Rahman II, the son of the Umayyad caliph. Here, Ziryab found prosperity and recognition of his art, becoming the court entertainer with a monthly salary of 200 golden dinars in addition to many privileges.

His accomplishments are many, including establishing the world's first conservatoire in Córdoba; teaching harmony and composition; introducing the Arab lute *(al-'ud)* to Europe and adding the fifth bass string to it; replacing the wooden plectrum with a quill feather from a vulture; and rearranging musical theory completely by setting free metrical and rhythmical parameters.

Henri Terrasse, the French 20th-century historian, said, "After the arrival of this oriental [Ziryab], a wind of pleasure and luxurious life blew through Córdoba. An atmosphere filled with poetry and exquisite delight surrounded Ziryab; he composed his songs at night in the company of two servants who played the lute. He gave his art an unprecedented value."

MILITARY MUSICAL BANDS

The Ottoman State was the first Euro-Asian state to have a permanent military musical band. Founded in 1299, the famous Mehterhane military band followed the sultan on his expeditions. It would arrive in the middle of battles to rouse the spirit of the soldiers while also terrifying the enemy. The Janissary, an elite army, also had a band of six to nine members with instruments like drums *(zurna)*, clarinets, triangles, cymbals *(zil)*, and kettledrums of war *(kös* and *naqqara)*. These were carried on the backs of camels.

Europeans met the Janissary bands in peace and war. On various ambassadorial receptions it became fashionable to have Ottoman-Turkish instruments, the "Turquerie" fashion, in Europe. The Janissaries were defeated at the gates of Vienna in 1683 and left behind their musical instruments. This was an event that led to the rise of European military bands. Even Napoleon Bonaparte's French military bands were equipped with Ottoman war musical instruments such as zil and the kettledrums. It is said that Napoleon's success in the battle of Austerlitz (1805) was due in part to the psychological impact of the noise of his fanfares.

A Janissary band of the Ottoman State

07 CLEANLINESS

Medieval times are often imagined as smelly, dark, rough, and unclean. In the tenth-century Islamic world though, the hygiene practices and products found in bathroom cabinets could compete with those we have today.

Cleanliness is vital in Islam, with ablutions known as *wudhu* carried out before prayers. In the 13th century there was an outstanding mechanical engineer called Al-Jazari who wrote a book called *The Book of Knowledge of Ingenious Mechanical Devices,* including robotic wudhu machines. One elaborate machine, resembling a peacock, would be brought to each guest, who would tap the peacock's head to make water pour in eight short spurts, providing just enough for ablution. Some machines could even hand you a towel.

Muslims made soap by mixing oil (usually olive oil) with *al-qali* (a saltlike substance). According to manuscripts, this was boiled to achieve the right consistency, left to harden, and used in the *hammams* or bathhouses.

"Allah is Beautiful and He loves beauty."

PROPHET MUHAMMAD
NARRATED BY MUSLIM (NO. 131)

An illustration from a manuscript shows Al-Jazari's wudhu, *or washing, machine.*

Medieval Muslims also went to great lengths with their appearance. One expert was Al-Zahrawi, a famous physician and surgeon about whom you can read more in the Hospital chapter. He included in his medical book, called *Al-Tasrif,* a chapter devoted completely to cosmetics, called *The Medicines of Beauty.*

He described the care and beautification of hair and skin, teeth whitening, and gum strengthening, all within the boundaries of Islam. He included nasal sprays, mouthwashes, and hand creams, and talked of perfumed sticks, rolled and pressed in special molds, a bit like today's roll-on deodorants. He also named medicated cosmetics like hair-removing sticks, as well as hair dyes that turned blond hair to black and lotions for straightening kinky or curly hair. The benefits of suntan lotions were also discussed as were their ingredients, all amazing considering this was a thousand years ago.

> *"When you rise up for prayer, wash your face, and your hands up to the elbows, and lightly rub your heads and [wash] your feet up to the ankles."*
>
> QURAN (5:6)

Born in Kufa, now in Iraq, Al-Kindi also wrote a book on perfumes called *Book of the Chemistry of Perfume and Distillations*. His book contained more than a hundred recipes for fragrant oils, salves, aromatic waters, and substitutes or imitations of costly drugs. Initially, the more affluent in society used these; later they became accessible to all.

Muslim chemists also distilled plants and flowers, making perfumes and substances for therapeutic pharmacy.

These processes and ideas of the Muslims filtered into Europe via merchants, travelers, and the Crusaders. The BBC documentary *What the Ancients Did for Us: The Islamic World* said that the knowledge of the Muslims eventually arrived at Haute Provence in the south of France, which has the perfect climate and the right kind of soil for the perfume industry, which still flourishes after 700 years.

Another important cosmetic in Islam is henna, known for its beautiful, intricate designs on elegant hands. With the spread of Islam, it reached different parts of the world, becoming an essential cosmetic ingredient.

Prophet Muhammad and his companions dyed their beards, while women decorated their hands and feet and also dyed their hair like women of today. Modern scientists have found henna to be antibacterial, antifungal, and anti-hemorrhagic. It is useful in healing athlete's foot, fungal skin infections, and local inflammation. The leaves and seeds of the plant possess medicinal properties, and both act as cooling agents for the head and body. Henna also contains natural ingredients that are used for hair nourishment.

SAKE DEAN MAHOMED

Sake Dean Mahomed opened the Indian Vapor Baths on the Brighton seafront in 1759.

In the 1770s and 1780s, Brighton, England, was a blossoming beach resort and it was at this scene that Sake (Sheikh, but because of accents this became Sake) Dean Mahomed arrived.

Sake Dean Mahomed was from a Muslim family in Patna, India, and in 1759 opened what was known as Mahomed's Indian Vapour Baths on the Brighton seafront, the site of what is now the Queen's Hotel. These baths were similar to Turkish baths, but clients were placed in a flannel tent and received an Indian treatment of *champi* (shampooing) or therapeutic massage from a person whose hands came through slits in the flannel. This remarkable "vaporing" and shampooing bath led him to receive the ultimate accolade of being appointed "Shampooing Surgeon" to both George IV and William IV.

08 Trick Devices

Maybe you can hear the click-clack of the metal balls swinging on wires as they knock each other rhythmically while you fiddle with a Rubik's Cube. Games and puzzles, whether for business or leisure, are a source of fascination to many.

This sense of human enjoyment was captured by three brothers in the ninth century. Muhammad ibn Musa ibn Shakir, Ahmed ibn Musa ibn Shakir, and Al-Hasan ibn Musa ibn Shakir were known as the Banu Musa brothers. They were part of the famous House of Wisdom, the intellectual academy of Baghdad in the ninth century, which you can read more about in the School chapter. As well as being great mathematicians and translators of Greek scientific treatises, they also invented fabulous trick devices, which, some would say, are a precursor to executive toys. The brothers fed their peers' obsession by designing and making trick inventions, and their *Book of Ingenious Devices* lists more than a hundred of them. These were the beginnings of mechanical technology.

Like toys today, they had little practical function, but these 1,100-year-old mechanisms displayed amazing craftsmanship and knowledge.

A drawing explains the Banu Musa brothers' ninth-century "Drinking Bull" robot.

Many of the mechanisms involved water, fake animals, and sounds. For example, the drinking bull made a sound of contentment when it finished, as if its thirst had been satisfied. It did this using a series of filling chambers, floats, vacuums, and plugs.

See if you can follow the Banu Musa brothers' thinking on the diagram below.

Initially water comes from the tap into compartment A and then it is closed off. The bowl is then filled with water, too. The float m (seen in the diagram opposite) rises with the level of water, pulling the plug out of the valve. Water drains from compartment A into compartment B. Float B rises with the water, pushing up plug B and allowing water to flow between the two compartments. When the air in compartment B is fully evacuated, a vacuum forms in compartment A since no air is allowed to flow into it. Water from the bowl is then drawn through the pipe and into A. Once all the water is gone from the bowl, air is sucked in so it appears that the

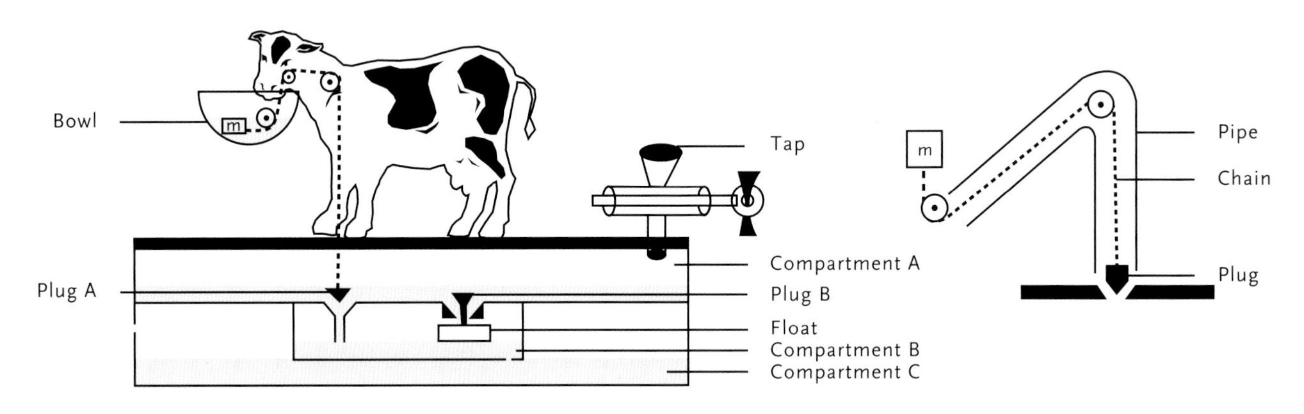

bull is making a sound of contentment. Since no water is left in the bowl to keep the plug afloat, that particular plug closes, so only plug B is open to empty compartment A. Compartment B empties via a small hole between B and C. Air is allowed to flow freely from a hole on the side of compartment C.

Highly complex and mind-bending, this must have kept people enthralled for hours.

Another of the Banu Musa brothers' trick devices was a flask with two spouts. Different-colored liquids were poured into each spout, but when it was time to pour, the "wrong" color came out of the "wrong" spout. Like the magician who can make orange juice come out of his elbow, the brothers had an even better, and simpler, intricate mechanism up their sleeves.

What they had done was to divide the jar into two vertically, with each section totally separate from the other. Liquid came into the right side from the right funnel and into the left side by the left funnel, but it could not leave this way. Instead, another pipe had been inserted for the outflow. Of course, people observing could not see any of this, and although it was simple, the trick still had impact and amazed them. The brothers' imagination for fun also moved into designing fountains. Take a look at the Fountains section in the Town chapter.

"A joke is not a thing but a process, a trick you play on the listener's mind. You start him off toward a plausible goal, and then by a sudden twist you land him nowhere at all or just where he didn't expect to go."

MAX EASTMAN, AMERICAN WRITER

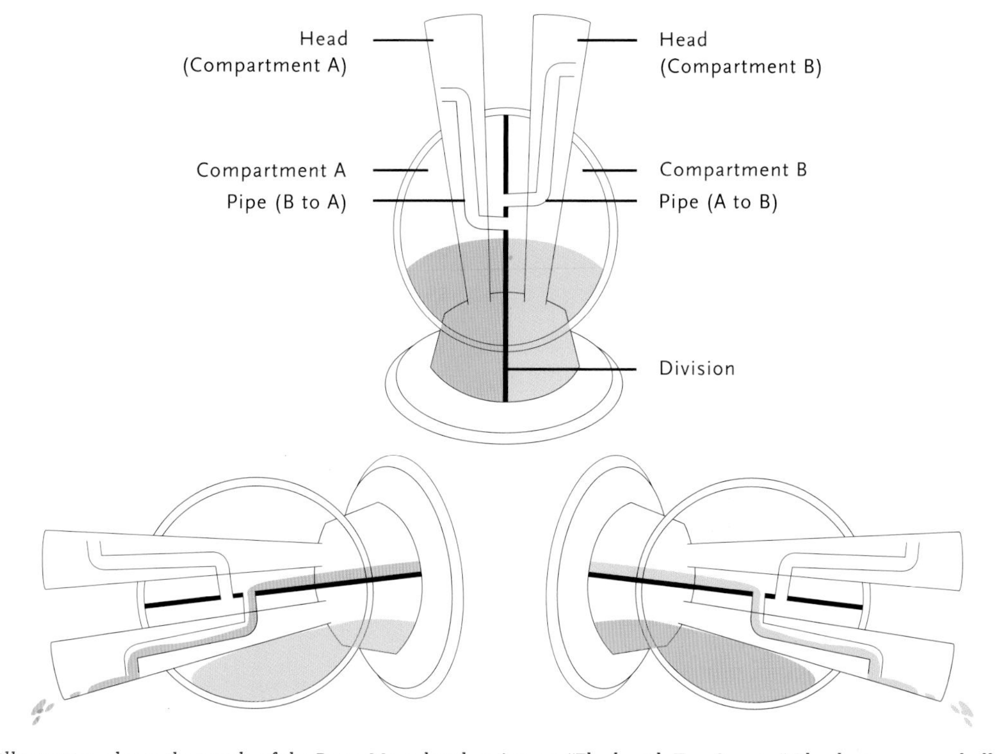

An illustration shows the inside of the Banu Musa brothers' game "Flask with Two Spouts." The device amazed all those who observed the trick developed by the inventive brothers.

09 VISION AND CAMERAS

As a child did you ever wonder how we see? Did you think that if you shut your eyes and you could not see anyone, then no one could see you? Some ancient Greek scholars had less than conventional ideas of sight as well, and the first understanding of optics consisted of two main theories.

The first maintained that rays came out from our eyes, a bit like laser technology today, and these rays were cut off by the objects in our vision. So, sight was carried out through the movement of the rays from the eye to the object.

The second idea said that we see because something is entering our eye that represents the object. Aristotle, Galen, and their followers believed in this model, but their theories were speculation and not backed up by experiments.

Ninth-century polymath Al-Kindi first laid down the foundations of modern-day optics by questioning the Greek theories of vision. He said that how we see, our visual cone, is not formed of discrete rays as Euclid had said, but appeared as a volume, in three dimensions, of continuous radiations.

Sixteenth-century Italian physician and mathematician Geronimo Cardano said Al-Kindi was "one of the twelve giant minds of history" because he discussed sight with and without a mirror, how light rays came in a straight line, and the effect of distance and angle on sight—including optical illusions. Al-Kindi wrote two treatises on geometrical and physiological optics, which were used by English scholar Roger Bacon and German physicist Witelo during the 13th century. According to Sebastian Vogel, a 20th-century Danish scholar, "Roger Bacon not merely counted al-Kindi one of the masters of perspective but in his own *Perspectiva* he and others in his field referred repeatedly to al-Kindi's optics."

The questioning originally begun by Al-Kindi was built upon by Al-Hasan Ibn al-Haytham in the tenth century, who eventually explained that vision was made possible because of the refraction of light rays. Distinguished 20th-century science historian George Sarton said that the leap forward made in optical science was due to Ibn al-Haytham's work, which scientifically explained much of what we know today about optics.

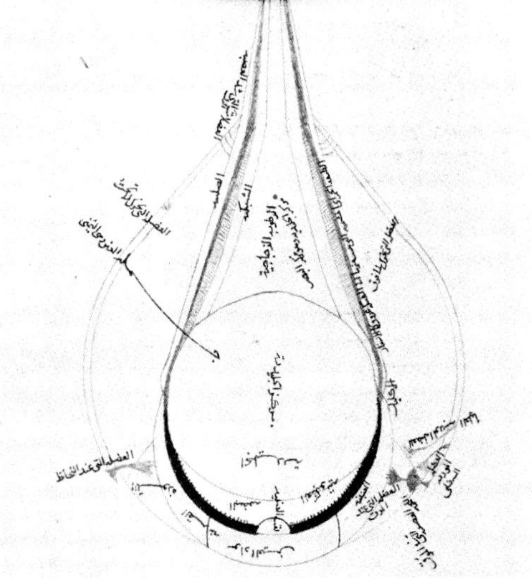

The anatomy of the eye by 13th-century Kamal al-Din al-Farisi is based on Ibn al-Haytham's ideas. The Arabic text refers to the role of the brain in interpreting the image on the retina of the eye.

In fact, a tenth-century physicist, Ibn Sahl from Baghdad, had worked on light refraction by lenses before Ibn al-Haytham, although we are not sure that Ibn al-Haytham knew of Ibn Sahl's work. Al-Hasan Ibn al-Haytham, usually called just Ibn al-Haytham and also known in the West as Alhazen, carried out meticulous experiments a thousand years ago, which enabled him to provide the scientific explanation that vision was caused by light reflecting off an object and entering the eye. Alhazen was the first to totally reject the theories of the Greeks.

"He, Ibn al-Haytham, was the greatest Muslim physicist and student of optics of all times. Whether it be in England or far away Persia, all drank from the same fountain. He exerted a great influence on European thought from Bacon to Kepler."

GEORGE SARTON IN HIS *HISTORY OF SCIENCE*

Born in Basra, Iraq, he moved to Egypt on the invitation of its ruler to help reduce the effects of the Nile's flooding, and was the first to combine the "mathematical" approach of Euclid and Ptolemy with the "physical" principle favored by natural philosophers. Ibn al-Haytham said, "The knowledge of optics demands a combination of physical and mathematical study."

He was also a mathematician, astronomer, physician, and chemist, but it is his *Book of Optics* that has formed the foundations for the science of optics. Generally known as *Magnum Opus,* it discussed the nature of light, the physiology and mechanism of sight, the structure and anatomy of the eye, reflection and refraction, and catoptrics.

This frontispiece is from the 1572 Latin edition of Book of Optics *by Alhazen, the Western name for Al-Hasan Ibn al-Haytham.*

Ibn al-Haytham also studied lenses, experimenting with different mirrors: flat, spherical, parabolic, cylindrical, concave, and convex. He also treated the eye as a dioptric system, by applying the geometry of refraction to it. He brilliantly investigated the phenomenon of atmospheric refraction, calculating the height of the atmosphere to be 16 kilometers (10 miles). This compares well with modern measurements of the troposphere, the lowest layer of the atmosphere, which measures 11 kilometers (7 miles).

Ibn al-Haytham used experimental evidence to check his theories. This was unusual for his time because physics had until then been more like philosophy, without experiment. He was the first to introduce experimental evidence as a requirement for accepting a theory, and his *Book of Optics* was actually a critique of Ptolemy's book *Almagest*. A thousand years later, this book of optics is still quoted by professors training research students to be factual. Some science historians believe that Snell's law, in optics, actually comes from the work of Ibn Sahl.

A NEW VIEW OF VISION

How One Man's Imprisonment Led to the Discovery of How We See

LEGACY: **Ibn al-Haytham overturned the ancient idea that our eyes emit rays to make objects visible**

LOCATION: **Cairo, Egypt**

DATE: **11th century**

KEY FIGURE: **Ibn al-Haytham**

How do we see? Arguments over this question went on between ancient Greek scholars for centuries. Some said rays came out of our eyes, while others thought something entered the eyes to represent an object.

Ibn al-Haytham was born in Basra, Iraq, in 965. In his lifetime, he completely changed our understanding of light and vision—although he paid a high price for his scientific honesty.

The story goes that Ibn al-Haytham was summoned to Cairo by the ruling caliph, to see if he could help to control the Nile's unpredictable floods. Confident of his own abilities, Ibn al-Haytham boasted that he would tame the great river by building a dam and reservoir. But when he saw the extent of the challenge he had taken on, he realized it was impossible.

To avoid the caliph's wrath at his failure, Ibn al-Haytham decided to pretend to be mad, and so, for his own protection, the caliph placed him under house arrest. It was while imprisoned that Ibn al-Haytham made the discovery for which he is best remembered.

One day, he saw light shining through a pinhole into his darkened room, projecting an image of the world outside onto the opposite wall. At the time, people believed the ancient view that our eyes see by sending out invisible rays. But Ibn al-Haytham realized that we see because light rays emitted from visible objects actually enter our eyes.

By experimenting with a dark room "camera obscura" he showed that the light enters our eyes in the shape of a geometric cone of vision. He also conducted experiments with lenses and mirrors. He laid out his new ideas about light and vision in the *Book of Optics,* written between 1011 and 1021.

Ibn al-Haytham realized that the senses were prone to error, and he devised scientific methods of verification, testing, and experimentation to uncover the truth about the natural phenomena he saw. Like many people before and since, Ibn al-Haytham wondered why the moon looks so big when it is low in the sky. Previously, scholars had thought it was an effect caused by the atmosphere. Ibn al-Haytham, however, was the first to state correctly that it is just an optical illusion—and scientists still are not quite sure why this happens.

Later, Kamal al-Din al-Farisi, a Persian mathematician born around 1260, built on Ibn al-Haytham's work. In an attempt to explain the colors of the rainbow, Al-Farisi experimented with a glass sphere filled with water as a model of a raindrop to show that sunlight is bent twice through a water droplet.

Ibn al-Haytham's search for evidence set the scene for the development of experimental science and the rational approach of later scholars.

■ *Ibn al-Haytham laid the foundation for optical devices like cameras, such as this one that dates from the mid-20th century.*

"Light issues in all directions opposite any body that is illuminated with any light [and of course, also opposite any self-luminous body]. Therefore when the eye is opposite a visible object and the object is illuminated with light of any sort, light comes to the surface of the eye from the light of the visible object."

11TH-CENTURY IBN AL-HAYTHAM FROM HIS *BOOK OF OPTICS*

Light shines through a tiny hole into a darkened room, projecting an image onto the opposite wall in this artist's re-creation of Ibn al-Haytham's camera obscura.

Many of his works, including the huge *Book of Optics,* were translated into Latin, and among those influenced by his work and methods were Roger Bacon, Leonardo da Vinci, and Johannes Kepler. Today, many modern inventions rely on the accurate understanding of light and vision, the foundations of which Ibn al-Haytham laid more than a thousand years ago.

10 FASHION AND STYLE

Fashions come and go, but good taste never goes out of style. So it may not be surprising to find out that many present European styles and ideas of dressing arrived 1,200 years ago when Spain was part of the Islamic world.

Ziryab, the musician and etiquette teacher, was also a trendsetter and style icon in ninth-century Córdoba, Spain. "He brought with him all the fashion. Baghdad was the Paris or New York of its day and . . . you have this influx of ideas from Baghdad to Córdoba, so he brought with him toothpaste and deodorant, and short hair. . . . This is the thing; Córdoba had street lighting, sewage works, running water," said author Jason Webster about Ziryab when speaking with Rageh Omaar on the BBC's *An Islamic History of Europe.*

Baghdad, in Iraq, was a great cultural and intellectual center of the Islamic world, from where Ziryab also brought new tableware, new sartorial fashions, and even the games of chess and polo. He was renowned as an eclectic man with good taste and his name was connected with elegance. With his refined and luxurious ways, he defined the court of the caliphs while the average Cordoban imitated his hairstyle, the new short look, and enjoyed the leather furniture he brought to Spain.

Henri Terrasse, the French historian, said 1,200 years later of Ziryab, "He introduced winter and summer dresses, setting exactly the dates when each fashion was to be worn. He also added dresses of half season for intervals between seasons. Through him, luxurious dresses of the Orient were introduced in Spain. Under his influence a fashion industry was set up, producing colored striped fabric and coats of transparent fabric, which is still found in Morocco today."

Ziryab's achievements gained him the respect of successive generations, even up to the present day. In the Muslim world, there is not a single country that does not have a street, a hotel, a club, or a café named after him. In the West, scholars and musicians still pay him tribute.

Muslims, especially in Andalusia, developed a sophisticated lifestyle pattern that was based on seasonal influences. The choice to eat particular

Bolts of cloth run the gamut of colors.

foods and wear certain types of clothing and material was crucial in providing comfort and well-being. In clothing terms, winter costumes were made essentially from warm cotton or wool items, usually in dark colors. Summer costumes were made of light materials such as cotton, silk, and flax that came in light and brilliant colors from local dye works.

> "Beauty of style and harmony and grace and good rhythm depend on simplicity."
>
> PLATO, GREEK PHILOSOPHER

Andalusian Muslims were also heirs to a number of oak-based industries developed by the Romans, including the making of cork-soled shoes. They intensified and diversified the production technique and cork-soled shoes became widespread in the country, and a staple of the export trade. The shoe was called *qurq*, the plural is *aqraq*, which subsequently returned to Castilian in the form *alcorque*. The artisan who made the product was a *qarraq*. Such an artisan was 'Abdullah, a Sufi mystic sandal maker of Seville, mentioned by Ibn 'Arabi. Artisans of this trade had living quarters called *qarraqin*, now Caraquin, in Granada.

Two medieval Muslim writers, Al-Saqati and Ibn 'Abdun, provide detailed specifications of the making of cork-soled shoes, notably that the leather stitched to the back should not be skimpy, and that leather should be sewn to leather, with no filler inserted in between. Some shoemakers put sand below the heel to make it higher, causing it to break when worn. The more sophisticated styles and methods were then adopted by Christians after the conquest of Al-Andalus.

An early 17th-century manuscript titled Album of the Sultan Ahmed I *by Kalandar Pasha depicts typical costumes of the time.*

So the next time you are out shopping for the latest fashions in fancy designer shops, remember the high heels of a thousand years ago. When you try on a light pair of summer trousers or a dress remember Ziryab, "the Blackbird" from 1,200 years ago, because this was the time such ideas were flying from Spain, Sicily, and the Middle East into Christian Europe.

11 CARPETS

Carpets were first made long before Islam by the Bedouin tribes of Arabia, Persia, and Anatolia. They used carpets as tents, sheltering them from sandstorms; floor coverings providing great comfort for the household; wall curtains providing privacy; and for items such as blankets, bags, and saddles.

For Muslims, carpets are held in special esteem and admired for being part of Paradise. Inspired by this, and new tinctures for tanning and textiles, they developed both the design and weaving technique, so that their carpets came in wonderful colors. A Tunisian scientist called Ibn Badis in the 11th century carried out pioneering work on inks and the coloring of dyes and mixtures to produce his *Staff of the Scribes*.

As well as colorful, Muslim carpets were renowned for their quality and rich geometric patterns of stars, octagons, triangles, and rosettes, all arranged around a large central medallion. Arabesque and floral patterns filled the areas around these shapes, pulling them all together with a sense of unity.

In Europe carpets caught on quickly and became status symbols. England's King Henry VIII (ruled 1509–1547) is known to have owned more than 400 Muslim carpets, and a portrait made of him in 1537 shows him standing on a Turkish carpet with its *Ushak* star. Muslim designs also decorate his robe and curtains.

But the earliest English contact with Muslim carpets was when the grandson of William the Conqueror, who lived in the Abbey of Cluny, gave a carpet to an English church in the 12th century. At this same time, Muslim geographer and philosopher Al-Idrisi said that woolen carpets were

Carpets are draped over camels on long journeys to add some comfort for the rider. They also serve as saddlebags for storing provisions.

Muslim carpets are known for their rich colors and geometric patterns.

produced in Chinchilla and Murcia, both now in Spain, and were exported all over the world.

Paintings made in the late medieval period show us how and where carpets were used and what people thought of them. In 14th- and 15th-century Europe, they were first used in Christian religious paintings. The 17th century saw decorative carpets covering tabletops and their bases. Cupboard and window carpets also made an appearance.

Belgian artists were also inspired. Van Eyck's painting "Virgin and Child with St. Donatian, St. George, and Canon Van der Paele," which he painted in 1436 at Bruges, shows the Virgin Mary seated on a carpet with geometrical shapes, mainly circles, drawn around rosettes combined with lozenges and eight-pointed star motifs.

Muslim carpets were so highly prized that a Victoria and Albert Museum publication quotes a chapter in 16th-century Hakluyt's *Voyages*, which talks of a plan to import Persian carpet makers into England. It says: "In Persia you shall find carpets of

"In Persia you shall find carpets . . . the best of the world."

RICHARD HAKLUYT, WRITER

course thrummed wool, the best of the world, and excellently coloured: those cities and towns you must repair to, and you must use means to learn all the order of the dyeing of those thrums, which are so dyed as neither rain, wine, nor yet vinegar can stain. . . . If before you return you could procure a singular good workman in the art of Turkish carpet making, you should bring the art into the Realm and also thereby increase work to your company."

Besides the Ottoman/Turkish carpet, no other carpet reached the status and popularity of the Persian carpet, which became a state enterprise in the Safavids' reign. These rulers developed trade relations with Europe under Shah Abbas I (1587-1629), and their export and the silk trade became the main sources of income and wealth for the Safavid state.

Carpet making was a huge industry, and manufacturers received orders from across Europe. Persian craftsmen from Tabriz, Kashan, Isfahan, and Kerman produced eye-dazzling and mesmerizing designs.

But by the early 19th century the carpet industry started to decline, partly due to historical events and conflicts, which lost Persia its stability and security, but also because Europeans had begun manufacturing their own carpets in the 18th century.

The first production of imitated Muslim carpets in Europe was under English patrons. The Royal Society of Arts promoted the establishment of successful carpet manufacturing "on the Principle of Turkish Carpets" through subsidies and awards. Between 1757 and 1759, the society gave £150 as awards for the best "imitated" Turkish carpets.

Today, the fame of the flying carpet of 'Al'a al-Din continues in films and stories, and Berber carpets from North Africa are once again increasing in popularity.

CHAPTER THREE

"What is learnt in youth is carved in stone."

ARABIC PROVERB

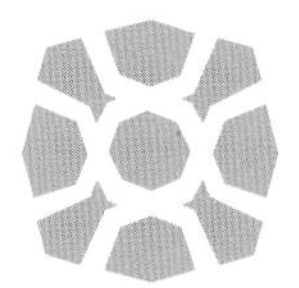

SCHOOL

SCHOOLS • UNIVERSITIES • HOUSE OF WISDOM • LIBRARIES AND BOOKSHOPS • TRANSLATING KNOWLEDGE • MATHEMATICS TRIGONOMETRY • CHEMISTRY • COMMERCIAL CHEMISTRY • GEOMETRY ART AND THE ARABESQUE • THE SCRIBE • WORD POWER

THIS CHAPTER REVEALS THAT WHICH MAKES ALL CIVILIZATIONS GREAT: EDUCATION, learning, and applying knowledge for a practical outcome such as bettering society. School is one of the institutions we learn in, and the medieval Muslims excelled in learning, from the primary-level mosque schools through to universities and the illustrious House of Wisdom, an intellectual academy in ninth-century Baghdad.

The ethos of learning was a culture where inquiring minds searched for truth based on scientific rigor and experimentation, where opinion and speculation were cast out as unworthy pupils. This system of learning embodied by medieval Islam formed the foundation from which came exceptional inventions and discoveries.

Here you will discover how a thousand-year-old chemical process helped form many of today's global industries; where the word "chairperson" came from; the origins of admiral, tabby, and sofa; as well as read that the 18th-century story of Robinson Crusoe was not the first tale of a person stranded on an island.

OPPOSITE: *Layla and Majnun, characters in a classical Islamic love story, are shown at school in a 15th-century Persian manuscript.*

01 SCHOOLS

After spending our formative years in schools, we have favorite teachers, hated subjects, and a bagful of memories from sports days to difficult exams. Our lives are molded by timetables until finally we emerge with a head full of some kind of knowledge.

In Muslim countries a thousand years ago, the school was the mosque. There was little distinction between religion and knowledge, as the mosque was both the place of prayer and the place of learning. Subjects included science, so religion and science sat side by side comfortably, which has not always been the case in other parts of the world. According to Danish historian Johannes Pedersen, learning "was intimately bound up with religion . . . to devote oneself to both, afforded . . . inner satisfaction and . . . service to God . . . it not only made men of letters willing to accept deprivation . . . it prompted others to lend them aid."

Prophet Muhammad made the mosque the main place of learning, traveling between them, teaching, and supervising schooling. Anywhere a mosque was established, basic instruction began. He also sent teachers of the Quran to the Arab tribes and they were known as *Ahl al-'ilm,* or "the people with knowledge." This meant that education spread everywhere and these traveling teachers lived lives of great contentment. In Palermo, Ibn Hawqal, a tenth-century geographer, merchant, and traveler, claimed to have counted about 300 primary teachers.

Young students attend an elementary Islamic school in Laem Pho, Thailand.

At the time of Prophet Muhammad in the seventh century, there were nine mosques in Medina, which is now in Saudi Arabia. The first school appeared here in 622 and the idea of schooling spread, so another sprang up in Damascus, Syria, in 744. Eighth-century Córdoba, Spain, had schools, and by the late ninth century nearly every mosque had an elementary school for the education of both boys and girls.

"Stand up for your teacher and honor him with praise. For the teacher is almost a prophet. Did you see greater or more honorable than he who creates, fosters, and develops personalities and brains?"

A VERSE FROM A POEM BY AHMED SHAWKI

At the age of six nearly all boys, except the rich (who had private tutors), and some girls began elementary school. Tuition was normally free or so inexpensive that it was accessible to all. One of the first lessons in writing was to learn how to write the 99 most beautiful names of God and simple verses from the Quran. This was followed by a thorough study of the Quran and instruction in arithmetic.

By the tenth century, teaching was moving away from the mosque and into the teacher's house, which meant that gradually schools developed; this occurred in Persia first. Then, by 1066, the Seljuks built the Nizamiyah school, named after its founder, Vizier Nizam al-Mulk of Baghdad. This was the first proper school that had a separate teaching building. However, schools were established and salaries were designated for teachers in the early days of Islam.

Like many Muslim buildings, schools were constructed with no expense spared, and beauty was

FOUR LEVELS OF SCHOOLING

As schools developed, they could be divided into four categories depending on what they taught and at what level. These were regular schools, high schools or houses of readers, houses of hadiths, and medical schools.

Regular schools taught general subjects and were the equivalent of primary schools. Students left with a primary level certificate so they could enter high schools. Regular schools were the most common, found in every village.

Dar-al-Qurra', or the house of readers, was a high school that taught proficiency in Arabic, and in reading and reciting the Quran. The school also trained imams and muezzins.

Dar-al-Hadith, or the house of hadith, specialized in teaching and researching the hadith, the sayings of Prophet Muhammad. Graduates from this school were awarded the equivalent to a university degree so they could work at religious institutions.

The first school dedicated solely to the teaching of medicine was founded in Damascus in 1231. Before this, medical teaching was carried out in hospitals and through apprenticeships. Medical schools were not widely established until the 16th century, during the reign of Suleyman the Magnificent, the Ottoman sultan.

A school in Baghdad, Iraq, in 1890

an important consideration. Each had a courtyard with one, two, three, or four *iwans* (large arched halls directly open to the courtyard), which were used for lessons, as well as a prayer hall, living accommodation (individual rooms), and an ablution complex. The state or ruling caliph exercised some supervision over teaching, and teachers had to have permission before they could teach.

A 14th-century Muslim educator Ibn al-Hajj had much to say about schools: "The schools should be in the bazaar or a busy street, not in a secluded place... It is a place for teaching, not an eating house, so the boys should not bring food or money... In the organization, a teacher must have a deputy to set the class in their places, also visitors according to their rank, to awaken the sleepers, to warn those who do what they ought not or omit what they ought to do, and bid them listen to the instruction. In class, conversation, laughing, and jokes are forbidden."

By the 15th century, the Ottomans had revolutionized schools by setting up learning complexes in towns like Istanbul and Edirne in Turkey. Their school system was called Kulliye, and constituted a campuslike education, with a mosque, hospital, school, public kitchen, and dining area. These made learning accessible to a wider public while also offering free meals, health care, and sometimes accommodation. The Fatih Kulliye in Istanbul was such a complex, with 16 schools teaching science and theology.

Where did the money come from for all these institutions? Not so much in taxes, but from public funds that were charitable donations from a foundation called *waqf*. Anyone could set up a school under a deed of foundation as long as he abided by the beliefs of Islam. Finance covered maintenance, teachers' salaries, accommodation, food for students, and scholarships for those in need.

Because education was held in such high esteem, money was given generously and learning flourished. Ibn Battuta, the 14th-century Muslim traveler, talks about the student, who was supported 100 percent: "anyone who wishes to pursue a course of studies or to devote himself to the religious life receives every aid to the execution of this purpose."

Many students in the 21st century would like such free education, and even though the 14th century may seem like a long time ago, the methods of organization, logistics, and system of institutions could offer a few pointers for today.

"It was this great liberality which they [Muslims] displayed in educating their people in the schools which was one of the most potent factors in the brilliant and rapid growth of their civilization. Education was so universally diffused that it was said to be difficult to find a Muslim who could not read or write."

EDUCATOR E. H. WILDS

The university complex of Bayazid II Külliye consists of a mosque, madrasa *(school), and hospital in Edirne, Turkey.*

A Turkish miniature from the 17th century by Mehmed ibn Amirshah shows the Ghazanfar Aga madrasa in Istanbul.

02 UNIVERSITIES

Today, more people than ever before are applying for a university education. This quest for knowledge was also close to the hearts of medieval Muslims as they were urged throughout the Quran to seek knowledge, observe, and reflect. This meant that all over the Muslim world, advanced subjects were taught in mosques, schools, hospitals, observatories, and the homes of scholars—developing knowledge and learning that eventually spread to Europe.

There was some overlap between school and university education. Both began in the mosque, but "university" in Arabic is *Jami'ah,* which is the feminine form of the Arabic word for mosque, *Jami'.* So in Arabic the place of religion and the place of advanced learning are conjoined. There is no equivalent in other cultures or languages, and some of the mosques of Islam are the oldest universities.

Famous mosque universities include Al-Azhar, which still exists today—1,030 years later. As the focal point of higher learning in Egypt, it attracted the cream of intellects. So it is known for its age and also for its illustrious alumni: Ibn al-Haytham, father of experimental optics, lived there for a long time, and Ibn Khaldun, a leading 14th-century sociologist, taught there.

Another grand college mosque complex was Al-Qarawiyin in Fez, Morocco. This university was originally built as a mosque in 859 during

LEFT: *The courtyard of Al-Qarawiyin shows a* Mihrab *behind the fountain; it was usually used for prayer on summer evenings.* RIGHT: *The timing room at Al-Qarawiyin houses a fully functioning water clock, which includes a series of brass jars and an astrolabe indicating days and months. On the white wall hangs one of the oldest European grandfather clocks. On the floor is the bed on which the timekeeper sleeps.*

UNIVERSITY OF SANKORE

Prayer mats lie in the courtyard of Sankore Mosque in Timbuktu, Mali.

At one of the most southerly points of the Muslim lands was the University of Sankore, in Timbuktu, and it was the intellectual institution of Mali, Ghana, and Songhay.

It developed out of the Sankore Mosque, founded in 989 by the erudite chief judge of Timbuktu, Al-Qadi Aqib ibn Mahmud ibn Umar. The inner court of the mosque was in the exact dimension of the *Ka'bah* in holy Mecca. A wealthy Mandika lady then financed Sankore University, making it a leading center of education. It prospered and by the 12th century, student numbers were at 25,000, an enormous amount in a city of 100,000 people.

The university had several independent colleges, each run by a single master. Subjects included the Quran, Islamic studies, law, literature, medicine and surgery, astronomy, mathematics, physics, chemistry, philosophy, language and linguistics, geography, history, and art.

It was not all cerebral, as the students also spent time learning a trade, the business code, and ethics. These trade shops offered classes in business, carpentry, farming, fishing, construction, shoemaking, tailoring, and navigation.

The highest "superior" degree, equivalent to a Ph.D., took about ten years, and produced world-class scholars who were recognized by their publications and for their erudition. The Ph.D. thesis was called *Risaleh* (literally meaning "letter"), and those graduating with this degree were named Ayatullah, as they still are in the theological Shiite centers Qum (Iran) and Najaf (Oraq).

the Idrisids' rule by Fatima al-Fihri, a devout and pious young woman. She was well educated, and after inheriting a large amount of money from her father, a successful businessman, she vowed to spend her entire inheritance on building a mosque-university suitable for her community in Fez. She put a design constraint on the building that all the building material should be from the same land. On launching the project she began a daily fast until the campus building was completed.

Like some of the great mosques, Al-Qarawiyin soon developed into a center for religious instruction and political discussion, gradually extending its education to all subjects, particularly the natural sciences, and so it earned its name as one of the first universities in history. The university was well equipped, especially with astronomy instruments, and the "timers room" had astrolabes, sand clocks, and other instruments to calculate time. As well as astronomy, studies were in the Quran and theology, law, rhetoric, prose and verse writing, logic, arithmetic, geography, and medicine. There were also courses on grammar, Muslim history, and elements of chemistry and mathematics. This variety of topics and the high quality of its teaching drew scholars and students from a widespread area.

These mosque "universities" not only took local students, but also those from neighboring countries. So in the famous Abbasid universities of Baghdad, Iraq, medicine, pharmacology, engineering, astronomy, and other subjects were taught to students from Syria, Persia, and India. Students at Al-Azhar University in Cairo included large numbers of foreigners, alongside Egyptians from areas outside Cairo.

> *"Books were presented and many a scholar bequeathed his library to the mosque of his city to ensure its preservation and to render the books accessible to the learned who frequented it. And so grew up the great universities of Córdoba and Toledo to which flocked Christians as well as Moslems from all over the world."*
>
> R. S. MACKENSEN, A CONTEMPORARY EUROPEAN HISTORIAN OF ISLAMIC LIBRARIANSHIP

At the Zaytuna Mosque in Tunisia, there were manuscripts on grammar, logic, documentation, the etiquette of research, cosmology, arithmetic, geometry, minerals, and vocational training. At the Tunisian Qayrawan's Atiqa Library, there was an Arabic translation of the *History of Ancient Nations* written by St. Jerome before 420.

Courses were difficult, and medicine was particularly grueling, as in universities today, with the department of medicine having a hard and long examination. Anything less than a pass meant that person could not practice medicine.

The students of law went through undergraduate training and, if they were successful, were chosen by their master as a fellow. Only then could they go on to graduate studies, which lasted an indefinite period of time. It could be up to 20 years before they acquired their own professorial chair. The law student had to get a certificate of authorization and a license before practicing.

These certificates, known as *ijazas,* could be the origin of the word "baccalareus," which is the lowest university degree. The term first appeared in the University of Paris degree system set up in 1231 by Pope Gregory IX. It could be a Westernized Arabic phrase that the Muslims used. *Bi-haqq al-riwayah* meant "the right to teach on the authority of another," and this phrase was used in the degree certificates, *ijazas,* for six centuries. When a student graduated he was given this license, and it literally meant he now had "the right to teach."

Now the International Baccalaureate is a qualification for students to prepare them for universities anywhere in the world.

THE PROFESSOR'S CHAIR

How did the tradition arise of awarding a professor a "chair," or addressing the person in charge of a meeting as "chair"? In schools and universities more than a thousand years ago, you would have seen a study circle or a *Halaqat al-'ilm* or *halaqa* gathered around a professor who was seated on a high chair to make him visible and audible to the students. The professor would have been either chosen by the caliph or by a committee of scholars—and once a professor was appointed by the caliph to a chair in one of the main mosques or Jami', he usually held it for the remainder of his life. Cases of lengthy tenure are frequent, like Abu 'Ali al-Kattani, who was in his 80s when he died in 1061 after occupying his chair for 50 years.

A lecturer sits in a chair, or minbar, while giving a sermon at a mosque.

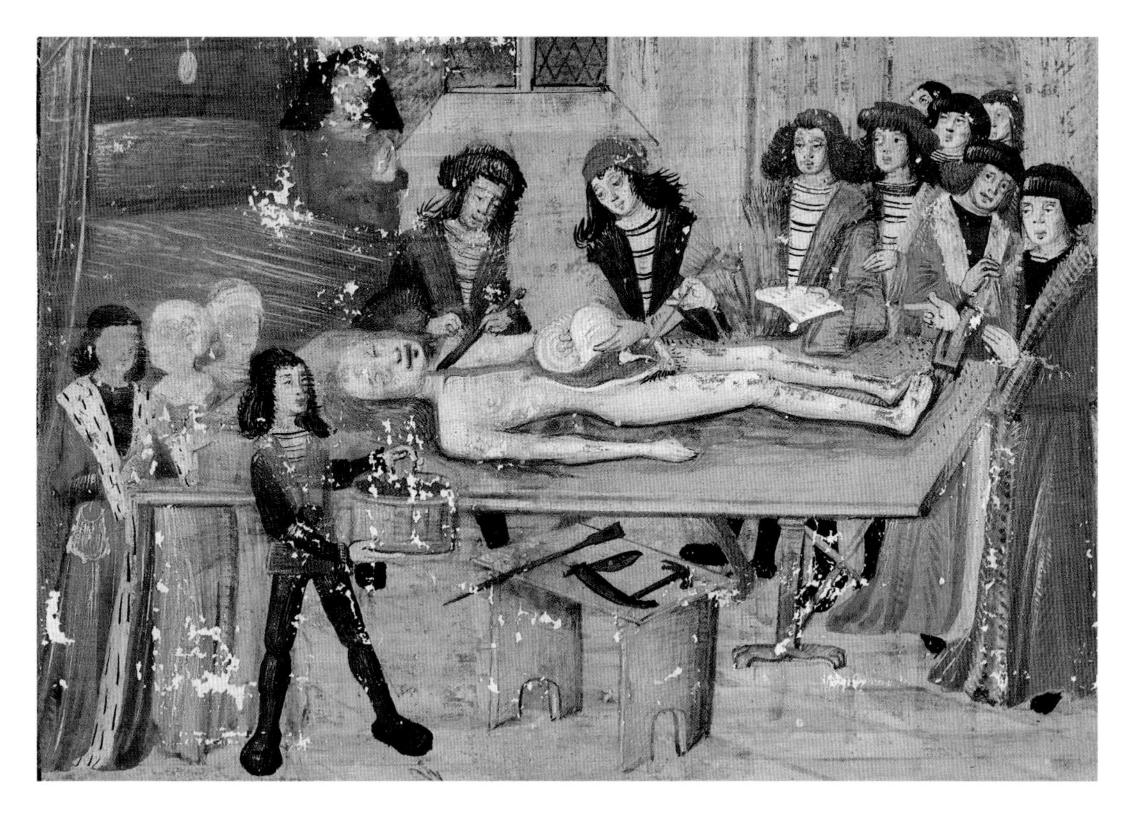

A 14th-century manuscript shows a dissection lesson at the faculty of medicine at Montpellier University, France.

Muslims institutionalized higher level education. There were entrance exams, challenging finals, degree certificates, study circles, international students, and grants. In fact, there is a remarkable similarity between the teaching procedures in medieval "universities" and the methods of the present day. They even had collegiate courses and prizes for proficiency in scholarship.

Muslim learning hit medieval Europe in the 12th century. A massive translation exercise began of Arabic works from the previous 500 years into medieval Latin, making available the rational ideas from experiments to a new audience. The availability of well-referenced material kick-started European tertiary education and questioned the idea that there had to be conflict between religion and science. At Chartres cathedral school in the 1140s, Thierry of Chartres taught that the scientific approach was compatible with the story of creation in the Bible, paving the way for the Renaissance.

The first university in western Europe was at Salerno in Italy, which burst into life in the late 11th century after the arrival of Constantine the African. The French city of Montpellier was an offshoot of Salerno and a major center for the study of Muslim medicine and astronomy. It was close to Muslim Spain, with its large population of Muslims and Jews.

By the beginning of the 12th century, the intellectual powerhouse of the Western world had shifted to Paris, "a city of teachers," as the knowledge of Arabic works continued to spread with traveling scholars. Indeed, many historians today say that the blueprints of the earliest English universities, like Oxford, came with these traveling, open-minded scholars and returning Crusaders who, as well as visiting Muslim universities in places like Córdoba, brought back translated books based on rational thought rather than confined to religious thought.

03 House of Wisdom

The heyday of Baghdad was 1,200 years ago when it was the thriving capital of the Islamic world. For about 500 years the city boasted the cream of intellectuals and culture, a reputation gained during the reigns of Caliphs Al-Rashid, Al-Ma'mun, Al-Mu'tadhid, and Al-Muktafi.

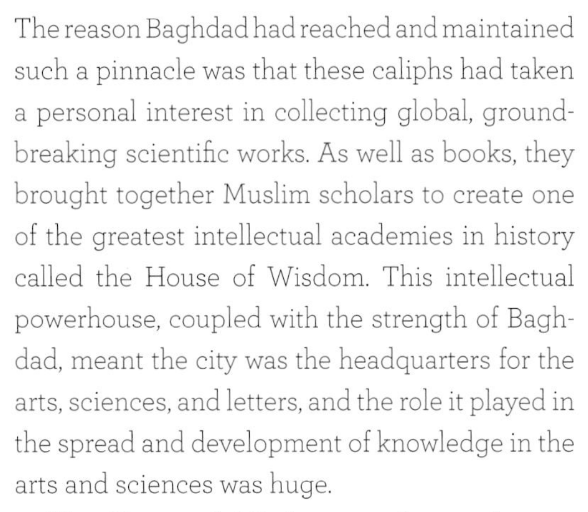

The reason Baghdad had reached and maintained such a pinnacle was that these caliphs had taken a personal interest in collecting global, groundbreaking scientific works. As well as books, they brought together Muslim scholars to create one of the greatest intellectual academies in history called the House of Wisdom. This intellectual powerhouse, coupled with the strength of Baghdad, meant the city was the headquarters for the arts, sciences, and letters, and the role it played in the spread and development of knowledge in the arts and sciences was huge.

The House of Wisdom was known by two names, according to its development stages. When it was a single hall in the time of Harun al-Rashid, it was named *Khizanat al-Hikmah* (Library of Wisdom), but later, as it grew into a large academy in the time of Al-Ma'mun, it was named *Bayt al-Hikmah* (House of Wisdom). It housed a large library, which held a huge collection of different scientific subjects in many languages, thus making it a scientific academy.

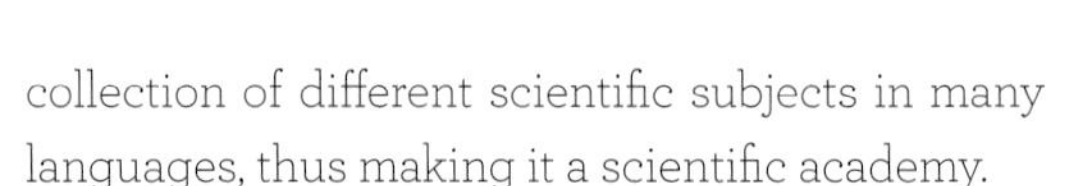

Caliph Mohammad al-Mahdi first began collecting manuscripts when he came across them during his war expeditions. His son, Caliph al-Hadi, carried on this work and later his son, Caliph Harun al-Rashid, who reigned from 786 to 809, built the scientific collection and Academy of Science. Caliph Al-Ma'mun, who reigned for 20 years from 813, extended the House of Wisdom and designated a section or wing for each branch of science, so the place was full to bursting with scholars or *'ulama*, art scholars, famous translators, authors, men of letters, poets, and professionals in the various arts and crafts.

These medieval brains met every day for translation, reading, writing, and discourse. The place was a cosmopolitan melting pot, and the languages that were spoken and written included Arabic (the lingua franca), Farsi, Hebrew, Syriac, Aramaic, Greek, Latin, and Sanskrit, which was used to translate the ancient Indian mathematics manuscripts.

Among the famous translators was Yuhanna ibn al-Bitriq al-Turjuman, known as "the Translator Jonah, son of the Patriarch." He was more at home with philosophy than medicine, and translated

TOP: *A Syrian stamp issued in 1994 shows Al-Kindi, a leading scholar in the House of Wisdom, who translated the works of Aristotle.* BOTTOM: *An archival photo shows Baghdad in 1932. Eleven centuries earlier, the city was the site of the House of Wisdom.*

from Latin *The Book of Animals* by Aristotle, which was in 19 chapters. Hunayn ibn Ishaq was also a renowned translator of the books by Greek physicians Hippocrates and Galen.

It is said that Caliph Al-Ma'mun wrote to the king of Sicily asking him for the entire contents of the Library of Sicily, which was rich in philosophical and scientific books. The king responded positively to the caliph by sending him copies from the Sicilian library.

The transportation of books varied. Without the availability of modern planes, it is said that Al-Ma'mun used a hundred camels to carry handwritten books and manuscripts from Khurasan in Iran to Baghdad.

The Byzantine emperor was also approached because Al-Ma'mun wanted to send some of his scientists to translate the useful books that were stored in his domain. The emperor agreed and the scientists went, also charged with bringing back any books of the Greek intellectuals.

Caliph Al-Ma'mun not only steered the organization of the House of Wisdom, but also participated with the scientists and scholars in their discussions. He built an astronomy center called Marsad Falaki, which was run by his personal astronomers, a Jew named Sanad ibn Ali al-Yahudi and a Muslim named Yahya ibn Abi Mansur.

Al-Ma'mun took after his father in establishing many higher institutes, observatories, and factories for textiles. It is said that the number of higher institutes during his reign reached 332.

He also apparently asked a group of wise men to prepare a map of the world for him, which they did. This was known as "*al-Ma'mun's* map," or *al-surah al-ma'muniyah*, which expanded upon those that were available during the lifetime of Ptolemy and other Greek geographers.

Among the House of Wisdom's luminaries of the time were the Banu Musa brothers, Muhammad, Ahmed, and Al-Hasan, known as mathematicians and inventors of trick devices; Al-Khwarizmi, the "father" of algebra; Al-Kindi, inventor of decryption and musical theory; Sa'id ibn Harun al-Katib, a scribe; Hunayn ibn Ishaq al-'Ibadi, physician and translator; and his son Ishaq. These names appear time and time again throughout this book because these individuals were researching, discovering, and building a vast edifice of knowledge, based on real experiments, which has provided a firm bedrock for much of what we know today.

However, we must distinguish between the Abbasid House of Wisdom above and the Fatimid House of Wisdom (*Dar al-Hikmah*), which was established in Cairo in 1005 by the Caliph Al-Hakim and lasted for 165 years. Other cities in the eastern provinces of the Islamic world also established Houses of Science (*Dar al-'Ilm*), or more accurately Houses of Knowledge, in the ninth and tenth centuries to emulate that of Baghdad.

AN UNRIVALED CENTER

Brian Whitaker wrote in the United Kingdom's *Guardian* newspaper in September 2004: "The House of Wisdom was an unrivalled centre for the study of humanities and for sciences, including mathematics, astronomy, medicine, chemistry, zoology and geography . . . Drawing on Persian, Indian and Greek texts—Aristotle, Plato, Hippocrates, Euclid, Pythagoras and others—the scholars accumulated the greatest collection of knowledge in the world, and built on it through their own discoveries."

AN INTELLECTUAL POWERHOUSE

Baghdad's World-Class Academy That Attracted Scholars from Far and Wide

LEGACY: **Contributed to algebra; and translations of medical, philosophical, and mathematical works**

LOCATION: **Baghdad, Iraq**

DATE: **Founded in the early ninth century**

KEY FIGURE: **Caliph Al-Ma'mun, building on his father's foundation**

A thousand years ago, Baghdad was one of the world's biggest and richest cities. But its wealth went far beyond money. For more than two centuries, it was home to the House of Wisdom, an academy of knowledge that attracted scholars from far and wide. From mathematics and astronomy to zoology, the academy was a major center of research, thought, and debate in Muslim civilization—the intellectual powerhouse of its day.

Scholars at the academy could draw on a vast collection of scientific, medical, and philosophical books. The caliphs who established the House of Wisdom assembled a world-class library that included the writings of many ancient civilizations. Caliph Harun al-Rashid, who reigned from the year 786, founded the library with manuscripts collected by his father and grandfather. Three decades later, the collection had grown so large that his son, Caliph Al-Ma'mun, built extensions to house different branches of knowledge. Later, he added numerous other study centers to allow more scholars to pursue their research, and an observatory in the year 829.

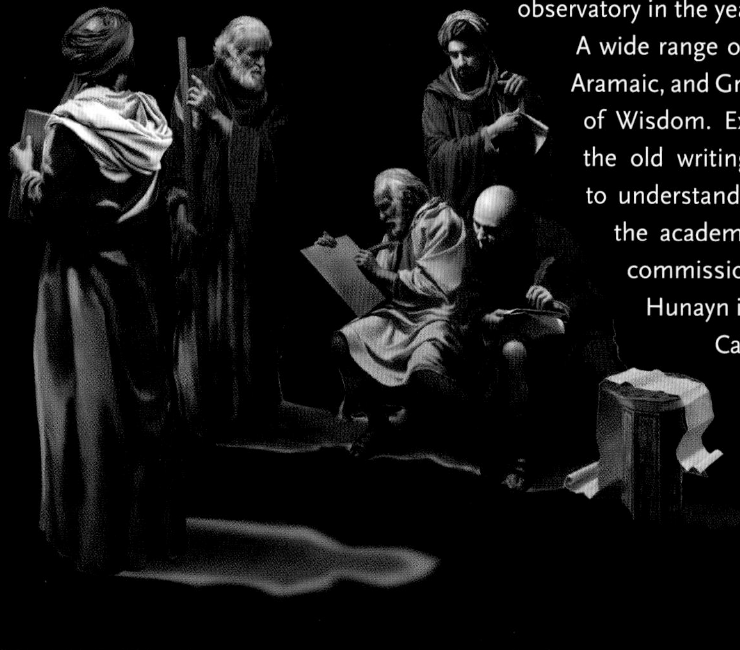

A wide range of languages including Arabic, Persian, Aramaic, and Greek were spoken and read in the House of Wisdom. Experts constantly worked to translate the old writings into Arabic to allow the scholars to understand, debate, and build on them. Among the academy's leading lights were Al-Kindi, who commissioned the translation of Aristotle, and Hunayn ibn Ishaq, who translated Hippocrates. Caliph Al-Ma'mun is said to have encouraged translators and scholars to add to the library in the House of Wisdom by paying them the weight of each completed book in gold.

This successful knowledge transfer was mirrored years later in

2th-century Spain. At that time, Toledo was the focus of another huge translation effort—this time from Arabic into Latin. Arabic works and translations of important ancient Greek texts came to light, and Christian, Jewish, and Muslim scholars flocked to the city. Together they worked to translate the writings of ancient Greece and the original Arabic treatises into Latin and then into European languages. Here, and elsewhere in southern Europe, some of Aristotle's works were translated from Arabic into Latin by Michael the Scot. Gerard of Cremona made translations of Al-Zahrawi's medical encyclopedia and Al-Razi's chemical writing.

Today, the advances made by scholars in the House of Wisdom are still having an impact. The word "algebra" comes from the title of the book *Al-Jabr wal Muqabalah* by the scholar Al-Khwarizmi. He put algebra on a secure footing in the early ninth century while working in the House of Wisdom. Al-Khwarizmi's successor, Al-Karaji, developed and refined these ideas, eventually starting an algebraic tradition that thrived for hundreds of years. These scholars and others developed the basis of abstract mathematical thought on which much of modern-day computing relies.

The vision of Caliph Al-Ma'mun, and his intellectual legacy are clear—and his name lives on, too, through a lunar crater named "Almanon."

OPPOSITE AND THIS PAGE: *People from all over the Muslim world flocked to the House of Wisdom, the intellectual powerhouse of ninth-century Baghdad. The illustrations depict scholars, both male and female and of many faiths, who came from far and wide to study various scientific disciplines at the academy.*

04 LIBRARIES AND BOOKSHOPS

It is said that the Abbasid Caliph Al-Ma'mun paid translators the weight in gold of each book that they translated from Greek into Arabic. This produced a vast supply of books, commanding the attention and respect of succeeding generations, Muslim and non-Muslim. During the Abbasid period, hundreds of libraries, many privately owned, were opened, making many thousands of books available to readers.

Before science books came the very first book in Islam in the seventh century. This was the Quran, which was revealed to Prophet Muhammad in the form of messages called *Ayats,* or verses. These were immediately memorized by several companions and written down by scribes on whatever material was available such as leaves, cloth, bones, and stones. The earliest full copy of the book was kept by Hafsah, the daughter of the second caliph, Omar. The arrangements of the verses were in chapters, or suras, and the location of each chapter was personally checked and revised by the Prophet himself. Several copies existed, but most of these contained personal explanatory notes by their owners.

All these copies needed to be collected to produce a single standard copy without additional comments, and that was also checked against the original version of Hafsah. This copy of the Quran was produced by 'Uthman ibn 'Affan, the third caliph, which led to the standardization of reading and writing styles, and made it easier for the Quran to spread. Copies of this 1,400-year-old Uthman manuscript are still available in major libraries of the world, and the present-day copy of the Quran is an authentic duplication of this original seventh-century manuscript.

Developing strong attachments to books meant Muslims also loved book collecting and establishing libraries. There were public and private libraries, with a large network of public libraries in mosques in most big cities, plus prestigious private collections, which attracted

The Zaytuna Mosque college complex was built in 732 in Tunis, Tunisia. In the 13th century, its library contained more than 100,000 volumes.

"The book is silent
as long as you need silence,
eloquent whenever
you want discourse.
It never interrupts you
if you are engaged, but if
you feel lonely it will
be a good companion.
It is a friend who never
deceives or flatters you,
and it is a companion
who does not grow tired of you."

AL-JAHIZ, MUSLIM PHILOSOPHER AND MAN OF LITERATURE, EIGHTH CENTURY, BASRA, IRAQ

scholars from all parts of the Muslim world. The books or manuscripts in them were of various sizes, containing good quality paper with writing on both sides, and bound in leather covers.

Public book collections were so widespread that it was impossible to find a mosque, the place of learning, without a collection of books. Before the Mongols decimated Baghdad in 1258, it had 36 libraries and more than a hundred book dealers, some of whom were also publishers, employing a corps of copyists. There were similar libraries in Cairo, Aleppo, and in major cities of Iran, Central Asia, and Mesopotamia.

Mosque libraries were called *dar al-kutub*, or the House of Books, and they were the focus of intellectual activity. Here writers and scholars dictated the results of their studies to mixed audiences of young people, other scholars, and interested laymen. Anyone and everyone could take part in the discussions. Professional *warraqs (nussakh)*, or scribes, then copied and turned them into books. Even when the books were specially commissioned, they would still be published in this way.

Aleppo in Syria probably had the largest and oldest mosque library, called the *Sayfiya*, at the city's grand Umayyad Mosque, with a collection of 10,000 volumes. These were reportedly bequeathed by the city's most famous ruler, Prince Sayf al-Dawla.

The *Sayfiya* was the oldest and largest, but the library at the Zaytuna Mosque college complex in Tunis was possibly the richest of all. It had tens of thousands of books, and it is said that most rulers of the Hafsid dynasty competed with each other for the prestige associated with maintaining and strengthening this library. At one point, the collection exceeded 100,000 volumes.

Al-Jahiz, an eighth-century Muslim philosopher and man of literature, returned to his home in Basra after spending more than 50 years in Baghdad studying and writing about 200 books. These included a seven-volume *Book of Animals*, which had observations on the social organization of ants, communication between animals, and the effects of diet and environment. Other books were *The Art of Keeping One's Mouth Shut* and *Against Civil Servants*. He died an appropriate death in his private library in 868, at the age of 92, when a pile of books fell on him.

When book lovers died, it was a tradition that they would donate their collected manuscripts, sometimes thousands of volumes, to the mosque libraries for all to enjoy. Historian Al-Jaburi says that Nayla Khatun, a wealthy widow of Turkish origin, founded a mosque in memory of her deceased husband, Murad Efandi, and attached a school and a library to it. Other books came from

UMAYYAD LIBRARY

The Umayyad rulers of Spain had a library of 600,000 volumes in their huge Córdoba library. So much better was the company of books for Al-Hakam II, caliph in Spain from 961 to 978, that he said they were "a more consuming passion than his throne."

traveling scholars as they showed their gratitude to mosques for giving them free accommodations, food, and stationery.

Libraries could be grand affairs. In Shiraz, Iran, these tenth-century complexes were described by the medieval historian Al-Muqaddasi, as "buildings surrounded by gardens with lakes and waterways . . . topped with domes, and comprised of an upper and a lower story with a total . . . of 360 rooms . . . In each department, catalogues were placed on a shelf . . . the rooms were furnished with carpets."

Some libraries, like those of Shiraz, Córdoba, and Cairo, were in buildings separate from the mosque. They were spacious, with many rooms for different uses: shelved galleries to store books, reading rooms, rooms for making copies of manuscripts, and rooms for literary assemblies. All these were adequately lit and comfortable, with carpets, mats, and seating cushions.

Like libraries today, those of a thousand years ago were highly ordered, with both public and private libraries having book classification systems, and accurate cataloguing to help readers. Librarians also had control over the quality and quantity of their resources.

In 1050, the book collection of Al-Azhar library in Cairo had more than 120,000 volumes recorded in a 60-volume catalog totaling about 3,500 pages. In Spain, the catalog for the works in Al-Hakam's library was alleged to have consisted of 44 volumes.

Librarians were appointed to take charge and this was an honored position, only for the most learned. Only those "of unusual attainment" were considered as custodians of the libraries, the guardians and protectors of knowledge. The management of the libraries of the Almohad dynasty, the rulers in North Africa in the 12th and 13th centuries, was one of the most privileged state positions.

All these libraries were the holders of vital knowledge and as Ralph Waldo Emerson, a 19th-century American writer, said: "Consider what you have in the smallest chosen library. A company of the wisest and wittiest men that could be picked out of all civil countries, in a thousand years, have set in best order the results of their learning and wisdom . . . [it] is here written out in transparent words to us, the strangers of another age."

"There can be no education without books."

ARABIC PROVERB

Bookshops, too, had their place in sharing knowledge. The celebrated bookshop of Ibn al-Nadim, the tenth-century bibliophile and bookseller, was said to be on an upper story of a large building where buyers came to examine manuscripts, enjoy refreshments, and exchange ideas. In the Muslim world, a thousand years ago, there were bookshops containing hundreds of titles as well as massive public and private libraries.

With paper, *waraq* in Arabic, came the profession of *Warraq*. The title *Warraq* has been used for paper dealers, writers, translators, copiers, booksellers, librarians, and illuminators. The profession of the *Warraqeen* is generally believed to have started shortly after the introduction of the art of papermaking into the Muslim world from China. Baghdad was probably the first major city where the *warraqi* bookshops first appeared, and as the manufacture of paper spread, the number of these bookshops increased dramatically throughout the Muslim world.

Kutubiyun is a Moroccan name for bookbinders or book merchants who set up their bookshops and libraries, copyists and scribes in a district of 12th-century Marrakech, Morocco. This district was a street with a hundred bookshops and libraries, 50 on each side. Such activity reached its zenith during the reign of Yaqub al-Mansur, who constantly encouraged the spread of book printing and promoted general reading activity.

A 13th-century manuscript of Maqamat al-Hariri *shows the public library of Hulwan in Baghdad.*

05 Translating Knowledge

What is striking about the discoveries, innovations, research, and writings of Muslim scientists and scholars during the European medieval period is their insatiable thirst for knowledge. This was not knowledge for the sake of knowledge; in most cases it had practical application—improving the quality of life of the people.

There was also a spiritual influence, as Prophet Muhammad had said: "When a man dies, his actions cease except for three things: a continuous charity, knowledge which continues to benefit people, or a righteous son who prays for him."

Amazing energy was shown by encyclopedic individuals who were writing down their findings at incredible rates, filling up enormous tomes with groundbreaking information. Books ran to thousands of pages and numerous volumes, and filled vast libraries. The golden age of this civilization, the eighth to the thirteenth centuries, saved ancient learning from extinction, modified it, added new discoveries, and spread knowledge in an enlarged and enriched form. To read more about great feats of learning and knowledge gathering see the House of Wisdom section in this chapter.

At the heart of this understanding was the idea of direct observation. In order to understand how something worked, you had to see it with your own eyes, and only then could you write it down. One man, Ibn al-Haytham, in the late tenth century, did his experiments in complete darkness. Ibn al-Haytham was one of the first people in the world to test his theories with experiments, establishing one of the keystones of all scientific method—prove what you believe. You can read more about him and his experiments in the Home chapter and Vision and Cameras section.

This thirst for knowledge was infectious and even reached its tentacles across oceans, touching non-Muslims who flocked to absorb the vast encyclopedias, based on experimentation that the Muslim polymaths had produced.

Ibn al-Haytham, left, and Galileo, right, explored the world through observation and rational thought. Both men are shown on the front of Johannes Hevelius's Selenographia, *a 1647 description of the moon. Al-Haytham holds a geometrical diagram, and Galileo clutches a telescope.*

A contemporary photo shows Toledo, Spain. In the 12th century, Muslim Toledo hosted people of at least three religions—Muslims, Jews, and Christians. They lived and worked side by side. This melting pot of people and ideas attracted scholars and translators from East and West.

"On the Day of Resurrection the feet of the son of Adam [man] will not move away till he is questioned about four matters: how he spent his lifetime, how he spent his youth, from where he acquired his wealth and how he spent it, and what he did with his knowledge."

PROPHET MUHAMMAD
NARRATED BY AL-TIRMIDHI, NO. 2417

Daniel of Morley was an English priest and scientist, born circa 1140 in a small, sleepy village in Norfolk, who went in search of knowledge. He is just one example of an outward-looking and forward-thinking European who opened his mind to Muslim knowledge.

Daniel was possibly a student of Adelard of Bath, who had written to the future King Henry II saying, "It happens that you not only read carefully and with understanding those things that the writings of the Latins contain, but you also . . . wish to understand the opinions of the Arabs concerning the sphere, and the circles and movements of the planets. For you say that whoever has been born and brought up in the hall of the world, if he does not bother to get to know the reason behind such wonderful beauty, is unworthy of that hall and should be thrown out . . . Therefore I shall write in Latin what I have learnt in Arabic about the world and its parts."

To further his education, Daniel, like many young students, left his native England and headed east, first stopping at the university in Paris. Unfortunately, according to him, it had become "stale and moribund" and he could hardly wait to leave. He said, "These masters [in Paris] were so ignorant that they stood as still as statues, pretending to show wisdom by remaining silent."

So where did he go? Well, in his own words, "since these days it is at Toledo that Arabic teachings are widely celebrated, I hurried there to listen to the world's wisest philosophers." In 12th-century Toledo at least three cultures lived side by side: Muslims, Jews, and Christians. This was a time of cultural richness where all shared the same, breathtaking desire for knowledge. Today, the way they worked and lived together is known by the Spanish word as *convivencia.*

Although thousands of Arabic manuscripts in the Toledo Cathedral archives were burned, about 2,500 translated manuscripts—from Arabic to Latin—remain, dating from Daniel of Morley's time in the 12th century.

It was in 12th-century Toledo that possibly the greatest translation effort in the history of science took place, from Arabic into Latin. This attracted numerous scholars and translators from the Christian West. Important works by Greek philosophers and mathematicians, which had been lost in the West, were turning up in Toledo, saved and enhanced by Muslims. The critique and commentary on Aristotle by

> *"The real jewel of Toledo is to be found in the city's libraries and involved all three communities [Muslims, Jews, and Christians] working in a particular field, translation. Teams of Muslims, Jews, and Christians translated texts into Arabic, then into Castellan Spanish and Latin. It required close cooperation and religious tolerance. The Andalusian word for this is* convivencia *and means living together."*
>
> RAGEH OMAAR, PRESENTING THE BBC'S *AN ISLAMIC HISTORY OF EUROPE*

Ibn Rushd, known in the West as Averroes, was the real start of Europe's classical revival, and this was 200 years before the start of the European Renaissance.

Many of the commentaries on and summaries of Aristotle's works by Ibn Rushd, writing in Arabic in Córdoba in the late 12th century, were translated into Latin by Michael Scott, a scholar from Scotland who died before 1236, and by his

Today, King Peter I's palace in Toledo is a center for teaching Arabic and Hebrew translation skills. It is a 14th-century Mudéjar—the name for Muslims who stayed under Spanish rule—building; at that time Jews, Christians, and Muslims lived and worked together translating scholarly works from Arabic and Hebrew into Latin and Spanish.

successor, Herman the German. These Latin translations, made both in Toledo and Sicily, were destined to set Europe ablaze: "He [Averroes] would launch Paris as the intellectual capital of Europe . . . Averroes was trying to defuse a conflict between science and religion because the truth revealed by science was often at odds with the truth of divine revelation. This attempt had the opposite effect when his ideas came to the attention of the Christian church. They immediately banned Averroes and Aristotle's works. The Paris intellectuals fought back and a debate raged for years," said Rageh Omaar in the BBC's *An Islamic History of Europe.*

As well as Michael Scott and Daniel of Morley, the city of Toledo was buzzing with contemporary translation scholars. There was Gerard of Cremona, who was translating into Latin important works like Al-Zahrawi's 30-volume medical encyclopedia; Ibn al-Haytham's voluminous *Book of Optics*; Al-Kindi's treatise on geometrical optics; Al-Razi's *De aluminibus et salibus* or *A Study and Classification of Salts and Alums [Sulfates];* and the book of geometry by the Banu Musa brothers. What is amazing about Gerard of Cremona is that he made more than 80 translations but never had a full grasp of Arabic. Instead, he had to work with and rely on Mozarab locals and the Christian Spanish, who did know the language.

The BBC's *Voices from the Dark* program said, "The process [of translation] varied from translation to translation. Sometimes it was a team helped by a local person with Arabic as their mother tongue. He read the text aloud to an intermediary who also knew Arabic and was expert in Romance, the language that preceded modern Spanish. Then the Romance translation would be put into Latin. Some translators could work alone as they had full command of all three languages."

Even though Alfonso VI had retaken Toledo into Christian hands, the city remained Muslim in that the lingua franca was still Arabic, spoken by Muslims, Jews, and Mozarabs alike; the culture and customs were Muslim; and the architecture was Islamic. Long winding narrow streets provided rooms for lodgings and study for all the translators and scholars who arrived. For all these Western scholars, Toledo was the place to be.

Manuscripts of the Latin translations made in Toledo are still in the Toledo Cathedral archives. About 2,500 manuscripts are there, including translations from Arabic dating from Daniel of Morley's day.

06 MATHEMATICS

There are quite a few mathematical ideas that were thought to have been brilliant conceptions of 16th-, 17th-, and 18th-century Europeans. From the studying and unearthing of manuscripts we now know that Muslim mathematicians, about 1,000 years earlier, were calculating with great intensity. Many of these mathematicians came from the Iran/Iraq region around 800, when the House of Wisdom was the leading intellectual academy in Baghdad. You can read more about the House of Wisdom in a section in this chapter.

This remarkable period in the history of mathematics began with Al-Khwarizmi's work, when he introduced the beginnings of algebra. It is important to understand just how significant this new idea was. In fact, it was a revolutionary move away from the Greek concept of mathematics, which was essentially based on geometry.

Algebra was a unifying theory that allowed rational numbers, irrational numbers, and geometrical magnitudes all to be treated as algebraic objects. It gave mathematics a whole new dimension and a development path, much broader in concept than before. It also enabled future development. Another important aspect of the introduction of algebraic ideas was that it allowed mathematics to be applied to itself in a way that had not been possible earlier.

The torch of algebra was taken up by the successor of Al-Khwarizmi, a man called Al-Karaji, born in 953. He is seen by many as the first person to free algebra completely from geometrical operations, and to replace it with the arithmetical type of operations, which are at the core of algebra today. He was first to define the monomials x, x^2, x^3, . . . and $1/x$, $1/x^2$, $1/x^3$, . . . and to give rules for products of any two of these. He started a school of algebra, which flourished for several hundred years.

Two hundred years later, 12th-century scholar Al-Samawal was an important member of Al-Karaji's school. He was the first to give algebra the precise description of "operating on unknowns using all the arithmetical tools, in the same way as the arithmetician operates on the known."

The next contributor to the algebraic story was the poet Umar al-Khayyam, known today as Omar Khayyám, who was born in 1048. He gave

A 1983 commemorative stamp issued by the former Soviet Union depicts Al-Khwarizmi, the "father of algebra."

The algebra studied today in school has as its basis Al-Khwarizmi's book Algebr wal Muqabala.

a complete classification of cubic equations, with geometric solutions found by means of intersecting conic sections. He hoped to give a full description of the algebraic solution of cubic equations and said: "If the opportunity arises and I can succeed, I shall give all these 14 forms with all their branches and cases, and how to distinguish whatever is possible or impossible so that a paper, containing elements which are greatly useful in this art, will be prepared."

In the mid-12th century, while Al-Samawal was studying in Al-Karaji's school, Sharaf al-Din al-Tusi was following Al-Khayyam's application of algebra to geometry. He wrote a treatise on cubic equations, and in it said that algebra "represents an essential contribution to another field, which aimed to study curves by means of equations," thus inaugurating the field of algebraic geometry.

Algebra is only one area where Muslim mathematicians significantly changed the course of development. In ninth-century Baghdad, in the House of Wisdom, was a group of three brothers called the Banu Musa brothers. You can read more about them in the Home chapter and how they developed their trick devices. They were gifted mathematicians, and one of their students was Thabit ibn Qurra, who was born in 836. He is probably best known for his contribution to number theory, where he discovered a beautiful theorem allowing pairs of amicable numbers to be found. This term refers to two numbers such that each is the sum of the proper divisors of the other.

Amicable numbers played a large role in Arabic mathematics, and in the 13th-century Al-Farisi gave new proof of Thabit's theorem, introducing important ideas concerning factorization and combinatorial methods. He also discovered the pair of amicable numbers 17,296 and 18,416, which have been attributed to Euler, an 18th-century Swiss mathematician. And many years before Euler, another Muslim mathematician, Muhammed Baqir Yazdi, in the 17th century discovered the pair of amicable numbers 9,363,584 and 9,437,056.

"[Algebra operates] on unknowns using all the arithmetical tools, in the same way as the arithmetician operates on the known."

AL-SAMAWAL, MATHEMATICIAN AND ASTRONOMER

Muslim mathematicians excelled in the tenth century in yet another area when Ibn al-Haytham was the first to attempt to classify all even perfect numbers (numbers equal to the sum of their proper divisors), such as those of the form $2^{k-1}(2^k-1)$ where 2^k-1 is prime. He was also the first person that we know of to state Wilson's theorem, namely that if p is prime, then the polynomial $1+(p-1)!$ is divisible by p, but it is unclear whether he knew how to prove this. It is called Wilson's theorem because its "discovery" is attributed to John Wilson, a Cambridge mathematician in 1770.

But again, we do not know whether he could prove it or whether it was just a guess. It was a year later when a mathematician named Lagrange gave the first proof, 750 years after its "first discovery."

Mathematics was also needed in business and everyday use, and in particular it was essential in counting systems. Today, most of us are only aware of one counting system, which begins with zero and carries on into the billions and trillions. But in tenth-century Muslim countries, there were three different types of arithmetic used, and by the end of the century, authors such as Al-Baghdadi were writing texts comparing them. These three systems were finger-reckoning arithmetic, the sexagesimal system, and the Arabic numeral system.

Finger-reckoning arithmetic came from counting on fingers with the numerals written entirely in words, and this was used by the business community. Mathematicians such as Abu al-Wafa' in Baghdad in the tenth century wrote several treatises using this system. He was actually an expert in the use of Arabic numerals but said these "did not find application in business circles and among the population of the Eastern Caliphate for a long time." The sexagesimal system had numerals denoted by letters of the Arabic alphabet. It came originally from the Babylonians and was most frequently used by Arabic mathematicians in astronomical work.

The arithmetic of the Arabic numerals and fractions with the decimal place-value system was developed from an Indian version. Muslims adapted the Indian numerals into the modern numbers, 1 to 9, known as Arabic numerals. They are believed to have been based on the number of angles each character carries, but the number 7 creates a challenge, as the medial horizontal line crossing the vertical leg is a recent 19th century development. These have become the numerals we use in Europe and North Africa today, as distinct from the Indian numerals that are still used in some eastern parts of the Muslim world. Number 1, for example, has one angle, numeral 2 has two angles, 3 has three, and so on. The arrival of these numerals resolved the problems caused by Latin numerals in use until then. Arabic numerals were referred to as *ghubari* numerals because Muslims used dust (*ghubar*) boards when making calculations, rather than an abacus.

"Mathematics is the door and key of the sciences and things of this world . . . It is evident that if we want to come to certitude without doubt and to truth without error, we must place the foundations of knowledge in mathematics."

ROGER BACON, ENGLISH SCHOLAR

A great refinement by Muslim mathematicians of the Indian system was the wider definition and application of zero. Muslims gave it a mathematical property, such that zero multiplied by a number equals zero. Previously zero defined a space or a "nothing." It was also used for decimalization, making it possible to know whether, for example, the writing down of 23 meant 230, 23, or 2,300. It is interesting to note that if we imagined the zero sitting inside a hexagon, the ratio of the diameter of the circle to the side of the hexagon would equal the golden ratio. To read more about the golden ratio see the Geometry section in this chapter.

Muslim scholars were also fascinated by the significance of some numbers, such as the link of zero and one to the "One" as one of the 99 attributes of God, "nothing before Him and nothing after Him." It is interesting to note that the numerals 0 and 1 are the only two digits used in the computer language of today.

Arabic numerals came into Europe from three sources. First, through Gerbert (Pope Sylvester I) in the late tenth century, who studied in Córdoba and returned to Rome. Then through Robert of Chester in the 12th century, who translated the second book of Al-Khwarizmi's, which contained the second *ghubari* (Arabic numerals). This route of Arabic numerals into Europe is mentioned by contemporary historian Karl Menniger in *Number Words and Number Symbols*. The third route was through Fibonacci in the 13th century, who inherited and delivered them to the population of Europe. Fibonacci learned of these methods when he was sent by his father to the city of Bougie, Algeria, to learn mathematics from a teacher called Sidi Omar, who taught the mathematics of the schools of Baghdad and Mosul, which included algebraic and simultaneous equations.

Fibonacci also visited the libraries of Alexandria, Cairo, and Damascus, after which he wrote his book *Liber Abaci*. The first chapter deals with Arabic numerals. He introduced the numerals in the following words, "The nine numerals of the Indians are these: 987654321. With them, and with this sign '0,' which in Arabic is called *cephirum* [cipher], any desired number can be written."

It was this system of calculating with Arabic numerals that allowed most of the advances in numerical methods by Muslim mathematicians. Now the extraction of roots became possible by mathematicians such as Abu al-Wafa' and Umar al-Khayyam. The discovery of the binomial theorem for integer exponents by Al-Karaji was a major factor in the development of numerical analysis based on the decimal system. In the 14th century, Al-Kashi contributed to the development of decimal fractions, not only for approximating algebraic numbers, but also for real numbers such as pi. His contribution to decimal fractions is so major that for many years he was considered their inventor. Although not the first to do so, Al-Kashi gave an algorithm for calculating "nth roots" that is a particular example of methods developed many centuries later by Ruffini and Horner, 19th-century mathematicians from Italy and England respectively.

Although Arab mathematicians are most well known for their work on algebra, number theory, and number systems, they also made considerable contributions to geometry, trigonometry, and mathematical astronomy.

LEFT: *An example of the Babylonian sexagesimal number system illustrates how the figure of 424,000 was written.* RIGHT: *The progression of Arabic numerals from the tenth to the fourteenth centuries shows how the Muslims devised modern numerals—the numbers 1 to 9 we use today—based on the use of angles.*

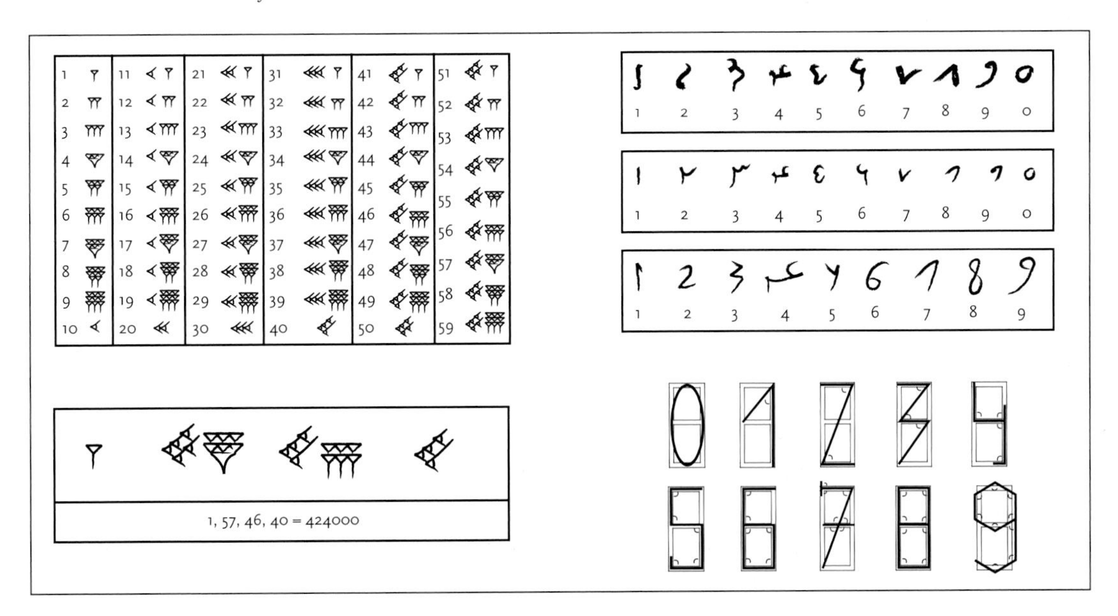

07 TRIGONOMETRY

The birth of trigonometry lies in astronomy, one of the sciences studied most vigorously by the Muslims, particularly due to its relevance in determining the exact times of the ritual prayer. But even before the Muslims, Greek astronomers were calculating the unknown sides and angles of certain triangles, given the value of the remaining sides or angles, in order to understand the motions of the sun, the moon, and the then-known five planets.

Motivated by questions such as the position of the sun, moon, and planets, the Greeks composed tables and rules that enabled geometric problems to be tackled. The most thorough treatment of the subject is found in the work *Almagest* by Ptolemy, who was an astronomer working in Alexandria in the early part of the second century C.E. Ptolemy's treatise reached European scholars via Muslim hands, who translated the original Greek title, which meant *The Great Arrangement,* into more succinct terms to produce *Al-Majisti,* simply meaning *The Greatest.*

Astronomers from late antiquity would draw principally upon a table found in Book I of *Almagest,* which was called *A Table of Chords in a Circle,* to solve all their plane trigonometric problems. For arcs at angles in increments of half a degree up to 180 degrees, the table gives the lengths of the chords subtending the angles in a circle of radius 60 units.

In his work *The Transversal Figure,* 13th-century Muslim astronomer Al-Tusi explains how this table of chord lengths was employed to solve problems relating to right-angled triangles. Al-Tusi made the crucial observation that established the link between triangles and arcs of circles: Any triangle may be inscribed in a circle; therefore, its sides may be viewed as the chords subtending the arcs opposite the angles of the triangle.

But there were two drawbacks to relying on these tables. First, considerable manipulation of the table and intermediate steps were required to solve all the variations that might arise in solving unknown lengths or angles of a right-angled triangle. This is in contrast to using the six familiar trigonometric functions—the sine, cosine, and tangent, and their reciprocals, the secant, cosecant, and cotangent—that are characteristic of modern techniques, which were first devised and arranged in a systematic way by Muslim mathematicians. The second inconvenience of the chord length tables is that they often required angles to be doubled in order to calculate the length of an arc.

Actually, a chain of Muslim scholars had already laid the foundations of trigonometry before the tenth century, paving the way for Al-Tusi to collect, organize, and elaborate on their contributions. It was Al-Battani, born in Harran, Turkey, who was one of the most influential figures in trigonometry. He is considered to be one of the greatest Muslim astronomers and mathematicians, eventually dying in Samarra, now in Iraq, in 929. His motivation for pioneering the study of trigonometry was his observation of the movements of planets. You can read more about him in the Astronomy section of the Universe chapter.

More crucially, Al-Battani explained his mathematical operations and urged others "to continue

A thousand years ago, Muslim scholars pioneered the study of trigonometry as they observed the movement of the planets, and predicted unknown lengths and angles. Today, trigonometry, including spherical trigonometry, is used in solving complex problems in astronomy, cartography, and navigation.

observation and to search" in order to perfect and expand his work. As well as Al-Battani, Abu al-Wafa', Ibn Yunus, and Ibn al-Haytham also developed spherical trigonometry and applied it to the solution of astronomical problems.

Al-Battani was the first to use the expressions sine and cosine, defining them as lengths, rather than the ratios we know them as today. The tangent was referred to by Al-Battani as the "extended shadow," the shadow of a notional horizontal rod mounted on a wall. In the 11th century, Al-Biruni defined the trigonometric functions of tangent and cotangent, which were inherited in a tentative form from the Indians.

Al-Biruni, born in 973, was among those who laid the foundation for modern trigonometry; Al-Khwarizmi, born in 780, developed the sine, cosine, and trigonometric tables, which were later translated to the West.

It would be another 500 years, though, before the trigonometry of tangents was discovered by modern mathematicians, and another 100 years before Nicolaus Copernicus was aware of it.

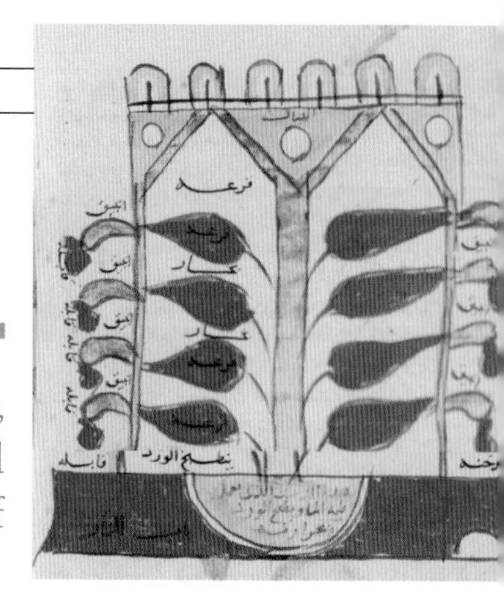

08 CHEMISTRY

Materials such as plastic, rayon, artificial rubber, and gasoline, and medicines such as insulin and penicillin, all stem from the chemical industry of the early Muslims, who were chemistry revolutionaries.

The word "chemistry" in Arabic is *kimia*, and with *"al"* as the definite article *kimia* becomes *alkimia* (the chemistry). In the West the last "a" was dropped and the word became "alchemy" in English. Alchemy, for the main Islamic medieval scientists, was not about folklore or occult practices but about the chemistry.

There are three people who stand out in Muslim chemistry from a golden era spanning 200 years.

Jabir ibn Hayyan, or Geber (722 to ca 815, Iran)
Jabir ibn Hayyan was known in the West as Geber, and all scholars agree that he is the founder of chemistry. The son of a druggist, he spent most of his life in Kufa, Iraq, where he scientifically systemized chemistry. Constantly in the laboratory, he devised and perfected sublimation, liquefaction, crystallization, distillation, purification, amalgamation, oxidation, evaporation, and filtration. He also wrote about how chemicals combined, without loss of character, to form a union of elements together that were too small for the naked eye to see. Now this may seem like common sense, but more than 1,250 years ago, he was a man ahead of the times.

Jabir vastly increased the possibilities of chemical experiments by discovering sulfuric, nitric, and nitromuriatic acids, all now vitally important in the chemical industry.

He also built a precise scale, which weighed items 6,480 times smaller than the *ratl* (ratl=1 kilogram or 2.20 pounds), and noticed in certain conditions of oxidation, the weight of a metal was lessened.

Some of Jabir ibn Hayyan's writings include the *Great Book of Chemical Properties, The Weights and Measures, The Chemical Combination,* and *The Dyes*. Among his greatest contributions to the theory of chemistry were his views on the constitution of metals, and these survived with slight alterations and additions until the beginning of modern chemistry in the 18th century.

All this research was carried out in his laboratory in Kufa, Iraq, which was rediscovered about two centuries after his death during the demolition of some houses in the quarter of the town known as the Damascus Gate. Found among the rubble were a mortar and a large piece of gold.

Al-Razi, or Rhazes (864 to 925, Iran)
Muhammad ibn Zakariya' al-Razi was known in the West as Rhazes, and he wrote *The Book of the Secret of the Secrets*. In this, he proved himself to be a greater expert than all his predecessors, including Jabir, in the exact classification of natural substances. He divided them into earthly, vegetable, and animal substances, while also adding a number of artificially obtained materials like lead oxide, caustic soda, and various alloys. Before him, Jabir had divided mineral substances into bodies (like gold and silver), souls (like sulfur and arsenic), and spirits (like mercury and sal-ammoniac).

TOP: *A 14th-century manuscript shows hemispherical vessels with a rose-and-water mixture resting on a fire—the red base. The vapors are collected and cooled in the eight vessels, which feed into eight external alembics.*

Al-Razi also excelled in writing up his experiments. From his *Secret of the Secrets* we know he was performing distillation, calcination, and crystallization more than 1,100 years ago.

He designed, described, and used more than 20 laboratory instruments, many of which are still in use, such as the crucible, cucurbit, or retort for distillation.

Al-Kindi (801 to 873, Iraq)

A lot of Al-Kindi's work was translated into Latin by men like Gerard of Cremona, so today there is more in Latin than Arabic. For instance, there is *De gradibus,* in which Al-Kindi explains that the complex of a compound medicine could be mathematically derived from the qualities and degrees of its component samples, and that there was a geometrical relationship between increasing quantity and degree of effectiveness.

Like much of the knowledge built up in the Muslim world, the work of Al-Kindi did not stay there; like all good ideas, it spread around the globe. It was translated into Latin and even into local languages, which explains its spread to Europe. Italian Gerard of Cremona made the more valuable translations like Al-Razi's *De aluminibus et salibus,* a study and classification of salts and alums (sulfates).

Important scientists of 13th-century Europe, like Albertus Magnus and Roger Bacon, came to know about these works. Roger Bacon particularly believed in the great importance of chemistry, which he discovered from the Latin translations of Arabic works.

This huge translation process, from Arabic to Latin, began in the middle of the 12th century. One work of Jabir's, *Liber Claritatis,* appeared in the last third of the 13th century and along with *Summa Perfectionis Magisterii* or *Sum of Perfection* was translated into Latin. Frequently printed together with other treatises in one volume between the 15th and 17th centuries, they were all known as *The Summa,* and the volume was so successful that it became the main chemistry textbook of medieval Europe. This manual on general chemical literature remained unrivaled for centuries.

Ninth-century chemist Al-Razi is shown in his Baghdad laboratory. Modern perfumes would not exist today without the distillation process.

EARLY CHEMISTS AND CHEMICALS

Muslim Civilization's Ingenious Experimenters and Their Legacy

LEGACY: **Influenced European chemistry textbooks up to the 17th century**

LOCATION: **Persia**

DATE: **Ninth century onward**

KEY FIGURES: **Al-Razi and Jabir ibn-Hayyan**

From rose water to hair dye, soap to paint, early chemists worked to create a panoply of useful substances. As early as the middle of the ninth century, experimenters in Muslim civilization were aware of the processes of crystallization, oxidation, evaporation, sublimation, and filtration. To make their experiments more accurate, they invented precise scales to use for weighing chemical samples. But alongside this experimental work, they came up with new theoretical ideas and chemical concepts, some of which survived for centuries.

Scientists of this period laid important foundations of the modern chemical industry. Jabir ibn Hayyan, and his successor, Muhammad ibn Zakariya al-Razi, developed new ways of classifying substances and organizing chemical knowledge. They wrote chemical textbooks and researched processes to improve ceramic glazes, formulate new hair dyes, and create varnishes for waterproofing fabrics. Other scholars worked on synthetic chemicals useful for pesticides, papermaking, paints, and medicines. Al-Razi, or Rhazes in Latin, made hundreds of discoveries in his chemical laboratory, writing up his findings in a book intriguingly entitled *The Book of the Secret of Secrets.*

Jabir, known as Geber in the West, carried out many ingenious experiments, including attempts to make paper that would not burn and ink that you could read in the dark. He is said to have used an alembic still for distillation, one of the most important techniques of the period. In this curiously shaped glass vessel, a liquid could be boiled down, allowing its separate pure parts to be collected as they condensed and trickled down the spout. Rose water was one of the first products of the distillation process, a delicately scented liquid vital for flavoring foods and drinks, and in perfumes and cosmetics. Al-Kindi wrote a book on the chemistry of perfumes, which contained 107 recipes for different scents.

The early chemists distilled wine, not to make a drink, but to use the pure alcohol as a disinfectant or ink mixed with ground silver filings. Perhaps most useful of all, they distilled the thick crude oil known as *naft* to produce the fuel kerosene, and in the 12th century made stronger acids by distilling vinegar. Today, distillation is still crucial for refining oil, and is used widely in the chemical industry.

During the 12th and 13th centuries, many Arabic textbooks and writings on chemistry were translated into Latin to reach a wider audience. One particular set of works said to be associated with Jabir was republished many times up to the 17th century, becoming the main chemistry textbook throughout medieval Europe.

■ *The distillation process is shown in an 18th-century Arabic treatise on chemistry. The Arabic text refers to the various vessels and the alembic, describing how the condensation is conveyed from the upper cooling vessel to the recipient flask.*

Works by other influential chemists were also keenly consulted. Al-Razi's writings showed how to prepare and use hundreds of chemicals. Al-Kindi's book *De gradibus* about the effects of compound medicines was translated by Gerard of Cremona. With sophisticated mathematical formulas for the qualities of the mixtures of the ingredients, it influenced Western pharmacology.

09 COMMERCIAL CHEMISTRY

The systematic approach of Muslim chemists more than 1,100 years ago led to the discovery of a process that today affects every person and every nation on Earth. And a product of this process, after water, is now considered one of life's essentials. Who would have thought that the black sludge known in Arabic as *naft* could have more than 4,000 uses? Without the process of distillation, and in this case of crude oil, we would have no gasoline, kerosene, asphalts, or plastics.

Distillation is a means of separating liquids through differences in their boiling points, and has been known to Muslim chemists since the eighth century. Its first and most renowned application was in the production of rose water and "essential oils." Pure alcohol was also obtained from the distillation of wine, which was produced and consumed mainly by non-Muslim communities, such as Christians living under Muslim rule. Jabir ibn Hayyan described a cooling technique that could be applied for its distillation. The distilled alcohol and alcoholic mashes were then used in chemical processes for the production of acids, medicines, perfumes, and inks for writing, but not drinks, as Islam prohibits the consumption of alcohol and other intoxicants.

Jabir was the first to develop the alembic still in the eighth century, which is still used today in distillation laboratories. It cooled and collected the necessary liquids in the distillation process.

Refineries, like this one in Malaysia, produce gasoline and kerosene through distillation from crude oil.

The word "alembic," like much chemical terminology, comes from the Arabic *al-anbiq,* which means "the head of the still." The alembic still has two retorts connected by a tube. It was in the alembic still that Jabir observed the flammable vapors coming from boiling wine and salt. In his chemistry book he wrote: "And fire which burns on the mouths of bottles due to boiled wine and salt, and similar things with nice characteristics, which are thought to be of little use, these are of great significance in these sciences."

The flammable property of alcohol was used extensively from Jabir's time. There are descriptions in military treatises from the 14th century of distilled old grape wine becoming an important ingredient in the production of military fires. These manuscripts also came with warnings that such distillates could ignite easily and that they should therefore be stored in containers buried in sand.

Al-Kindi was famous for his perfume distillations, which he wrote about in the *Book of the Chemistry of Perfume and Distillations* in the ninth century. In this, he described a distillation process: "and so one can distill wine using a water-bath it comes out like rose water in color. Also vinegar is distilled and it comes out like rose water in color."

Nine centuries ago, Ibn Badis from Tunisia described how silver filings were pulverized with distilled wine to provide a means of writing with silver. He said, "Take silver filings and grind them with distilled wine for three days; then dry them and grind them again with distilled wine until they become like mud, then rinse them with water."

As we have said, alcoholic drinks are *haram,* or forbidden, for Muslims, but their interest and discovery of it through distillation was intended to use its beneficial and harmless elements. Its discovery has given rise to a huge number of products in industries from pharmaceuticals to cosmetics. Much of their work a thousand years ago had practical application, and with their research and acquired knowledge from other cultures, new items could be manufactured, such as ink, lacquers, solders, cements, and imitation pearls.

A diagram warns of the presence of acid. The ancient world knew no stronger acids than vinegar, but the acids introduced by Jabir ibn Hayyan vastly increased the range of experiments that could be carried out.

Among the key experiments that marked the beginning of synthetic chemistry were those of Al-Razi, when he described how to obtain mercuric chloride as "corrosive sublimate" in *On Alums and Salts.* This, coupled with the discovery of chloride of mercury, today used in pesticides, inspired the discovery of other synthetic substances. The discovery of corrosive sublimate, and the fact that it was capable of chlorinating other materials, began the unearthing of mineral acids. Corrosive sublimate today has important applications in medicines as an astringent, stimulant, caustic, and antiseptic.

In the field of industrial chemistry and heavy chemicals, one of the greatest advances of medieval times was the isolation and manufacture of alum from "aluminous" rocks, through artificial weathering of alunite. Alum was used in paper-making, paint production, and the production of sulfuric acid. It was Jabir who discovered acids like sulfuric and hydrochloric.

10 GEOMETRY

Muslims are famous for intricate and elegant geometrical designs decorating their buildings, which you can read more about in the Art and Arabesque section of this chapter. These wonderful designs could not have happened without leaps made in geometry, or the measurement, properties, and relationships of points, lines, angles, and two-dimensional and three-dimensional figures.

Scholars inherited, developed, and extended geometry from the Greeks, who took a keen interest, and Euclid spent a lot of time on it in the *Elements*. For most avid mathematicians, their starting point into geometry is through Euclid's monumental and timeless work.

The investigations Muslims undertook in geometry rested on three Hellenist pillars. The first was Euclid's *Elements*, which was translated in Baghdad in the eighth-century House of Wisdom. The second was two works of Archimedes: *On the Sphere and Cylinder* and *The Heptagon in the Circle*. The second one is now unavailable in Greek and reaches us through the Arabic translation by Thabit ibn Qurra. The third and final pillar is the difficult work of Apollonios of Perga, called *The Conics*. This appeared in eight books, written around 200 B.C.E. Only four of these survive in Greek, while seven came to us in Arabic.

Most of the geometrical constructions of both the Greek and Islamic worlds were unified under the theory of conic sections, which were used in geometrical constructions, the design of mirrors for focusing light, and the theory of sundials. The surface of a solid double cone is formed by extending out straight lines (generators) that radiate out of the circumference of a circle, called the base, and pass through a fixed point, denoted the vertex, not on the plane of the base. Conic sections are generated by cutting the double cone by a plane intersecting the generators. The shape of the plane section that remains is determined by the angle of the plane to the generators. Apollonios successfully argued that, other than the circle, only three kinds of conic sections could be generated: the ellipse, the parabola, and the hyperbola.

Abu Sahl al-Kuhi used the theory of conic sections to develop a remarkable procedure for the construction of a regular seven-sided polygon, the heptagon. Abu Sahl al-Kuhi was one of a group of gifted scientists who were brought together from all over the eastern segment of the Muslim world under the auspices of key members of the influential Buyid family in Baghdad. Emerging from the mountainous regions south of the Caspian Sea, and originally a juggler of glass bottles in the *souk*, or market, of Baghdad, Abu Sahl al-Kuhi turned his attention to the study of the sciences. He was interested in the work of Archimedes, writing a commentary on Book II of *On the Sphere and Cylinder*. His main focus lay in conic sections and their use in solving problems related to the construction of complex geometric objects.

> *"Let no one ignorant of geometry enter."*
>
> INSCRIPTION ABOVE PLATO'S ACADEMY

For instance, he explained how it was possible to construct, with conic sections, a sphere with a segment similar to a segment of one sphere and possessing a surface area equal to a segment of a second sphere. He elaborated on a new instrument that could be used for drawing conic sections, "the complete compass." But Abu Sahl al-Kuhi had set his sights on even greater ambitions: detailed instructions for the construction of the regular heptagon. Archimedes had supplied a proof relating to a regular heptagon inscribed within a circle that suggests that it ought to be possible to construct a heptagon, but this did not go quite far enough to provide an actual procedure. This is quite common in the abstract universe of mathematics. Occasionally, it is very difficult to derive a step-by-step procedure for the construction of certain mathematical objects. In such situations, mathematicians concern themselves with proving that at least such a procedure exists, leaving the discovery of the detailed procedure to others.

Even though Archimedes gave proof of the existence, the actual construction of the heptagon eluded the best Greek and Muslim mathematicians for centuries, so much so that the tenth-century Muslim scholar Abu al-Jud remarked that "perhaps its execution is more difficult and its proof more remote than that for which it serves as a premise." Cue Abu Sahl al-Kuhi to take up this challenge. Through deft manipulation, Abu Sahl al-Kuhi was able to tame the beast, reducing the problem to three steps, which, if reversed, would lead to the construction. He said to start with the construction of a relevant conic section based on the length of the side of the heptagon. Then generate a divided line segment according to given proportions, and from the divided line segment, form a triangle possessing certain properties. Finally, produce the heptagon from the constructed triangle.

Abu Sahl is also known for his discovery of a method for trisecting a given angle. This was

Tiles from Alhambra Palace, Granada, Spain, make a colorful design. Most Islamic tile designs have geometrical and mathematical codes.

referred to as "the lemma of Abu Sahl al-Kuhi" by Abd ul-Jalil al-Sijzi, a younger contemporary of Abu Sahl, and used in the construction of a regular nine-sided polygon, the nonagon.

Knowledge of conic sections was required by instrument makers to engrave them on the surfaces of sundials. The Greeks knew "that as the sun traces its circular path across the sky during the day, the rays that pass over the tip of a vertical rod set in the earth form a double cone, and, because the plane of the horizon cuts both parts of this cone, the section of the cone by the horizon plane must be a hyperbola." This motivated Ibrahim ibn Sinan, the grandson of Thabit ibn Qurra, to make a study of the subject. His life was cut short due to a liver tumor and he met his demise at the early age of 37 in 946. Yet "his surviving works ensure his reputation as an important figure in the history of mathematics," said J. L. Berggren, a contemporary writer.

Berggren then summarizes Ibrahim ibn Sinan's achievements: "His treatment of the area of

Measurements in nature often follow mathematical patterns, which inspired Muslim scholars. The golden ratio is where the relative sizes of consecutive features form a ratio such that the ratio of the larger size to the smaller one is equal to the ratio of the sum of the two sizes to the larger one, as seen here in the chambers of a mollusk shell and the arrangement of the spiny center of a coneflower

a segment of a parabola is the simplest that has come down to us from the period prior to the Renaissance . . . in his work on sundials, he treats the design of all possible kinds of dials according to a single, unified procedure, and it represents a fresh, successful attack on problems that had often defeated his predecessors."

In relation to practical geometric design, which would be used to embellish public buildings as mosques, palaces, and libraries, Muslim geometers were interested in justifying the craft of the artisans and exploring the limits of their art. Abu Nasr al-Farabi—who died in 950 and is better known for his work on music, philosophy, and his commentaries on Aristotle—is credited with a treatise of geometric constructions from tools with various restrictions. His work was titled, rather exotically, *A Book of Spiritual Crafts and Natural Secrets in the Details of Geometrical Figures*. This contribution of Al-Farabi was later incorporated by Abu al-Wafa', in his youth when Al-Farabi died, in his book *On Those Parts of Geometry Needed by Craftsmen*, providing full constructional details and justifications.

The kind of problems to which Abu al-Wafa' devoted his attention included constructing a perpendicular to a given segment at its endpoint; dividing a line segment into any number of equal parts; and constructing a square in a given circle and various regular polygons (with 3, 4, 5, 6, 8, 10 sides). All these constructions were to be carried out with nothing more than a straight edge and a "rusty compass," a compass with one fixed opening.

Geometry had special significance also for Muslim artists, architects, and calligraphers. They had a keen awareness of the affinity between measurements in nature and mathematical expressions, and they were constantly inspired by these deep connections.

Such measurements included the golden ratio, a ratio of measurements that is pleasing to the eye and appears a lot in nature, such as in mollusk shells and plant leaves. In layman's terms, it means the width of an object is roughly two-thirds that of its height, or approximately 1.618. It is also called a golden section or line, so that if a line is divided, the smaller part of the line is to the larger part of the line as the larger part is to the whole line. This turns out to be approximately the ratio of 8:13 and is visible in many works of art and architecture.

As well as being fascinated by these geometrical occurrences, artists were also looking for the center of any system of "chaos," so this concept of center in terms of proportion remained their focus, too.

The *Ikhwan al-Safa'*, or Brothers of Purity, were a group of scholars in the tenth century, who recorded their ideas about proportion in their *Epistles* or *Rasa'il*. They knew of the Roman canon of Vitruvius, a first-century B.C.E. architect and writer, who measured the human body as a system of proportion. It was this idea the *Ikhwan* considered to be defective, as it was centered on the sacrum or the groin, instead of the navel.

Vitruvius's findings were based on a Greek canon, and this was founded on an ancient Egyptian rule of proportion, which related to the backbone of the god Osiris. The "sacred backbone," or *Djet* pillar, was a predynastic representation of Osiris and it represented stability, endurance, and goodness.

After painstaking research, the *Ikhwan*'s epistles came to a different conclusion. They established that when the human body was stretched and extended out, the fingertips and the tips of the toes touched the circumference of an imagined circle. The center of this circle was then the navel and not the groin, if the body was that of a child under age seven. This perfect proportion, with the navel being the center, begins to be disproportionately placed after the age of seven—the age of innocence. At birth, the midpoint of the body is at the navel. As the individual grows the midpoint drops until it reaches the groin or sacrum.

The proportional ratio produces an ideal figure for religious painting. The width is eight spans, the height is ten spans, and the midpoint is on the navel. The division of the figure is a body eight heads long, a foot is an eighth of a body, the face is an eighth of a body, the forehead is a third of the face, and a face is four noses or four ears.

With the navel as the center of the circle, which represented the Earth and the place of life sustenance, this demonstrated a divine manifestation. These divine proportions were reflected in cosmology, musicology, and calligraphy, and in all arts from the tenth century. They were seen as the key to finding harmony and, for the mystics, closeness to God.

For example, the natural harmony of the figure of eight was seen by Muslim scholars as the basic number, which motivated them to make measurements in music scale, poetry, calligraphy, and artistic themes.

There is, of course, the whole fascinating area of the algebraic geometry of Umar al-Khayyam, and the geometric theory of lenses by Al-Tusi, which were both new fields of geometry. To read more about these go to the sections Mathematics in this chapter, and Vision and Cameras in the Home chapter.

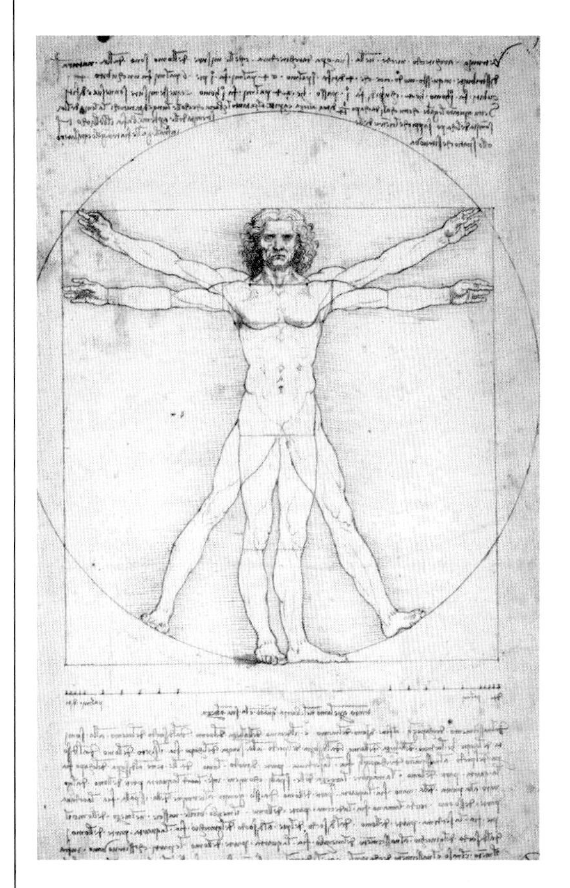

Leonardo da Vinci's "Vitruvian Man" shows the proportions of the human body, which were discussed in the tenth-century Epistles of the Ikhwan al-Safa'.

11 ART AND THE ARABESQUE

You can gaze into some art designs, and with each blink of an eye you see different shapes and forms. This type of geometric art is a fusion of pure mathematics and the art of space, an interplay of shapes and repeating patterns. It has no figures of people, but is made of flowing lines in complex designs. These designs seem to change as they are looked at, encouraging deep thought and spiritual contemplation, and, because of this, they fit well into mosques.

Prophet Muhammad spoke out against the portrayal of human or animal forms in art. He did not want Muslims at the time to revert to worshipping idols, figures, or the material world, a pre-Islamic practice, which would take attention away from God.

Geometry became central to the art of the Muslim world. Artists freed their imagination and creativity to produce a totally new art form called the arabesque, a development of geometric art. Arabesque is a pattern of many units joined and interlaced together, all flowing from the others in all directions. Each independent unit is complete and capable of standing alone, but all are interlinked and form a part of the whole design. These two-dimensional designs were mostly used to decorate surfaces like ceilings, walls, carpets, furniture, and textiles.

Outstanding examples of this sophisticated art form were recently discovered when the Topkapi scroll was uncovered in Istanbul. The scroll, with its 114 individual geometric patterns for wall surfaces and vaulting, is the work of a master builder who worked in Persia during the late 15th or 16th century. It is the earliest of its kind to have been found intact. Before its discovery, the earliest known Islamic architectural scrolls were fragments from the 16th century around Bukhara, Uzbekistan.

Arabesque can also be floral, using a stalk, leaf, or flower, or a combination of floral and geometric patterns, and these designs equally fascinated European artists. Works from the Renaissance, Baroque, Rococo, modern art (particularly in the grotesque), and strapwork all featured the patterns.

Leonardo da Vinci found arabesque fascinating, and used to spend considerable time working out complicated patterns. The famous knot design was worn by England's King Henry VIII, and it appears in his portrait on the border of his cloak and the curtains. Dürer used geometric patterns, as did Raphael. The grotesque designs of 17th-century French

The Lotfollah Mosque in Isfahan, Iran, shows arabesque and interwoven monumental cursive-style calligraphy of Quranic verses above and below the pointed arch.

> *"The staggering array of geometric patterns shows the way the Muslim craftsmen explored the concept of infinity through mathematical repetition."*
>
> REPORTER RAGEH OMAAR TALKING ABOUT THE ALHAMBRA IN GRANADA IN THE BBC'S *AN ISLAMIC HISTORY OF EUROPE*

artist Jean Bérain show it, and 16th-century Italian artists called it a*rabeschi*.

One of the best known 20th-century artists inspired by geometric art was Dutchman M. C. Escher. He created unique and fascinating works of art that explored a wide range of mathematical ideas, and not surprisingly he drew his inspiration from the tile patterns used in the Alhambra, which he visited in 1936. He spent many days sketching these, and later said that this "was the richest source of inspiration that I have ever tapped."

It was not only the arabesque that came to Europe; in the 14th century an important breakthrough for European artists took place. From the Muslim world they imported oil paint. In the past they had only used tempera paint on wood panels, which was a substance made of a combination of egg, water, honey, and dye. The expensive linseed oil paint had a dramatic effect on European paintings, as it enhanced the color saturation of Flemish and Venetian pictures.

CLOCKWISE FROM LEFT: *A portrait of King Henry VIII, who ruled 1509-1547, shows him with the Islamic knot-style pattern on the border of his cloak and on the curtain; he stands on a Turkish carpet with its* Ushak *star. Ceramic tiles display the* Iznik *blue patterns at Topkapi Palace in Turkey. The Topkapi scroll from the late 15th or 16th century shows individual geometric patterns for wall surfaces and vaulting compiled by a Persian master builder.*

12 The Scribe

There are many types of decorative writing, like Egyptian hieroglyphics, or Chinese or Japanese scripts, but Arabic calligraphy developed independently from all of these. It existed in Arabia even before Islam in the seventh century, but Muslims significantly developed it. They used it in art, sometimes combined with geometrical and natural figures, but it was also a form of worship, as the Quran promises divine blessings to those who read and write it. With the pen as a symbol of knowledge, the art of calligraphy was an art in the remembrance of God.

With this great impetus to write artistically, a final ingredient gave calligraphy another popularity boost. This was the mystical power attributed to some words, names, and sentences as protections against evil.

The language of Arabic calligraphy belongs to the family of ancient Semitic languages, and it comes in many scripts, the most famous of which are *Kufic* and *Naskh*.

The *Kufic* script comes from the city of Kufa, Iraq, where it was used by scribes transcribing the Quran in the Kufa school of writing. The letters of this script are angular.

The *Naskh* script is older than *Kufic*, but it resembles the characters used by modern Arabic writing and printing. It is joined up, a cursive script, and round, and has a few semi-styles. As early as the tenth century, famous calligrapher Abu-'Ali ibn Muqla devised a systematic classification of the script according to geometric principles, establishing a unit of measurement for letters and creating a balance among them. He counted six cursive scripts, which became known as *al-aqlam al-sitta*. *Naskh* calligraphy became more popular than *Kufic* style, which was developed by the Ottomans.

Before paper was introduced, parchment and papyrus were the main materials for copying the Quran, writing manuscripts, and correspondence.

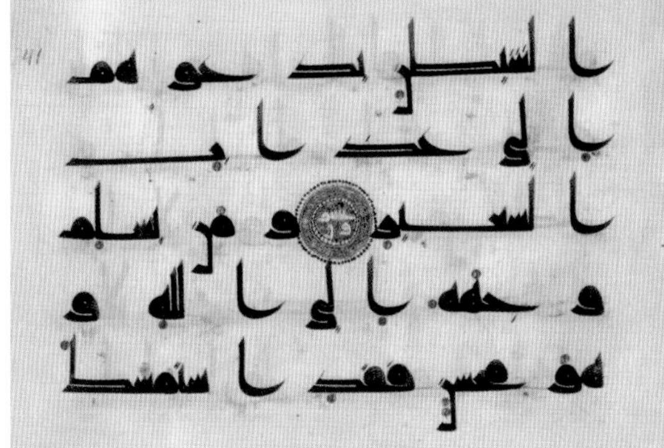

LEFT: *World-famous Turkish calligrapher Hasan Celebi instructs a student in the art of calligraphy.* RIGHT: *An ancient Kufic script displays two Quranic verses (21 and 22, chapter 31); the circle in the center signifies the separation between the verses.*

THE STORY CORNER: THE MYSTERY OF HAYY IBN YAQZAN

An illustration by Karima Solberg from Ibn Tufayl's Story of Hayy ibn Yaqzan *shows him with his adoptive "mother," a doe. Daniel Defoe's 18th-century* Robinson Crusoe *is very similar to Ibn Tufayl's 12th-century* Hayy ibn Yaqzan.

In early 12th-century Muslim Spain, a gifted philosopher, mathematician, poet, and medical doctor was born. Ibn Tufayl, or Abu Bakr ibn Abd al-Malik ibn Muhammad ibn Muhammad ibn Tufayl al-Qaysi, to give his full name, became known in the West as Abubacer. He held royal posts as an adviser and court physician to Abu Yaqub Yusef, the Almohad ruler of Al-Andalus, and he is remembered today for *The Story of Hayy ibn Yaqzan*, the original manuscript of which is now in the Bodleian Library at Oxford. This tale was inspired by an earlier story from the 11th-century physician-philosopher Ibn Sina, who also wrote a narrative called *Hayy ibn Yaqzan* about a century earlier. But was the story itself the inspiration for the book *Life and Strange Adventures of Robinson Crusoe*, by Daniel Defoe?

Hayy ibn Yaqzan means "Alive, son of Awake," so this is "The Story of Alive, son of Awake," which describes Hayy's character passing from sleepy childhood to knowledge by means of which he can fully contemplate the world and his surroundings.

It begins with Hayy as a child, a princess's son whose birth was a secret. He is cast upon the shore of an equatorial island where he is suckled by a doe and spends the first 50 years of his life without contact with any other human beings. His isolation is in seven stages of seven years. During each seven-year stage he is his own teacher and learns about himself and his surroundings.

The first English translations of Hayy ibn Yaqzan appeared in 1709. Eleven years later, Defoe's famous book was published. Many of Defoe's contemporaries said his inspiration lay in the experiences of Alexander Selkirk, a Scottish mariner who passed more than four years in solitude on one of the Juan Fernández Islands. But the similarities between *Robinson Crusoe* and *Hayy ibn Yaqzan* are enough to make it probable that Defoe knew the Muslim work. From the island shipwreck to the anguish of isolation and struggle for survival, *Robinson Crusoe* bears many similarities with the older work.

Parchment was durable, lustrous, and luxurious, though only one side could be used. Papyrus was brittle and could not be erased, which made it useful especially for government records. Both were expensive, so when the cheaper alternative, paper, was mass-produced in the late eighth century by the Muslims, after learning the art from China, the art of writing boomed.

Europe came into contact with Arabic calligraphy through trade and gift exchange between European and Muslim royal courts. At first, Europeans imitated Arabic calligraphy without knowing what it said, and *Kufic* inscriptions from the Ibn Tulun Mosque, built in Cairo in 879, were reproduced in Gothic art, first in France, then in other parts of Europe. Works such as the carved wooden doors by master carver Gan Fredus in a chapel of the under porch of the Cathedral of Le Puy in France, and another carved door in the church of la Vaute Chillac near Le Puy, are also attributed to the influence of the Ibn Tulun Mosque. Traders from Amalfi in Italy who visited

"Read: In the name of your Lord Who creates, creates man from a clot. Read: And your Lord is the Most Bounteous, Who teaches by the use of the pen, teaches man that which he knew not."

QURAN (96:1-5)—THE FIRST VERSE OF THE QURAN REVEALED TO PROPHET MUHAMMAD

Cairo are believed to be responsible for the transmission of these designs into Europe as they had special relations with Fatimid Cairo at that time.

In his book *Legacy of Islam*, Professor Thomas Arnold said a cross that probably dates back to the ninth century was found in Ireland with the phrase *Basmalah* (*bi-ism Allah*), or "In the name of God," inscribed in *Kufic* calligraphy on it. In other art forms, especially painting, *Kufic* inscription was added for style. People were drawn to calligraphy; even Italian Renaissance painter Gentile da Fabriano used it in the decorative edging of clothes in his painting "Adoration of the Magi."

Before pens, as we know them today, came other writing instruments including the *qalam*, or reed pen. The most sought-after reeds came from the coastal lands of the Arabian Gulf and they were valuable trading commodities. Their length varied between 24 and 30 centimeters (9.5 and 11.8 inches) and their diameter generally measured one centimeter (0.4 inch). Each style of script required a different reed, cut at a specific angle.

Inks were of different types and colors, with black and dark brown inks most often used; all differed in intensity and consistency. Calligraphers usually made their own inks, and sometimes the recipes were closely guarded secrets. Silver and gold inks were used on blue vellum, in frontispieces, for illustrations, and for title pages. Colored inks such as reds, whites, and blues were sometimes used in illuminated headings. Ink pots, polishing stones, and sand for drying the ink were additional accessories used by calligraphers.

MAKING OF A PEN

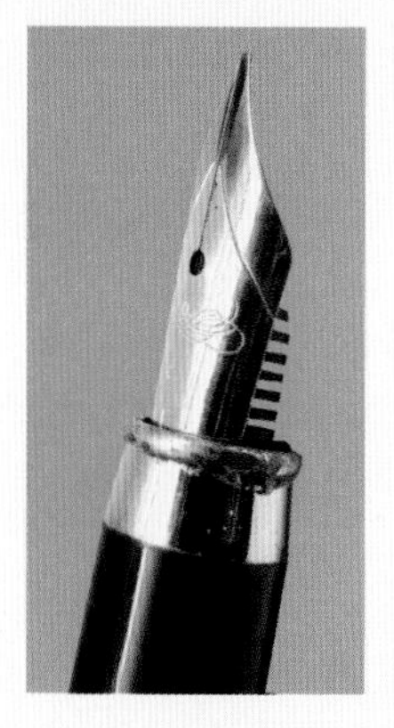

An outstanding publicist, confidant, and companion of Al-Mu'izz, the Egyptian sultan in 953, wrote a book called *The Book of Audiences and Concurrence*. His name was Qadi abu Hanifah al-Nu'man ibn Muhammad, and here he recounts Al-Mu'izz commissioning the construction of a fountain pen:

"We wish to construct a pen which can be used for writing without having recourse to an ink-holder and whose ink will be contained inside it. A person can fill it with ink and write whatever he likes. The writer can put it in his sleeve or anywhere he wishes and it will not stain nor will any drop of ink leak out of it. The ink will flow only when there is an intention to write. We are unaware of anyone previously ever constructing [a pen such as this] and an indication of 'penetrating wisdom' to whoever contemplates it and realises its exact significance and purpose." I exclaimed, "Is this possible?" He replied, "It is possible if God so wills."

The story continues that a few days later the craftsman brought a pen that wrote when it was filled with ink. The pen could be turned upside down and tipped from side to side without any ink being spilled. The pen did not release the ink except in writing and it did not leave stains on hands or clothes. Lastly, it did not need an ink pot because it had its own, hidden away.

The first chapter of the Quran in Jali Diwani *style was gilded by master gilder Mamure Oz from the Topkapi Palace in Istanbul.*

13 WORD POWER

This alphabet lists just some of the words that have come from sources in Muslim civilization and have passed into the English language with their original meaning intact. It is only a small selection. The actual list carries on into the thousands.

A is for **admiral**, from *amir-al-* "commander of . . . ," like *amir al-bahr*, or "commander of the seas." When the Europeans adapted *amir-al*, they added their own Latin prefix *ad-*, producing admiral. When this reached English, via Old French, it still meant "commander," and it was not until the time of England's Edward III that a strong naval link began to emerge. "A" is also for **arsenal**, from *dar al-sina'ah*, meaning "house of making/industry" like a factory. This was borrowed into Venetian Italian, where the initial "d" was not pronounced and became *arzaná*, which was applied to the large naval dockyard in Venice. The dockyard is known to this day as the Arzenale. English acquired the word either from Italian or from the French *arsenale*, using it only for dockyards. By the end of the 16th century, it was coming into more general use as an "ammunition storehouse."

B is for **barbican**, from the Arabic *bab al-baqarah*, or gate with holes.

C is for **crimson**, from *qirmizi*, which is related to the *qirmiz*, the insect that produced the red dye *qirmizi*. "C" is also for **caviar**, which may come from Farsi *kaya-dar*, meaning "having eggs" or from *chav-jar*, meaning "a cake of strength or power" or "bread of lovers," an allusion to its reputed aphrodisiac qualities. Others think it came from *havyar* in Turkish, which means "fish eggs."

D is for **dragoman**, an interpreter or guide in countries where Arabic, Turkish, or Persian is spoken; from the Arabic *tarjuman* and the verb, *tarjama*, to interpret.

E is for **El-Cid**, the hero of a Spanish epic poem from the 12th century, from *al-Sayyid*, meaning "the lord."

F is for **Fomalhaut**, the brightest star in the constellation Piscis Austrinus, the Southern Fish, 24 light-years from Earth; *fam al-hut* means "mouth of the fish."

G is for **ghoul**, from the Arabic *ghul*, meaning the demon. "G" is also for **giraffe** from the Arabic *Zarafa*.

H is for **hazard**, from *yasara*, which means "play at dice."

I is for **Izar**, name of a star in the constellation Andromeda, from the Arabic *al-'izar*, meaning the "veil or covering."

J is for **jar**, from *jarrah*, a large earthen vase. "J" is also for **jasmine**, from the Persian *yasmin*.

K is for **kohl**, from *kuhl*, meaning a fine powder, often of antimony, used in eye decoration or as eyeliner.

L is for **lilac**, from the Arabic *lilak*, which was taken from the Persian *nilak*, meaning "indigo." "L" is also for **lemon**, from the Persian *limun,* meaning "lemon."

M is for **mafia**, from an Arabic word *mahiyah,* meaning "boasting" or "flashy", i.e., "the swank set." In Sicily an unusually ornate and demonstrative cockerel is described as Mafioso. "M" is also for **mattress**, coming from old French *materas,* which was taken from *matrah,* a "place where something is thrown" and *taraha,* meaning "to throw."

N is for **nadir**, a point on the celestial sphere directly below the observer and diametrically opposite the zenith. It comes from *nadir as-samt,* meaning "opposite the zenith."

O is for **orange**, from the Persian *naranj* or *narang,* meaning "orange."

P is for **Pherkad**, a star in the constellation Ursa Minor, from the Arabic *al-farqad,* meaning "the calf."

Q is for **qanun**, the ancestor to the harp and zither, introduced by Al-Farabi in the tenth century, but used in Roman times as a freestanding instrument.

R is for the chess piece **rook**, from the Persian *rukh.*

S is for **sofa**; the seat was originally an Arabian ruler's throne and has been in existence since antiquity. Originally *suffah,* meaning "long bench" or "divan." "S" is also for **sugar** from the Arabic *sukkar,* meaning "sugar"; and for **so long**, from *salam,* a greeting and goodbye meaning "peace."

T is for **tabby**, which meant "silk cloth with striped pattern" and was borrowed in 1638 from the French. They used *tabis,* or rich-watered silk, from Arabic *'attabi,* originally meaning "manufactured at al-'Attabiyah," a suburb of Baghdad. By 1695 the phrase tabby cat was in use, and tabby as a noun meaning "striped cat" developed by 1774. "T" is also for **talcum** powder, which is from the Latin *talcum,* from the Arabic *talq*. It was first used in medieval Latin as *talc* around 1317, and in Spanish Talco and French as *talc* in 1582. In German, it is *Talkum.*

U is for **Unukalhai**, a star in the constellation of the Serpent, from the Arabic *'unuq al-hayyah,* meaning the "neck of the snake."

V is for **vizier**, from *wazir,* meaning "porter, public servant" from the verb *wazara,* to carry. "V" is also for **Vega**, the brightest star in the constellation Lyra from the Arabic *al-nisr al-waqi',* meaning "the falling vulture."

W is for **wadi**, a valley or gully that remains dry except during the rainy season, coming from the Arabic *wadi,* which means "valley."

X in algebra, meaning "a thing," is an Arabic invention to solve mathematic equations.

Y is for **yoghurt**. The original Turkish word was *yogurut,* but it had become yoghurt by the 11th century. The "g" is soft in the Turkish pronunciation but hard in English. *Yog* is said to mean, roughly, "to condense" in Turkish, while *yogur* means "to knead."

Z is for **zenith**, the point of culmination or the peak, coming from the Old Spanish *zenit,* which was from the Arabic *samt,* meaning "path," part of the Arabic phrase *samt al-ra's,* meaning "the road overhead," or directly above a person's head.

CHAPTER FOUR

"It is not permissible to sell an article without making everything [about it] clear, nor is it permissible for anyone who knows [about its defects] to refrain from mentioning them."

PROPHET MUHAMMAD, NARRATED BY AL-HAKIM AND AL-BAYHAQI

MARKET

AGRICULTURAL REVOLUTION • FARMING MANUALS
WATER MANAGEMENT • WATER SUPPLY • DAMS
WINDMILLS • TRADE • TEXTILES • PAPER • POTTERY
GLASS INDUSTRY • JEWELS • CURRENCY

WHETHER THROUGH BARTERING GOODS, GOLD, PAPER CURRENCY, OR DIGITAL TRANSFER, PEOPLE have been making deals and acquiring produce in the marketplace for many millennia. For 1,200 years, the Islamic world was a powerhouse of knowledge, influence, and innovation, all driven by a massive economy that bought and sold across three continents. Enterprising Muslims were producing goods at a fast rate, and great leaps in technology across many industries from textiles to chemicals meant that vast numbers of people were employed in these flourishing sectors.

Agricultural techniques, accompanied by research, improved irrigation, and landownership rights, meant that the standard of living was raised as people ate abundant food. Farming innovations included using pigeon manure for fertilization, a technique mastered in Iran where towers 18 to 21 meters high (60 to 70 feet) were dotted around the fields to house the birds. As these practices and knowledge drifted west they were accompanied by coinage, checks, and paper, while treasures of the world drifted back into the hustle and bustle of dynamic cities like Cairo. In this chapter you will peel back the layers of commerce to uncover the marketplace that was not so different from ours today.

OPPOSITE: *Muslim men and women bought and sold textiles, ceramics, and glass—as shown in this 13th-century painting.*

01 Agricultural Revolution

Today, we are more detached from our food sources than we were a thousand years ago. Few of us work the land or raise our own animals. We visit the local shops or supermarkets to sample delights from around the world, and can savor mangoes from Pakistan, strawberries from America, mushrooms from Holland, lamb from New Zealand, and beef from Argentina. No longer do we have to wait for summer apples or rely on pickled vegetables in winter; instead we just move along to the next shelf. But this concept of global food, not linked to local seasons and climates, is not new. What is new is that today it is mostly flown in, and not grown on local farms.

In the ninth century, Muslim farmers were making innovations: introducing new crops from all around the world, developing intensive irrigation systems, using global knowledge for local conditions in a scientific way, and promoting practical farming that included individual landownership. This all meant they could have a diversity of food previously unavailable.

Global Knowledge and Scientific Methods

Being from a civilization of travelers, Muslims combed the known world for knowledge and information, journeying in the harshest of environments from the steppes of Asia to the Pyrenees, detailing all they saw to produce huge agricultural manuals. These were a "spectacular cultural union of scientific knowledge from the past and the present, from the Near East, the Maghrib, and Andalusia," said American historian S. P. Scott in 1904.

Cotton, originally from India, was introduced as a major crop in Sicily and Al-Andalus.

As Professor Andrew Watson from the University of Toronto said, the Muslim world was "a large unified region which for three or four centuries . . . was unusually receptive to all that was new. It was also unusually able to diffuse novelties . . . Attitudes, social structure, institutions, infrastructure, scientific progress and economic development all played a part . . . And not only agriculture but also other spheres of the economy—and many areas of life that lay outside the economy—were touched by this capacity to absorb and to transmit."

With this vast array of knowledge coming from a diversity of geographic areas, Muslims could rear the finest horses and sheep, and cultivate the best orchards and vegetable gardens. They knew how to fight insect pests, use fertilizers, and were experts at grafting trees and crossing plants to produce new varieties.

Today, fresh food from around the world is readily available in local shops and markets.

> *"It is a blessed act to plant a tree even if it be the day the world ends."*
>
> PROPHET MUHAMMAD,
> NARRATED BY AL-BUKHARI AND AHMED

New Crops

In the ancient Mediterranean world, generally speaking, mainly winter crops were grown, and each field would give one harvest every two years. That was before the Andalusian Muslims arrived with crop rotation techniques as well as new crops, many from India. These needed warm or hot weather, which was provided by the long summer days, although there were also dry months with little rain. With the Muslim introduction of irrigation, though, four harvests each year could now be produced for numerous crops.

Subtropical crops, like bananas, were grown in the coastal parts of the country, and the new crops included rice, citrus fruit, peaches, plums, silk, apricots, cotton, artichokes, aubergines, saffron, and sugarcane. As well as introducing sugarcane to Spain where it had a huge impact, Muslims took it to Ethiopia, and to Zanzibar, now famous for its high-quality sugar.

A silk industry flourished, flax was cultivated, and linen exported. Esparto grass, which grew wild in the more arid parts of Spain, was collected and turned into products like baskets and floor coverings.

Al-Masudi, a tenth-century Muslim traveler and historian, wrote about the introduction of orange and citron trees: "The orange tree, *shajar al-naranj,* and the citron tree, *al-utrujj al-mudawwar,* were brought from India around 300 A.H. [912 C.E.] and were first planted in Oman. From here they were carried via al-Basra into Iraq and Syria. In a very short time they became numerous in the houses of the people of Tartus and other Syrian frontier and coastal towns. Very quickly the trees were sprouting up over Antioch, Palestine and Egypt where but a short time ago they were unknown."

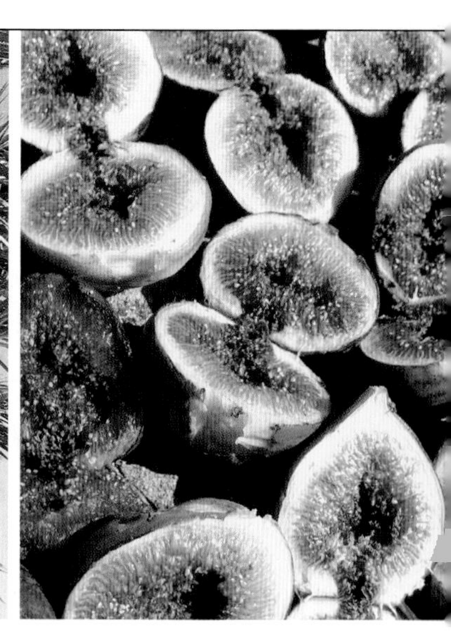

FROM LEFT: *Some of the crops that Muslims brought to and cultivated in southern Europe include citrus fruits, dates, and figs.*

The transfer of such crops was often due to the enthusiasm of individual people, like 'Abd al-Rahman I, who, out of nostalgia for his Syrian landscape, was personally responsible for the introduction of several species, including the date palm, to make himself feel more at home in this new land of Al-Andalus, or Spain. A variety of pomegranate was introduced from Damascus by the chief judge of Córdoba, Mu'awiya ibn Salih, and a Jordanian soldier named Safar took a fig cutting and planted it on his estate in the Málaga region. This species, called *safri* after the soldier, spread over the land.

Irrigation

As we have seen, crops were grown in the hot summers, and some of the new crops needed more water than was available, such as sugarcane, which had to be watered every four to eight days. Rice had to be continually submerged. Cotton was grown from the end of the 11th century and, according to medieval historian Ibn Bassal, had to be watered every two weeks from the time it sprouted until August. The Andalusis were self-sufficient in cotton, and exported to Sijilmasa (Morocco) and as far south as Ifriqiya, a region of Africa. Oranges and other citrus plants were also irrigated, as were many other fruit trees and dry-farming crops.

So how were these water demands met? Muslims were experts in raising water by several meters, guaranteeing a constant flow by using both pumps and waterwheels, or *norias*. In the Valencia area alone about 8,000 *norias* were built to take water to the rice plantations.

Muslims built upon the traditional use of animals to power machinery, and also devised advanced gearing mechanisms, and dug underground canals, or *qanats,* to take water through harsh, barren deserts like the Sahara. You can read more about irrigation and water practices in the Water Management and Water Supply sections in this chapter.

For the water to make it to the fields, the level of the irrigation systems had to be correctly calculated, and Muslims had the advantage of the advances they had made in mathematics. By using triangulation they could accurately make measurements of height.

It was not only mathematics that helped agriculture, because major advances in astronomy in 11th-century Toledo, Spain, were also having an impact. As reporter Rageh Omaar says in the BBC's *An Islamic History of Europe,* "astronomical tables were used in agriculture . . . the tables showed times for planting and harvesting."

New Landownership Approach

The last important factor for this boom in food production was the development of a new vigorous system of ownership. Farmers could now work more for themselves and the community, rather than in misery, suffering exploitation at the whim of big landowners. It was a revolutionary social transformation in landownership when laborers' rights were introduced. Any individual had the right to buy, sell, mortgage, inherit, and farm the land, or have it farmed according to his preferences.

The Spanish Muslims' agricultural system was "the most complex, the most scientific, the most perfect, ever devised by the ingenuity of man."

19TH-CENTURY AMERICAN HISTORIAN S. P. SCOTT

Every important transaction concerning agriculture, industry, commerce, and employment of a servant involved the signing of a contract and each party keeping a copy. Those who physically worked the land received a reasonable proportion of the fruits of their labor, and detailed records of contracts between landlords and cultivators have survived, showing that the landlord retained anything up to one-half.

With these farming innovations, the quality of life increased dramatically, and an enriched diet for all was possible with the introduction of year-round fresh fruit and vegetables. Thus, less needed to be dried for winter.

Citrus and olive plantations became a common sight, and market gardens and orchards sprang up around every city. All this involved intensive cropping, which could have led to decreased soil fertility, but the technique of intensive irrigation coupled with fertilization techniques, using mainly pigeon dung, had been mastered.

Animal husbandry and selective breeding using animals from different areas meant horse stocks improved and strong camels could carry the goods of the Saharan caravans.

Animal products such as meat and wool became plentiful in places where they had been a luxury. This included the use of animal manure. The fine-quality products from the Maghreb region of Tunisia, Algeria, and Morocco soon became known throughout the world.

Not only wool, but now silk and cotton were being produced. Cotton, originally from India, became a major crop in Sicily and Al-Andalus, making previously rare luxury goods available. Within a relatively short period, people had access to a wider range of textiles for clothing, which now also came in a greater variety of colors.

MAKING SUGAR CANE

"The cultivation of sugar cane in the West spread from Khurzistan in Persia, and throughout the middle ages Shuster [the ancient Susa] was renowned for its manufacture on a large scale. The art of sugar refining was practised extensively by the Arabs. Under Muslim rule the growth and manufacture of the cane spread far and wide, from India to Morocco. Through the Muslim dominions in Spain and Sicily it reached southern Europe."

GUY LE STRANGE, 20TH-CENTURY ORIENTALIST

02 FARMING MANUALS

For a garden or crop to bloom there has to be an ecological balance between nurture and nature. The elements of soil, water, and human intervention have to be in distinctly proportional amounts to ensure a good flowering and harvest. In their bid to achieve maximum output without destroying the things they relied upon, namely the soil and plants, Spanish Muslims started a systematic study of agriculture, including soil chemistry and soil erosion, hundreds of years ago.

Muslim agriculture was a sophisticated affair, which resulted in an ecologically friendly and very productive system. They had farming books that explained just about everything in detail, like how to enrich the soil by plowing, normal and deep hoeing, digging, and harrowing. Soil was classified, and so was water, according to its quality. Ibn Bassal, gardener to the emir of Toledo, wrote a *Book of Agriculture* in 1085. This classed ten types of soil, assigning each with different life-sustaining capabilities according to the season of the year. He insisted that fallow land should be plowed four times between January and May and, in certain cases, he recommended as many as ten plowings—for example, for cotton crops that were planted in heavy Mediterranean coastal soils.

A 14th-century Persian manuscript from Al-Biruni's Chronology of Ancient Nations *displays men at work in the field; one uses a spade.*

FROM LEFT: *The* Calendar of Córdoba of 961 *had tasks and timetables for each month. March noted that roses bloomed and quails appeared.*

Ibn al-Awwam, a 12th-century botanist from Seville in Muslim Spain, gathered together previous studies of Greek, Egyptian, and Persian scholars into another *Book of Agriculture*, which had 34 chapters on agriculture and animal husbandry, and also gave farmers precise instructions. It included 585 plants, explained the cultivation of more than 50 fruit trees, made observations on grafting, soil properties and preparation, manure, plant diseases and their treatments, gardening, irrigation, affinities between trees, and beekeeping. It covered all you could want to know about olives, from how to grow the trees, the treatment of their diseases, grafting, and harvesting olives, to the properties of olives, refining olive oil, and their conditioning. Then there was a section on plowing techniques, their frequency, times for sowing and how to sow, watering after sowing and during growth, maintenance of plants, and harvesting. So, with all this information an avid farmer could not go wrong, and most of this was published in Spanish and French between the end of the 18th and the middle of the 19th centuries.

Then there was the remarkable technical accuracy of the famous *Calendar of Córdoba of 961*. Each month of the year had tasks and timetables. For instance, March was when fig trees were grafted and early cereals began to rise. It was the time to plant sugarcane, and when early season roses and lilacs began to bud. Quails appeared, silkworms hatched, and mullet began to journey up rivers. This was also the time to plant cucumbers, and sow cotton, saffron, and aubergines. During this month mail orders to purchase horses for the government were sent to provincial tax officials. Locusts began to appear and their destruction was ordered. It was the time to plant lime and marjoram, and was also the mating season of many birds.

There was no agricultural stone left unturned; even individual crops were ruthlessly scrutinized. Rice, for example, had Ibn Bassal advising the use of plots that faced the rising sun; then the thorough preparation of the soil by adding manure was recommended. Sowing was advised between February and March. Ibn al-Awwam gave the specific amount of rice that needed to be sown on any given surface, and how that should be carried out. He also spoke at length of the watering process, specifying that land should be submerged with water up to a given height before the rice was planted. Once the soil had absorbed the water, the seeds were covered with earth, and the land submerged with water again.

"With a deep love for nature, and a relaxed way of life, classical Islamic society achieved ecological balance, a successful average economy of operation, based . . . on the acquired knowledge of many civilized traditions. A culmination more subtle than a simple accumulation of techniques, it has been an enduring ecological success, proven by the course of human history."

LUCIE BOLENS, AUTHOR OF *THE USE OF PLANTS FOR DYEING AND CLOTHING*

Rice experts also focused on fighting parasites, clearing weeds, and ways of harvesting and safe storage. The use of rice as a food took many forms, and Ibn al-Awwam specified that the best way to cook and eat rice was with butter, oil, fat, and milk. An anonymous author of the Almohad dynasty also wrote a recipe book called *The Cookery Book of Maghrib and Andalusia*, which included many recipes, five of them with rice, all sounding most appetizing.

A very important part of farming was ensuring field fertility to achieve a perfect balance. This was thoroughly explored, and interestingly, has not changed much in a thousand years, as medieval Muslims were also liberally applying manure to their fields. Ibn al-Awwam states that the best manure is from pigeons, and by today's standards it was definitely environmentally friendly and organic.

Dotted across the land in Iran were pigeon keeps—large circular towers made from mudbrick, with smaller turrets projecting from their summits. They stood at 18 to 21 meters (60 to 70 feet) in height and were constructed for collecting manure and breeding more pigeons.

Inside, the towers were made up of small cell-like compartments, like a honeycomb. The guano or dung accumulating over time would be spread on the surrounding fields after the pigeon towers were cleaned once a year. It is said that at one time there were as many as 3,000 of these pigeon towers outside Isfahan in Iran.

LEFT: *Ruins of a pigeon tower still stand near Isfahan, Iran. Muslims believed that the best organic fertilizer was manure from pigeon droppings, and they used it liberally on their fields.* RIGHT: *Pigeons were bred primarily to be used in the postal network for carrying messages.*

An artist's re-creation shows Muslim farmers at work. Innovative farmers in the ninth century were planting new crops, developing state-of-the-art irrigation techniques, using organic fertilizers, harnessing global knowledge in local areas, and basing their agronomy on scientific findings. This all led to an agricultural revolution, making fresh food available to more people.

03 WATER MANAGEMENT

Whether it is Andalusia or Afghanistan, Chicago or Cairo, water is essential for agriculture and sustenance. It is the source of all life. Muslims inherited existing techniques of irrigation, preserving some while modifying, improving, and constructing others.

Their engineering advances were partly down to progress in mathematics, which meant hydrology and the machinery for building irrigation devices were constantly being revolutionized. Eleventh-century Persian mathematician and engineer Muhammad al-Karaji talked about "the bringing to the surface of hidden waters." He also covered surveying instruments, methods of detecting sources of water, and instructions for the excavation of underground conduits.

These underground conduits or tunnels were dug to prevent water loss by evaporation. Called *qanats*, the oldest were in Persia and with the development of agriculture, and with more crops being planted, they became essential and *qanat* building became a necessity, especially in the dry environment of the Middle East. Later they came to Córdoba, Spain, making water available for urban domestic use.

Persia and today's Afghanistan had thousands of wells, all connected by these underground canals. They were constructed to withstand problems of silting and roof cavings, ensuring a continuous flow of water through miles and miles of formidable deserts and hostile terrain. In some areas of solid rock the *qanat* appeared as an overland stream, and then disappeared again as the geology changed. In the Algerian Sahara, there were also networks of underground tunnels, called *Foggaras*.

The Nilometer in Rawada Island at Fustat, near Cairo, was completed in 861-862. The octagonal column in the center is used to measure the height of the water in the Nile in cubits, an ancient method of measurement.

Here farmers also used a water clock, a clepsydra, to control water use for everyone in the area as it timed, night and day, the amount going to each farmer.

In parts of Iran, despite the existence of hydroelectric dams and modern irrigation systems, *qanats* are still a farmer's lifeline. Northeast of Shiraz, the precious commodity of water is still obtained from wells supplied by underground canals.

Given the scarcity of water in these hot, arid environments, it had to be controlled and regulated, just as it is today. The authorities of the time played a crucial role, too. In Iraq, hydraulic works

Muslims were able to transport water over long distances using a series of L-shaped wells connected to one another. Forming an underground tunnel, called a qanat—*these are near Isfahan, Iran—they had "manhole" covers for air circulation, which helped the water flow through the tunnel.* Qanats *are still used today.*

of a vast nature, like dams, were left to the state, while the local population focused its efforts on lesser ones, like local water-raising machines.

In Egypt, the management of the Nile waters was crucial to every single aspect of life. Both Al-Nuwayri and Al-Maqrizi, early Egyptian 14th-century historians, stressed the role of dam and waterway maintenance of the Nile. It was the responsibility of both sultans and large landholders, under both Ayyubids and Mamluks, to dig and clean canals and maintain dams. As in Iraq, the sultan took over the larger structures and the people the lesser ones. Most distinguished emirs and officials were made chief supervisors of such works. Under the Mamluks there was even an officer called the *Kashif al-Jusur,* whose job was to inspect dams for each province of Egypt.

Waste was banned and all disputes and violations of the water laws were dealt with by a court whose judges were chosen by the farmers themselves. This court was called "the Tribunal of the Waters," which sat on Thursdays at the door of the principal mosque. Ten centuries later, the same tribunal still sits in Valencia but now at the door of the cathedral.

Ibn al-Awwam, a 12th-century botanist, refers to a drip irrigation technique in his *Book of Agriculture,* saying that it conserves water and prevents overwatering of some species. He partially buried water-filled pots at the base of trees, with specific-sized holes for controlling the dripping rate. This technique is widely used around the world now.

As Muslims were accomplished civil and mechanical engineers, nothing came in the way of their extracting water. Even if the water source was in a gorge, the use of sophisticated machinery like water-raising machines and pumps revolutionized the society.

04 WATER SUPPLY

Imagine your life today without running water, where you have to walk for miles to a river or well and then contemplate how to get it into your bucket since you cannot get near the fast flow. This was the situation for Muslims before their groundbreaking inventions of water-raising machines and pumps, introduced about 800 years ago.

They devised new techniques to catch, channel, store, and lift the water, and made ingenious combinations of available devices, drawing on their own knowledge and that of other civilizations.

The ancient Egyptians already had the *shadoof*, a simple but effective contraption that took water from the river in a bucket tied to a long, pivoted pole. The bucket had a counterweight, and it was all supported between two pillars on a wooden horizontal bar. It is still used in Egypt today.

Large waterwheels, or *norias*, have raised water from fast-flowing waterways to higher land since 100 B.C.E. Vitruvius, the Roman writer, architect, and engineer, mentioned this simple yet powerful device. Like any waterwheel, it was turned by the force of flowing water against paddle compartments on its rim. These filled with water and took it to the top, where they emptied into a head tank connected to an aqueduct. Already used by

These norias, *which raise water from the Orontes River, are in Hama, Syria.*

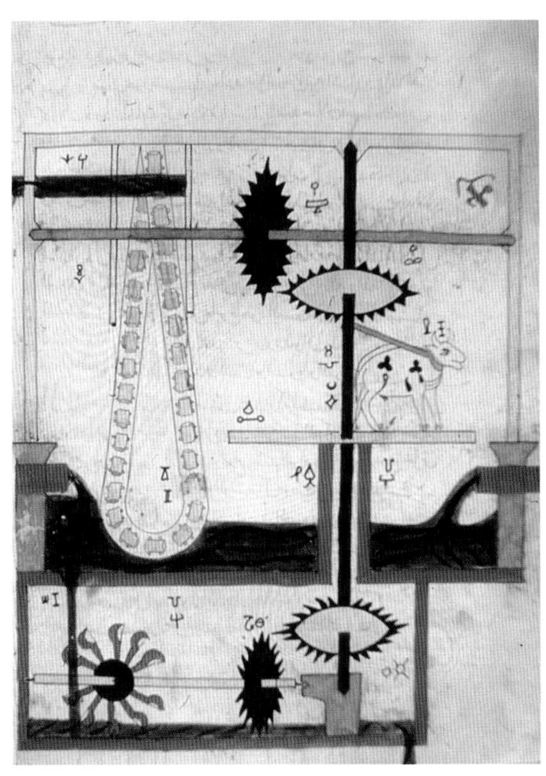

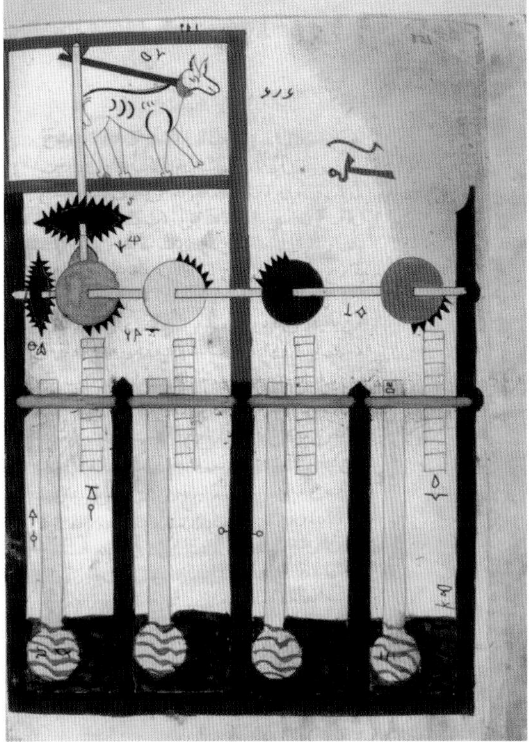

Pages from a 13th-century manuscript depict two water-raising machines designed by Al-Jazari. Water-raising machines are driven by a water turbine through geared shafts, which turn a sindi *wheel carrying a long belt of buckets. Al-Jazari made a wooden animal and placed it on the rotating disc on the machine so people would not think the automatic machine was driven by "magic"; they thought it was driven by an animal, operating multiple gears with partial teeth to produce a sequence of motion in four scoops that took water from the river one scoop at a time. This design included the first known appearance of a camshaft, which controled the mechanism.*

> *"It is impossible to over emphasize the importance of al-Jazari's work in the history of engineering. It provides a wealth of instructions for design, manufacture and assembly of machines."*
>
> BRITISH CHARTERED ENGINEER DONALD HILL, 1974

the Romans and Persians, they were adapted and redeveloped by the Muslims.

The first Muslim mention of *norias* refers to the excavation of a canal in the Basra region in the late seventh century. The wheels at Hama, on the river Orontes in Syria, still exist, although they are no longer in use. They were big wheels and the largest was about 20 meters (65.6 feet) in diameter, with its rim being divided into 120 compartments. The *noria* at Murcia in Spain, La Nora, is still in operation, although the original wheel has been replaced by a steel one. Apart from this, the Moorish system is otherwise virtually unchanged. There are still lots of norias in various parts of the world, and they are often able to compete successfully with modern pumps.

Many Muslim technologists recognized that harnessing power from both water and animals could increase the amount of work done. Two great innovators and Muslim engineers were Al-Jazari and Taqi al-Din. Both carried out a number of experiments, building remarkable machines that have led to automated machinery, which has made such an enormous impact on civilization today.

Al-Jazari lived in southeast Turkey in the late 12th and early 13th centuries, and was employed by the Artuqid king of Diyarbakir around 1180. As a skilled draftsman, he came up with an ingenious device for lifting huge amounts of water without lifting a finger. He was the first person to use the crank in his crank-connecting rod system. The crank is considered one of the most important mechanical discoveries made, because it translates rotary motion into linear motion. Today, cranks are in all kinds of things from toys to serious machinery like car engines and locomotives.

Al-Jazari used a machine powered by an animal with a flume-beam, which was moved up and down by an intricate system involving gears and a crank known as a slider-crank mechanism. The crank, as part of a machine, did not appear in Europe until the 15th century when it started a revolution in engineering.

An engraving shows the use of Egyptian shadoofs *for irrigation.*

Al-Jazari's Reciprocating Pump

Al-Jazari designed five water-raising machines. Two of them were improvements on the shadoof, and one replaced animal power with gears and water power. After the introduction of the crankshaft, his other radical breakthrough came when he made a water-driven pump. This involved cogwheels, copper pistons, suction and delivery pipes, and one-way clack valves. The pump sucked water, to be used in irrigation and sanitation, up 12 meters (39.4 feet) into the supply system. It is a very early example of the double-acting principle of one piston sucking while the other delivers, and Al-Jazari perfected the seals on the pistons and the one-way valve to make it all work.

If you ever felt like making your own 13th-century water-raising machine with reciprocating pump, here are details of how it worked.

Similar to a water mill, it would be built next to a flowing river with half of its paddle in the forceful current. This paddle wheel drove an internal gearing mechanism, powering pistons, which moved with the motion of the lever arm, and a reciprocating pump was created.

Clack valves helped to draw and expel the water through the pipes. The inlet pipe was submerged in water, and when the piston was pulled along the length of its cylinder, water would be sucked in through the inlet valve. The outlet valve remained closed during this time, because of gravity and the position of its pivot point.

When the piston was on its push stroke, the water in the cylinder was forced through the outlet valve and through an outlet pipe that was narrower than the inlet pipe. The inlet valve remained closed during this time, because of gravity and the position of its pivot point.

This motion was alternated between either side of the device, and so when one side was on its push stroke, the other was on its pull stroke. Therefore two "quantities" of water were being raised per one complete revolution of the

LEFT: *Artwork shows Taqi al-Din's six-cylinder water pump. The camshaft controls the motion of the connecting rods to produce a progressive motion of the six pistons, so water is raised continuously.* TOP RIGHT: *A view of the camshaft and waterwheel.* BOTTOM RIGHT: *A close-up of the pistons and cylinder block.*

waterwheel, and this carried on as long as there was flowing water to drive it.

Taqi al-Din's Six-Cylinder Pump

The other technological whiz was 16th-century Ottoman engineer Taqi al-Din ibn Ma'rouf al-Rasid, who wrote a book on mechanical engineering called *The Sublime Methods of Spiritual Machines*. As well as talking about water pumps, he also discussed the workings of a rudimentary steam engine, about a hundred years before the "discovery" of steam power.

His six-cylinder pump and water-raising machine form part of the study of the history of papermaking and metal works, as the pistons were similar to drop hammers, and they could have been used to either create wood pulp for paper or to beat long strips of metal in a single pass.

Taqi al-Din explained how the pump worked in his manuscript. The six-cylinder pump had a waterwheel attached to a long horizontal axle, or camshaft, which had six cams spaced along its length. The river drove the waterwheel, which rotated and turned the camshaft. Each cam on the camshaft pushed a connecting rod downward, and all connecting rods were pivoted at the center. At the other end of the connecting rod was a lead weight, which lifted upward and pulled a piston up with it. Now a vacuum was created, and water was sucked through a nonreturn clack valve into a piston cylinder. After the camshaft had rotated through a certain angle, the cam released the connecting rod and the piston's stroke ended. Through gravity, the lead weight pushed the piston down, forcing water against the clack valve, but the clack valve closed, so the water had to go through another hole and into the delivery pipes. The beauty of the mechanism was in the synchronization and control sequence of all the pistons, which were provided by the angular arrangement of the cams around the shaft.

In a time before dependence on machinery, when we were not surrounded by cars, bicycles, or electric pumps, these discoveries really changed society. These machines would not be mass-manufactured, but many towns would have a water pump. No longer were people heaving water containers around, or waiting their turn to use the shadoof. Instead they stood by pumps or aqueducts, waiting to catch the precious liquid gathered by their waterwheels, just as we wait for the water to flow from our faucets.

SUPPLYING WATER

New Technology for Irrigating a Thirsty Agricultural System

LEGACY: **Irrigation enabled crops like peaches, aubergines, and rice to spread west to Spain**

LOCATION: **Diyarbakir, modern-day Turkey**

DATE: **Late 12th century**

KEY FIGURES: **Al-Jazari, mechanical engineer**

Centuries ago in Muslim Spain, you could have seen hundreds of waterwheels busily irrigating the rice crops. But they did not have the capacity to supply every town and village.

In the 12th century, engineer Al-Jazari experimented with water-raising machinery to bring the supply directly to the local people. Of his five designs, the most advanced was the double-action suction pump. By using copper pistons, effective seals, and flapper valves, Al-Jazari made a pump that sucked up water from the river using two cylinders that emptied their contents into a single shared outlet above the machine.

Sixteenth-century scientist and engineer Taqi al-Din harnessed surging river water in his designs for an advanced six-cylinder pump, publishing his idea in a book called *The Sublime Methods of Spiritual Machines*. He also described a type of basic steam turbine, an idea that only resurfaced about a century later.

Taqi al-Din's six-cylinder pump was powered by a waterwheel on a shaft with six spiral-spaced studs, or cams. The cams on the rotating shaft activated a series of connecting rods, which would in turn each pull a piston up to suck water through a valve into the cylinder. As the shaft rotated, releasing the rod, a weight pushed the piston down again, forcing the water to leave the cylinder through a pipe that was connected to the supply system.

With such a synchronized action, the pump could deliver water smoothly—and may also have powered hammers that beat paper pulp and metal. High-capacity machines like Taqi al-Din's six-cylinder water pump meant people no longer had to wait in line for water.

Water was a significant factor in the agricultural revolution that occurred during Muslim civilization. Traders brought rice, peaches, apricots, and aubergines to Spain, where they were planted and watered using novel irrigation techniques. New ways of breeding and farming animals made meat and wool plentiful in places where before they had been a luxury. Men and women worked and traded on farms supplied by water from distant water-raising machines and a network of canals.

Modern farming methods are widespread today; nonetheless, you can still see ancient *shadoofs* for lifting water manually from rivers in use in Egypt.

A manuscript shows Al-Jazari's reciprocating pump. This was the first time an illustration of a crank appeared in a manuscript.

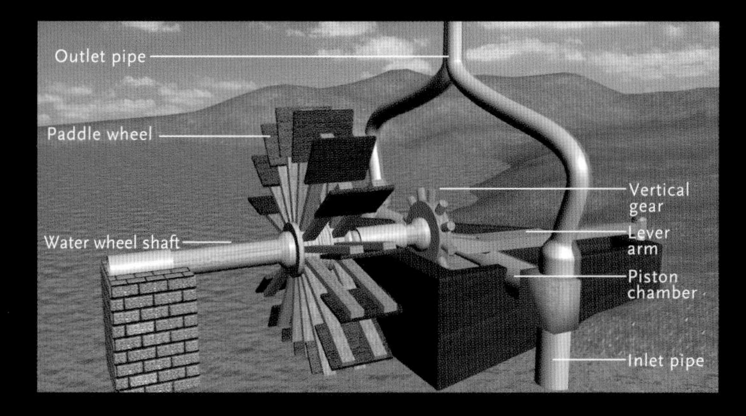

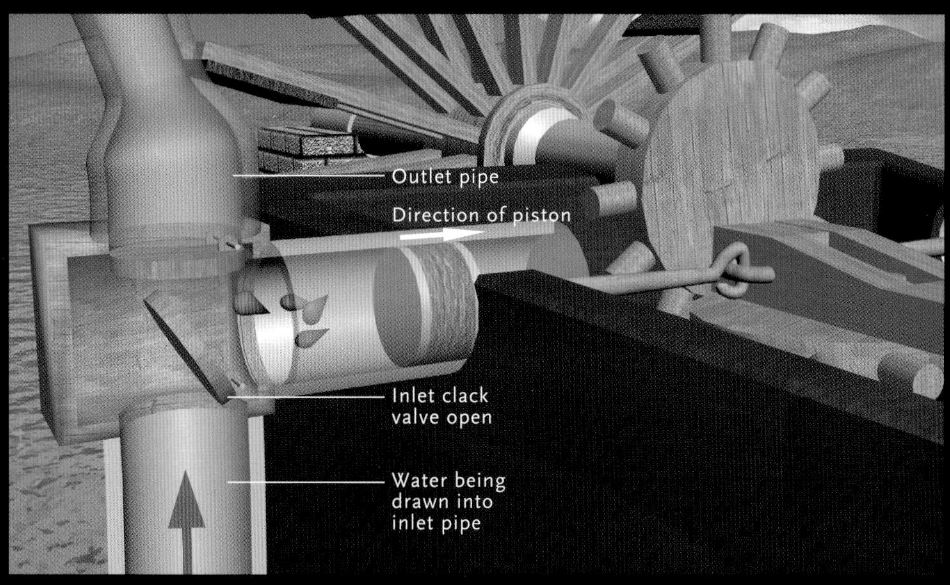

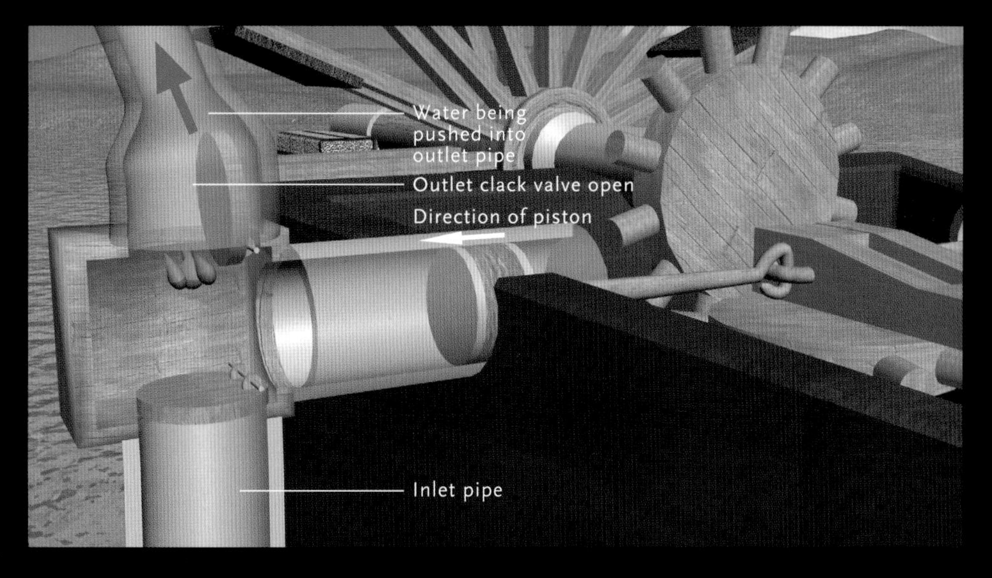

■ TOP: *A three-dimensional rendering shows Al-Jazari's reciprocating pump. It was designed to raise water to a height of 12 meters (39.4 feet) to deliver it into a supply system.* CENTER AND BOTTOM: *The piston movement causes the water to be pulled into the piston chamber, and pushed into the outlet pipe.*

05 DAMS

Dams are some of the largest civil engineering structures ever built and they play a vital role in civilization. Without dams, more floods would ravage lands, irrigation would not have been as large-scale, and we would not have hydroelectric plants pumping out power today.

Arch dams, buttress dams, embankment dams, Muslims built many dams in a rich variety of structures and forms centuries ago. The design and aesthetics of the most impressive of these dams were produced by the Aghlabids of Tunisia near their capital, Al-Qayrawan, in the ninth century. Al-Bakri, an 11th-century geographer and historian from southern Spain, described one as "circular in form and of enormous size. In the center rises an octagonal tower covered by a pavilion with four doors. A long series of arcades of arches resting one upon the other ends on the south side of the reservoir."

In Iran is the Kebar dam, the oldest known arched dam, which is about 700 years old. This dam, like many of its design, had a core of rubble masonry set in mortar. The mortar was made from lime crushed with the ash of a local desert plant, making it strong, hard, and impervious to cracking. Then there was the impressively curved Qusaybah dam, which was 30 meters high (98.4 feet) and 205 meters long (673 feet). It was built near Medina, now in Saudi Arabia.

In today's Afghanistan, three dams were completed by Sultan Mahmud of Ghazni in the 11th century near his capital city. One, named

Khaju Bridge, also a dam, on the Zayandeh River in Isfahan, Iran, was built in the mid-17th century by Shah Abbas II of the Safavid dynasty. The bridge was set on a stone platform and divided by sluices, which regulated the flow of the river.

"Historians of civil engineering have almost totally ignored the Muslim period, and in particular historians of dam building, such as there have been, either make no reference to Muslim work at all or, even worse, claim that during Umayyad and Abbasid times dam building, irrigation and other engineering activities suffered sharp decline and eventual extinction. Such a view is both unjust and untrue."

NORMAN SMITH, *HISTORY OF DAMS,* 1971

after him, was located a hundred kilometers (62 miles) southwest of Kabul. It was 32 meters high (105 feet) and 220 meters long (722 feet).

Dam construction in Muslim Spain was immense, and the masonry they used was a type of cement that was harder than stone itself. Each of the eight dams on the Turia River have foundations that go 15 feet into the riverbed, with further support provided by rows of wooden piles. The solid foundation was needed due to the river's erratic behavior: In times of flooding its flow was a hundred times greater than normal. More than ten centuries later, these dams still continue to meet the irrigation needs of Valencia, requiring no addition to the system.

The city of Córdoba, on the river Guadalquivir, probably has the oldest surviving Islamic dam in the country. According to 12th-century geographer Al-Idrisi, it was built of qibtiyya stone and included marble pillars. The dam follows a zigzag course across the river, a shape that shows that the builders were aiming at a long crest in order to increase its overflow

The reservoir on the wadi *Marj al-Lil near Al-Qayrawan in Tunisia, built by the Aghlabids in the ninth century, is one of the oldest surviving reservoirs in the Muslim world.*

capacity. Remains of the dam can still be seen today a few feet above the riverbed.

To build such immense structures, Muslim engineers used sophisticated land-surveying methods and instruments, like astrolabes and also trigonometric calculations. Dams were built of carefully cut stone blocks, joined together by iron dowels, while the holes in which the dowels fitted were filled by pouring in molten lead. The level of craftsmanship and superiority of design attained means that a third of all seventh- and eighth-century dams are still intact. The other two-thirds were destroyed by centuries of warfare, from the armies of Genghis Khan and the Mongols to Timur the Lame's hordes.

Muslims were also investing in "green energy" when they used stored water for mill power. In Khuzistan, at the Pul-i-Bulaiti dam on the Ab-i-Gargar, the mills were installed in tunnels cut through the rock on each side of the channel, constituting one of the earliest examples of a hydropower dam, and it was not the only one in the Muslim world. Another example was the bridge-dam at Dizful, which was used to provide power to operate a *noria,* a huge waterwheel, which was 50 cubits in diameter, and supplied water to all the houses of the town. Many such hydraulic works can still be seen today.

06 WINDMILLS

To produce anything, energy is needed, and before oil-powered machines, energy came from sustainable sources. Some energy in the Islamic world more than a thousand years ago came from water, and was harnessed in machines like the crank-rod system, which took water to higher levels and into aqueducts to quench the thirst of towns. Water drove mills to grind wheat, but in drier parts of the Islamic world there was not enough water, so alternative power supplies were sought.

One thing the vast deserts of Arabia had when the seasonal streams ran dry was wind, and these desert winds had a constant direction and blew regularly from the same place. The windmill was so simple yet effective that it quickly spread all over the world from its seventh-century Persian origins. Most historians believe that it was the Crusaders who introduced windmills to Europe in the 12th century.

A Persian had come to the second Caliph Umar, who reigned for ten years from 634, and claimed he could build a mill operated by wind, so the caliph ordered him to have one built. After this, wind power became widely used to run millstones for grinding corn, and also to draw up water for irrigation. This was done first in the Persian province of Sistan, and Al-Masudi, an Arab geographer who lived in the tenth century, described the region as a "country of wind and sand." He also wrote, "[A] characteristic of the area is that the power of the wind is used to drive pumps for watering gardens."

Early windmills were two-story buildings and were built on the towers of castles, hilltops, or platforms. On the upper story were the millstones, and in the lower one was a wheel, driven by the six or twelve sails that were covered with fabric. These turned the upper millstone. These lower chamber walls were pierced by four vents, with the narrower end toward the interior, which directed the wind onto the sails and increased its speed.

Windmills from that time were described as containing a millstone attached to the end of a wooden cylinder. This was half a meter wide (1.6 feet), and 3.5 to 4 meters high (11.5 to 13.1 feet), standing vertically in a tower open on the northeast side to catch the wind blowing from this direction. The cylinder had sails made of bundles of bush or palm leaves, attached to the shaft of the axle. The wind, blowing into the tower, pushed the sails and turned the shaft and millstone.

The introduction of the windmill had a great effect on the science of mechanical engineering and meant that new trades were born.

"Behold! a giant am I!
Aloft here in my tower,
With my granite jaws I devour
The maize, and the wheat, and the rye,
And grind them into flour.
I look down over the farms;
In the fields of grain I see
The harvest that is to be,
And I fling to the air my arms,
For I know it is all for me."

EXCERPT FROM "THE WINDMILL,"
BY HENRY WADSWORTH LONGFELLOW

Wind turbines turn in a field of rapeseed in France. Attempts to use environmentally friendly energy have revived the call for the use of wind power.

AIR POWER

Ancient and Modern Ways to Harness the Wind

LEGACY: Wind power is more popular than ever today as an alternative source of energy

LOCATION: Sistan, Persia

DATE: Tenth century to today

KEY FIGURES: Written and illustrated records by geographers Al-Masudi (tenth century) and Al-Dimashqi (14th century)

A thousand years ago, geographer Al-Masudi wrote of seeing windmills used to pump water for irrigating gardens in the Iranian province of Sistan, and in the western part of modern Afghanistan. When the streams ran dry each year in the region, the water mills would stop working. So, as the story goes, the Muslim rulers ordered a windmill built to take advantage of the desert winds that blew for four months of the year—and soon windmills appeared everywhere.

Unlike the traditional European design, Central Asian windmills had vertical vanes to catch the wind. Built on top of castles or at the crest of hills, the windmills had two stories. In one story were the millstones, one connected to a vertical wooden shaft. This shaft extended into the other story, where six to twelve windmill sails were mounted vertically, covered in cloth or palm leaves. The structure of the windmill was open to catch the wind on the northeast side.

Farmers used the windmills Al-Masudi saw to grind corn or pump water to irrigate their fields. Clean and green, it was a good solution to the power problem. And there have been many other ways that wind power was used in the past—including harnessing fast-moving air for natural ventilation.

Traditional courtyard houses were developed more than 4,500 years ago as private and energy-efficient family homes. Built with shared walls, these courtyard houses incorporated natural cooling elements in design. These included wind towers, which would catch moving air and channel it down from the roof terrace to pass through party walls to subterranean rooms.

Houses would have a variety of inward-facing rooms and spaces at different floor levels around a planted courtyard to suit different seasons and to enhance privacy.

Today, electrically powered air-conditioning is welcome in hot climates—but it may be that wind power holds the answer to making it more environmentally friendly. New windmills springing up across the landscape are now wind turbines, generating electricity to run a host of different machines.

■ *An old windmill still stands in Herat, Afghanistan.*

A 14th-century manuscript by Al-Dimashqi shows a cross-section of a typical windmill whose vertical vanes rotate around a vertical shaft.

07 TRADE

Trade has a long tradition in Islam and Prophet Muhammad, and many of his companions, were tradesmen. Because it played a major part of Islamic life, trade was governed by a well-developed body of legislation covering contracts, exchanges, loans, and market conduct.

The vast network of trade stretched over an empire that coursed with an eclectic collection of merchants and goods. Gold and white gold, as salt was known, traveled north and east from the African Sahara into Morocco, Spain, and France, with lesser quantities making their way into Greece, Turkey, Egypt, and Syria. Cowrie shells (they were a currency in the 14th century) went from the Maldives to West Africa. Pottery and paper money came west from China, but the paper currency did not catch on in Cairo. Travelers also flowed along with the goods: sheikhs and sultans, wise men and pilgrims.

Seljuk caravansaries still stand in Konya, Turkey. Caravansaries were charitable foundations and provided facilities, such as food and shelter, to travelers for free. They were the "highway service stations" of their time.

The land trade passing along the Silk Route was the heartbeat of the Muslim economy. The sea trade was mainly along the Mediterranean shores of Africa and Europe. The port of Málaga in southern Spain was a center of immense traffic, visited by traders from all countries, especially those from the mercantile republics of Italy, like the Genoese. Ibn Battuta sailed to Anatolia on a Genoese boat because they dominated this part of the trade routes, and he said, "The Christians treated us honorably and took no passage money from us."

On the crowded quays of Málaga, traders bartered the commodities of every country from silks, weapons, jewelry, and gilded pottery, to the delicious fruits of Spain.

Alexandria was a major port at the mouth of the Nile Delta, spilling into the Mediterranean Sea. On the Spice Route, it was the gateway into Europe for goods coming from the Indian Ocean, through the Red Sea and down the Nile. It had two harbors, a Muslim one in the west and a Christian one in the east, which were separated by the island of Pharos and its enormous lighthouse, known at this time as a wonder of the world.

Rest stops along the roads, called caravansaries, facilitated trade. Caravansaries were charitable foundations, providing travelers with three days of free shelter, food, and, in some cases, entertainment at regular intervals of about 30 kilometers (18.6 miles) along important trade routes.

As the merchants carried their wares across the world, they also took Islam with them. Up the Chinese coast in Guangzhou, now Canton, a colony of Muslim and Jewish merchants was well established in the eighth century. Muslim merchants penetrated Africa, and it was initially Berber merchants who carried Islam across the Sahara. All nomads in northeast Africa, where trade routes linked the Red Sea with the Nile, quickly became Muslims.

Some centers in the Islamic world constituted thriving communities due to their important place in commercial exchanges. Al-Qayrawan in Tunisia and Sijilmasa in Morocco were described by the tenth-century traveler Ibn Hawqal in his *Book of the Routes of the Kingdoms*: "Al-Qayrawan, the largest town in the Maghreb, surpasses all others in its commerce, its riches, and the beauty of its bazaars. I heard from Abu al-Hasa, the head of the public treasury that the income of all provinces and localities of the Maghreb was between seven hundred and eight hundred million dinars."

"The Arabs, masters of an empire extending from the Gulf of Gascony to beyond the Indus, involved in commercial enterprises reaching into Africa and Baltic Europe, brought East and West together, as never before."

ROBERT LOPEZ, HISTORIAN OF THE COMMERCIAL EXPANSION OF THE LATE MEDIEVAL PERIOD

Europe, Asia, and Africa imported vast amounts from Islamic lands, including enameled glassware, tooled leatherwork of all sorts, tiles, pottery, paper, carpets, carved ivories, illustrated manuscripts, metalwork including Damascene swords and vessels, fine cotton cloth, and rich silk fabrics.

Muslim textiles, metal, and glass pieces were highly prized, as were soaps. Mamluk gilt and enameled glass, a labor-intensive luxury product, has been uncovered by archaeologists on the northern shores of the Black Sea, as well as in Scandinavia, the Hanseatic ports, and Maastricht in Holland.

MUSLIM CARAVANS

Camel caravans carried traders across the Muslim world.

Muslim caravans were huge processions of people, their goods, and animals that traveled enormous distances and reached the farthest horizons. Their objective was either pilgrimage or trade, and it was these tradesmen who went as far as China in their caravans that bound this distant land to India, Persia, Syria, and Egypt.

Some of the camel caravans were so big that if you left your place you would not be able to find it again because of the vast number of people. Food was cooked in great brass cauldrons and given to the poorer pilgrims, and the spare camels took those who could not walk. Sheep and goats went with the caravans, providing milk, cheese, and meat. Camel milk and meat was also eaten, and the dried dung of these animals was used as fuel for the campfires. Flat bread, or pita bread, was made along the way from flour, salt, and water. Water was carried in goat and buffalo skin bags, and water points were welcome sights. The intense heat of the day in the deserts meant that caravans traveled by night with torches to light the way, making the desert glow with light, and turning night into day.

08 TEXTILES

Textiles drove much of medieval trade, and they were an exceptionally important part of the economy worldwide. It is estimated that textile manufacture and trade at this time would have kept the majority of the working population busy.

By the mid-ninth century the textile fabric of Muslim Spain had earned an international reputation, and even three centuries later Spanish silks with golden borders and ornamentation were used at the marriage of Queen Beatrice of Portugal.

The Spanish Muslims had as much delicacy and craftsmanship in their work as the famous Chinese artisans. In Córdoba alone, there were 3,000 weavers making carpets, cushions, silk curtains, shawls, divans, and Cordovan leather for the shoemakers of Europe, all of which found eager buyers everywhere. They were also producing superb woolen stuffs, especially rugs and tapestries, made in Cuenca, Spain. These were used as prayer mats as well as table and floor decorations in their beautiful houses.

In Al-Andalus, the production of Eastern-style cloth was concentrated in the towns of Málaga and Almeria, and because they were ports they were also the first to receive the new styles and techniques. From Muslim Spain the fine textile industry spread widely up into Europe.

Farther east and along the Mediterranean shores, textiles were made into clothing and the bulk of household furnishings. Nomad women wove tent bands, saddlebags, cradles, and other trappings for their mobile lives. Even in the urban centers and palaces, furnishings were mainly of carpets, covers, curtains, and hangings of various kinds. Instead of chairs, people sat on cushions and leaned against bolsters, covered with cloth whose quality and richness reflected their owners' financial status.

Textiles were important political tools as well. They made lavish diplomatic gifts, and it was customary to reward high officials and other favorites, at regular intervals and on special occasions,

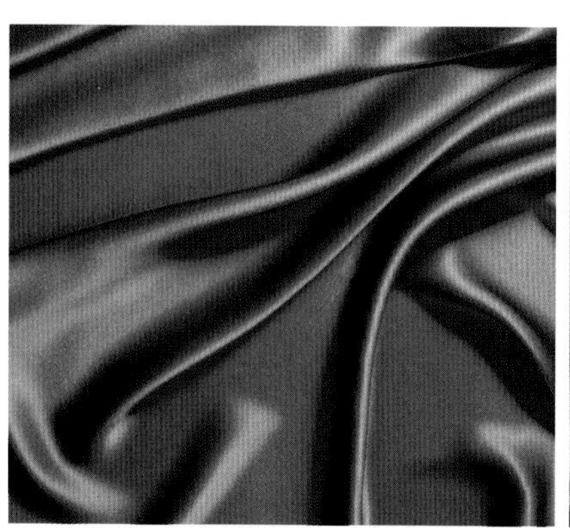

LEFT: *Muslim silk was so popular with the new bourgeois European society that local industry was threatened; so in 1700 the British government introduced a mandatory act restricting its import.* RIGHT: *A silkworm perches on a mulberry branch in a modern-day Turkish carpet factory. In the 1620s, King James I of England was so fascinated with Persian silk that he tried to establish his own silk industry.*

A 19th-century manuscript from the book Album of Kashmiri Trades *shows a dyer at work dyeing cloth.*

with robes of honor, turbans, and other garments woven in the rulers' own houses. It was also the caliphs' prerogative, and after 1250 that of the Mamluk sultans, to provide each year the new *kiswa*, the richly ornamented garment that veiled the *Ka'bah* at Mecca.

A full array of textiles was available in the Islamic world. Wool and linen were produced in quantity from Iran to Spain, and additional supplies of linen were imported since it was so popular. Cotton, native to India, was probably first produced on a large scale in the Mediterranean after the Muslim advance. It grew in Syria and Palestine as well, and from southern Spain it entered into Europe. Leather was also an important industry, and in the reign of Al-Mansur in the 12th-century Almohad dynasty in Fez, there were 86 tanneries and 116 dye works.

Some towns and cities were internationally recognized for their products. Shiraz was famous for its woolen cloths; Baghdad for its baldachin hangings and tabby silks; Khuzistan for fabrics of camel's or goat's hair; Khurasan for its sofa covers; Tyre for its carpets; Bukhara for its prayer rugs; and Herat for its gold brocades. No samples of these products from this period have survived the wear and tear of time though, although textile pieces from other periods can be found in Western museums and collections of Eastern art. One of the most precious fragments is the silk cape of an Egyptian Mamluk sultan, on which was inscribed "the learned Sultan," dating from the 14th century. This was found in St. Mary's Church, Danzig.

The Ka'bah *in Mecca, Saudi Arabia, is covered by kiswa, a cloth richly decorated with golden calligraphy. This is the place that Muslims around the world face when they do their daily prayers. It is a four-walled room normally covered by kiswa. The entrance to the room is on the left. The* Ka'bah *was originally built by Abraham and Ismail on the location believed to be the first place of worship in history by Adam. Before Mohammad, Arabs used it to house their statue gods, but these were destroyed by the advent of Islam; now there is nothing inside. The room is cleansed every year by the king and his guests, and nobody else is allowed inside. The lantern on the left houses a footprint believed to be that of Abraham. As part of the rituals, Muslims circle the* Ka'bah *seven times hailing God's Oneness.*

Europe's fascination with Muslim textiles goes back to the Middle Ages, when they were imported by Crusaders and traders. They were so valued that Pope Sylvester II was buried in luxurious Persian silk cloth. Queen Eleanor, the Castilian bride of King Edward I, brought Andalusian carpets to England as precious items of her dowry in 1255.

By the 17th century, trade relations with England were booming, coinciding with the peak of Persian textiles. In 1616, the Persian shah credited England with 3,000 bales to encourage trade, and after this Persian silk was at the top of the list of imports. Three years later, the ship *Royal Anne* brought in 11 bales of Persian silk, which came via Surat to England. The king at the time, James I, was so fascinated with Persian silk that he considered establishing a silk industry in England. He acquired silkworms and made special arrangements for their nursery at his country estates and Whitehall gardens. He also ordered Frenchman John Bonoeil, the manager of the royal silk works, to compile a treatise dealing with techniques of silk production, which was published in 1622.

Around the same time, trade with India was prolific, thanks to the active role of the East India Company in introducing Indian chintz to England. This fabric was cotton painted with Muslim elements, which provided a model for European cotton as well as wallpaper production.

By the 17th century, textiles imported from the Muslim world were all the rage with the new bourgeois European society, and local industry was threatened. Local silk weavers complained in 1685, while French and British silk and wool merchants sought bans on the East India Company, unwilling to suffer competition from the foreign textiles.

The British government reacted in 1700 by introducing a mandatory act restricting the import of silk from Muslim lands, which also prohibited the importation of Indian chintz, and Persian and Chinese fabrics.

An illustration from Maqamat al-Hariri *shows a girl working at either a spinning wheel or a spool-winding machine in Baghdad, Iraq.*

Fine silk did not come only from Persia, as the Turkish textile industry produced it as well. It was found in outstanding quality in Bursa, where silk weavers produced stunning pieces decorated with Iznik floral motifs. You can read more about these in the Pottery section. From here, silk and velvet reached the sultans' households, and were used in the Ottoman household on sofas, divans, and curtains, becoming essential for the interior decor. Lady Montagu, about whom you can read in the Inoculation section of the Hospital chapter, mentions the fame of Turkish textiles and admired Turkish dress style by wearing it herself.

Another 18th-century enthusiast of Turkish fabric and dress was the influential Swiss artist Jean-Etienne Liotard, who lived in Istanbul for five years and dressed like a native Turk. His female portraits of sitters "en Sultane" greatly helped to spread the fashion of Turkish dress throughout Europe.

We have products today that still bear their Muslim names, like muslin from the city of Mosul, where it was originally made; damask from Damascus; baldachin ("made in Baghdad"); gauze from Gaza; cotton from the Arabic *qutn,* meaning raw cotton; and satin from Zaytuni, named by Muslims after the Chinese port of Tseutung from where they imported it.

09 PAPER

Paper seems such an ordinary product today, but it has been fundamental to modern civilization. Think of all the pieces of paper you use every day—from magazines, TV guides, and newspapers, to paper towels and greeting cards.

Eleven hundred years ago Muslims were manufacturing paper in Baghdad after the capture of Chinese prisoners in the battle of Tallas in 751. The secrets of Chinese papermaking were passed to their captors, and papermaking was quickly refined and transformed into mass production by the mills of Baghdad and spread westward to Damascus, Tiberias, and Syrian Tripoli. As production increased, paper became cheaper and of better quality, and it was the mills of Damascus that were the major sources of supply to Europe.

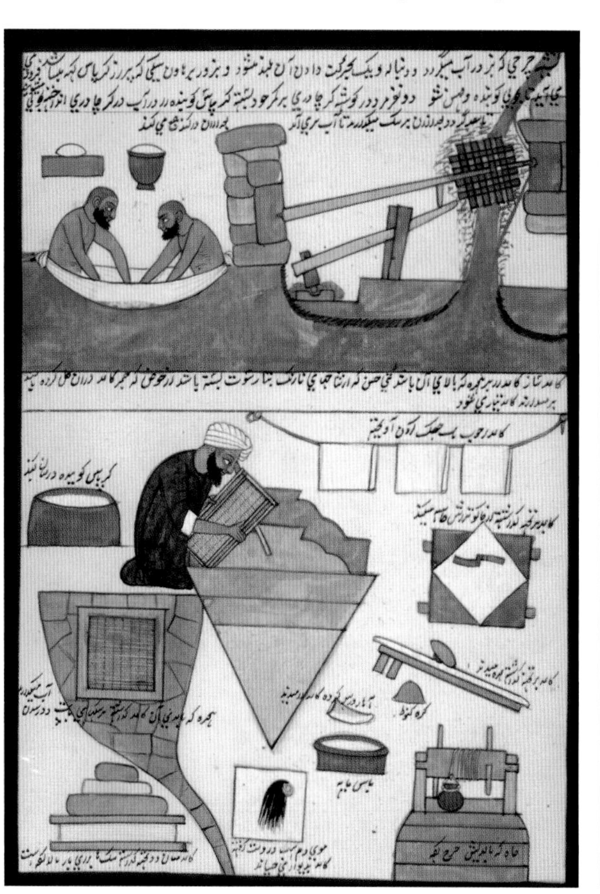

The Syrian factories benefited greatly from being able to grow hemp, a raw material whose fiber length and strength meant it produced high-quality paper. Today, hemp paper is considered renewable and environmentally friendly; it also costs less than half as much to process as wood-based paper.

As well as hemp, Muslims also introduced linen as a substitute for the bark of the mulberry, a raw material used by the Chinese. The linen rags were broken up, soaked in water, and fermented. They were then boiled and cleared of alkaline residue and dirt. The clean rags were beaten to a pulp by a trip hammer, a method pioneered by Muslims.

They also experimented with raw materials, making cotton paper. A Muslim manuscript on this dating from the 11th century was discovered in the library of the Escorial in Madrid.

By 800, paper production had reached Egypt, and possibly the earliest copy of the Quran on paper was recorded here in the tenth century. From Egypt, it traveled farther west, across North Africa to Morocco. Like much else, from there it crossed the straits into Muslim Spain around

TOP: *A vatman forms a sheet of paper in Kashmir, India, 1917—an ancient method of making sheet paper.* BOTTOM: *A 17th-century manuscript shows the papermaking process.*

950, where the Andalusians soon took it up, and the town of Jativa, near Valencia, became famed for its manufacture of thick, glossy paper, called *Shatibi*. Within 200 years of it being produced in Baghdad's mills, paper was in general use throughout the Islamic world.

This meant that producing books became easier and more cost effective because paper replaced the expensive and rare materials of papyrus and parchment, so mass book production was triggered. Before this, production had been complex and highly sophisticated: complex in that it was done through the labor of copyists, and sophisticated because of the skilled hands involved. The amount of labor in production decreased but the sophistication of the craftsmanship remained, so in the Muslim world hundreds, even thousands, of copies of reference materials were made available, stimulating a flourishing book trade and learning. Obviously the revolution in bookmaking was to happen much later after the use of printing machines in Europe.

The expansion of paper manufacturing engendered other professions, like those of dyers, ink makers, manuscript craftsmen, and calligraphists; the sciences also benefited. The pioneering Tunisian Ibn Badis, from the 11th century, described this in his *Staff of the Scribes*, writing about the excellence of the pen, the preparation of types of colored inks, the coloring of dyes and mixtures, secret writing, and the making of paper.

The first paper mill in Christian Europe was established in Bologna in 1293, and by 1309 the first use of paper in England was recorded. With all this paper and more cheaply produced books, the diffusion of knowledge into and around Europe sped up.

Danish historian Johannes Pedersen said that by manufacturing paper on a large scale, the Muslims "accomplished a feat of crucial significance not only to the history of Islamic books but also to the whole world of books."

DECORATING PAPER

Muslims developed techniques for decorating paper that are still used today in writing paper and books. One was marbling, which gave the paper a veined fabric look, and was used to cover important manuscripts.

The word for marbling in Turkish is *ebru*, which means cloud or cloudy, or *abru*, which means water face. Ebru comes from one of the older Central Asia languages, which means "veined fabric or paper." Its origin might ultimately go back to China, and it was through the Silk Route that marbling came first to Iran and then moved toward Anatolia, picking up the ebru name.

At the end of the 16th century, tradesmen, diplomats, and travelers coming from Anatolia brought the marbling art to Europe and after the 1550s it was prized by European book lovers, and became known as "Turkish paper" or "Turkish marbled papermaking." Later, it was widely used in Italy, Germany, France, and England.

Texts about ebru, like *Discourse on Decorating Paper in the Turkish Manner*, published in 1664 by Athanasius Kircher, a 17th-century German scholar in Rome, also spread the knowledge of marbling art.

Marbled paper was prized by 16th-century Europeans, who called it Turkish paper.

10 POTTERY

For more than a thousand years, Muslim lands produced some of the world's finest ceramics and pottery. They were traded, bought as ornaments, and used domestically in cooking, lighting, and washing. A millennium later, these pots have been turning up in European archaeological digs.

Pot making was a serious business and trade. The late 14th-century historian Al-Maqrizi said in Cairo: "Daily there is thrown on to the refuse heaps . . . to a value of some thousand dinars—the discarded remains of the red-baked clay in which milk-sellers put their milk, cheese-sellers their cheese, and the poor the rations they eat on the spot in the cook-shops."

In the east, pottery centers developed at Baghdad and Samarra, Iraq. Excavations at Samarra, the residence of the caliphs from 838 to 883, show us they had glazed and unglazed pots, incised and stamped, and that there were three main types. One was white, decorated with spots or pseudo-calligraphic motifs in cobalt blue. The second was decorated in polychrome, two-tone stripes, inspired by Chinese stonewares of the Tang period in the seventh and eighth centuries. The third pot type had a special luster, a decoration that looked metallic.

These pots were skillfully made in a similar way to the modern potter's wheel today, then dried and fired in kilns. They became collectors' items and icons of beauty and art, because what the Muslims did better than those before them was to improve and introduce new ways of glazing, coloring, and decorating their pottery.

The Romans had spread mostly red earthenware with shiny green or yellowish brown glazes to Mediterranean countries including Egypt. By adding more lead, the early Muslims produced a smoother, more brilliant finish to the pot and made it impermeable so it could hold liquids.

The Abbasid potters then took the lead glaze and added tin oxide to it, because they were trying to find a way of making pure white porcelain, like the expensive Chinese variety. The raw materials of Iraq and China were totally different, so the resourceful Muslim potters introduced a dash of tin oxide instead. This caused greater opacity and the exact white finish they were looking for.

Not satisfied, the potters made innovations in the design, producing the "blue-on-white" decoration, which was later re-exported to China, where it became hugely popular and spread to their porcelain. The blue-on-white ceramics were a source of pride for the Abbasid potters, who added their signature to much of their work. In one of these signatures a potter named Abawayh referred to himself as *"sani' amir al-mu'minin',"* telling us that he was the craftsman of the

TOP: *Andalusian geometric tile segments.*
BOTTOM: *This 14th-century Mamluk luster fritware jar was found at Trapani in Sicily.*

caliph, a reference to the caliphal promotion and patronage of crafts and pottery in particular.

In the eighth century, potters working in Iraq developed a mysterious process called luster. This was described as an "extraordinary metallic sheen, which rivals even precious metals in its effects, all but turning objects of clay to gold," by TV presenter Amani Zain in the BBC's *What the Ancients Did for Us: The Islamic World.*

Luster provided the right ingredients for producing these in a cheap and acceptable way, as Islam prohibits the use of gold and silver vessels.

The technique involved mixing silver or copper oxides with an earthy vehicle, such as ocher, and then vinegar or grape juice were added as a medium. The eighth-century Iraqi potters discovered that if they painted patterns with this mixture on the glazed coating of the clay, then put the wet pot into a kiln for a smoky and subdued second firing, a thin layer of metal was left. After wiping off the ash and dust, an amazing iridescent glow came through.

What was happening was that the copper and silver oxides separated out in the firing, leaving metal as a thin film on the surface as tin glaze. Silver left a pale yellow or golden and silvery effect, and copper produced a darker, redder, ruby color. The iridescence of these tones varied according to the fall of light. Exquisite monochromes and polychromes, in gold, green, brown, yellow, and red, in a hundred almost fluid tints, were possible.

Decorated tiles were also made in this way. The rich colors of these squares, and their harmonious combinations, gave mosques and palaces a regal splendor.

This luster technique from Baghdad passed through the Muslim world, and ninth-century Qayrawan in Tunisia starting producing luster tiles as well. Another century later it reached Spain. Archaeological finds at Madinat al-Zahra', the caliphate's city near Córdoba, uncovered a huge amount of pottery with patterns that

In Baghdad and Samarra, skilled craftspeople made pottery in a similar way to modern potters.

have been made with manganese brown for the painted lines and copper green for the colored surfaces. A few centuries later, Al-Andalus had its own centers of production such as Málaga, producing gold luster dishes and large jars like the "Alhambra Jar."

BBC presenter Amani Zain, on *What the Ancients Did for Us,* said, "These amazing vases [the Alhambra Jar] were originally used for storing oil and grains. But in the palaces of the caliphs their designs took on an extraordinary beauty. And for those who saw them, they must have thought they'd been made from precious metals."

Ordinary people needed practical pots, and in Spain the most popular pot was a *qâdûs,* which carried water on the *noria,* a waterwheel, which you can read about in the Water Supply section of this chapter. It became the universal unglazed pot, and must have been the mainstay of the rural pottery industry until it was replaced by tin fairly recently.

> *"Arabs invented the technique that makes these clay pots into art."*
>
> BBC PRESENTER AMANI ZAIN TALKING ABOUT LUSTER GLAZE ON *WHAT THE ANCIENTS DID FOR US*

As well as producing the necessary water-carrying pots, Spanish Muslims at the beginning of the 12th century were replacing Byzantine mosaics with tiles and *azulejos.* These were beautiful tiles in blue and white, covered with geometric, floral, and calligraphic patterns. These glazed faience tiles of Málaga are still famous. We know that the blue glaze of cobalt oxide, which the *azulejos* are decorated with, came from the east to Málaga, from where it spread to Murcia, then to Christian Spain and Valencia at beginning of the 14th century, and then to Barcelona.

Turkey was also a thriving pottery center because craftsmen crowded here, to the city of Konya, as they fled from invading Mongols. The collapse of the sultanate of Konya at the beginning of the 14th century brought the ceramic production of Anatolia to a standstill, but it was to have a brilliant revival when the Ottoman Turks made Bursa their capital in 1326. The city rose again, with fine buildings covered in ceramic tiles.

Even busier in production than Bursa was Iznik, which was the real center of the industry, and it flourished for two centuries from the end of the 14th century. A typical Iznik decoration was

MUSLIM POTTERY

"Along with the Chinese, the potters of the Arab world produced some of the most sophisticated and beautiful pottery known to the medieval world. When these Arab wares managed to reach the Christian west, they were highly prized and regarded as luxury items."

JOHN COTTER, LEADING ARCHAEOLOGIST, WRITING ABOUT THE ISLAMIC POTTERY FOUND AT THE LONGMARKET EXCAVATION IN CANTERBURY, ENGLAND

A market stall displays pottery in North Africa.

Iznik potters in Turkey made this serpent dish in the early 16th century.

painted on slip, in cobalt blue, turquoise, and green (from copper), outlined in black with an amazing tomato red in low relief. The patterns, made up from rectangular tiles, were all floral motives, with four flowers traditionally being used. These were the rose, jasmine, carnation, and tulip.

Muslim potters monopolized the skills of glaze and luster decoration for more than ten centuries, and the pottery of today is indebted to them. From the potters of Spain and Sicily, new modes and methods of pottery production, materials, and colors entered Europe. Tin glaze was not known in Europe until Muslims introduced it to Spain in the ninth century.

There is plenty of evidence today that Muslim pottery did travel outside Spain, as Malagan pottery has been found in England; 44 pieces of Moorish luster were discovered dating back to the late 13th and 14th centuries, and another 22 pieces were from the 15th century. More recently, in 1990, excavations in Longmarket in the center of Canterbury uncovered a large amount of Islamic luster and turquoise blue pottery.

John Cotter, who worked on the Canterbury find, wrote about how Muslim pots came to be in England: "Some pots may have made their way back to England in the baggage of Crusaders returning from the Holy Land . . . Another possibility is that medieval pilgrims either to the Holy Land or to the famous shrine of St James at Compostella in Spain might have brought back the occasional Islamic pot as a souvenir." On rare occasions the route was more direct. We know, for instance, that in 1289 Eleanor of Castile, the Spanish wife of England's Edward I, ordered four thousand pots of *Malik* for the royal household. In this case *Malik* almost certainly refers to Málaga—the main center for Andalusian lusterwares. (Malika is the Arabic name of Málaga.)

A 14th-century luster dish was found at a site called Blossom's Inn in London, decorated with the tree of life and with Kufic inscriptions. These were popular in Andalusia and North Africa at that time, and were copied widely in Europe. Amazingly, the dish's entry into England was recorded in 1303 in the accounts of the New Custom on goods imported and exported at the port of Sandwich, Kent. The dish is now at the Guildhall Museum of London.

Another famous ceramic brand left to us by Muslim potters is the so-called Maiolica ware. The story started at Majorca and other Balearic Islands, which were under Muslim rule until 1230. Italian ships, mainly Genoese and Venetian, often called there to collect tin-glazed pottery and recruit Moorish potters, who brought to Sicily the Majorcan pottery style. This was gradually established as a leading style, becoming renowned as "Majolica" or "Maiolica."

Since the 15th century, Maiolica has reached an astonishing degree of perfection, using the same production and decorative techniques as the Andalusians and Egyptians. Later, Italian artists developed it into new varieties, like Gubbio luster, which came in greenish yellow, strawberry pink, and a ruby red. This Maiolica pattern has dominated the ceramic industry in Italy till the present time.

11 GLASS INDUSTRY

What we know today about glass in the past has come from archaeological digs and the writings of travelers of the time. So we know that 13th- and 14th-century Syria was a great center of this fine material, in the cities of Aleppo and Damascus. Ibn Battuta described Damascus as a glassmaking center when he traveled through there in the 1300s. Not only Syria, but Egypt, Iraq, and Andalusia were all producing it in vast quantities from the eighth century onward, and it was either cut from crystal or blown in molds.

Muslims had inherited the famous Roman glass industry based in Syria and Egypt, developing it with double stamping (in which a stamp with decorative designs was pressed onto hot glass); free-form glassblowing with thread decoration (continuing from Roman and Byzantine traditions); mold blowing (where the glassmaker blows the liquid glass into a prepared mold); and engraving and cutting glass either by hand or with a wheel. They also perfected glass decoration and expanded the variety of products to include bottles, flasks, vases, and cups.

By the 13th century, Syrian glass was so fine that merchants and buyers all over the world were after samples, and recent digging has uncovered 700-year-old Syrian enameled glass in Sweden and southern Russia, and even China.

Samarra, in Iraq, was also famed for its glass. Among the most stunning finds was *millefiori*, or mosaic glass, which was different from earlier types in its peculiar coloring and design. Alongside this, another of the most beautiful finds at Samarra was a ninth-century straight-sided bowl in whitish glass.

Samarra's glassmakers were also renowned for making small bottles for things like perfumes. Some were pear-shaped, in blue and green glass, with four sides and a nearly cylindrical neck. These were heavier and frequently decorated

TOP: *The glassware industry flourished in the Muslim world.* BOTTOM: *The inscription on this 14th-century gilded and enameled Egyptian glass bottle reads, "Glory to our Master, the Wise, the Just King."*

with cutting. At Samarra, outstandingly beautiful fragments of ninth-century cut-glass bowls with strongly stylized decoration have also been found.

So much glass has been uncovered in excavations at Al-Fustat, "Old Cairo," founded in the 640s C.E., that from the eighth century to the later Middle Ages this area must have been a center of production. The earliest dated items, from 708, are coinlike weights, stamped with the names of rulers or government officials. They came in a variety of colors, from dark green, light green, and turquoise, to white and purple. Some of the most sophisticated Egyptian glass vessels were decorated with luster, the shiny, sometimes metallic effect made by painting copper or silver oxide on the surface of the object, which then was fired at a temperature of about 600°C (1112°F).

The glass industry was not restricted to the east, because in Al-Andalus glass was in the same great demand as the pottery. Jars with two, four, or eight handles, and bowls with handles and ribs have all been found. The chief centers for glassworks were in Almeria, Murcia, and Málaga, and it was Almeria that had a worldwide reputation. Glass goblets blown in these three cities imitating Eastern wares were found on the tables of nobles in tenth-century León.

The technique of cutting crystal was said to have been introduced by 'Abbas ibn Firnas in ninth-century Córdoba, Al-Andalus. He was a scholar and inventor in the courts of 'Abd al-Rahman II and Muhammad I, who could also decipher the most complex writing and also attempted to fly by building artificial wings. With glass, he understood its scientific properties and contributed to the early experimentation with lenses and the idea of magnifying scripts, having established Andalusia's crystal industry based on mined rocks.

So, glass had a colorful history as it traveled from the furnaces of Syria, Egypt, Iraq, and Andalusia around the known world, adorning people's tables and houses as a status symbol as well as practical necessity.

LEFT: *Glass from Samarra in Iraq was particularly famous in Muslim civilization. This mosaic glass vessel shows a colorful design.* RIGHT: *A glass blower in Venice, Italy.*

12 JEWELS

As you have read, the glass, textile, pottery, and paper industries formed the backbone of a successful empire whose goods were traded as far as China. Other vital industries included goods from mines and the sea, like jewels and pearls. Emeralds were extracted in upper Egypt, turquoise taken in Farghana, rubies found in Badakhshan, and carnelian and onyx obtained in Yemen and Spain.

The cinnabar mines of Almadén in Spain had a workforce of somewhere near a thousand; some cutting the stone down in the pit, others transporting wood for smelting, making the vessels for melting and refining the mercury, and manning the furnaces.

A surprisingly precious mined item was salt, or white gold, at Hadramawt (in Yemen), Isfahan, Armenia, and North Africa, which was carried by great camel caravans. "Throughout the greater part of Africa," wrote Leo the African, a historian and geographer who roamed Africa and the Mediterranean lands in the 16th century, "salt is entirely of the mined variety, taken from underground workings like those for marble or gypsum."

Precious stones were dressed and polished with emery, found in Nubia and Ceylon. Egypt and

TOP: *A 17th-century Indian gold pendant is inlaid with flat-cut rubies and emeralds and a large faceted diamond in the pattern of a flying bird against a leafy background of rubies.* BOTTOM: *These beads have an inscription of the name of God on each one. In the Quran there are 99 names of Allah.*

"The earth is like a beautiful bride who needs no man-made jewels to heighten her loveliness."

KHALIL GIBRAN, LEBANESE WRITER

Sudan both had alum, and parts of western Egypt, notably the famous Natron Valley, had *natron*, which was used for whitening copper, thread and linen, and also for curing leather. It was in demand with dyers, glassmakers, and goldsmiths; bakers even mixed it in with their dough and cooks used it as a tenderizer for meat.

From the sea came the beautifully smooth pearls that decorated many necks across the world. Pearl diving was carried out on both sides of the Persian Gulf, in the Arabian Sea, near Shiraf and the island of Kish, along the Bahrain coast toward the island of Dahlak and in Ceylon.

From the 14th century, Ibn Battuta refers to pearl-diving methods: "The diver attaches a cord to his waist and dives," he said. "On the bottom, he finds shells embedded in the sand among small stones. He dislodges them with his hand, or a knife brought down with him for the purpose, and collects them in a leather bag slung round his neck. When breath fails, he tugs at the cord, the sign for the man holding it in the boat to pull him up again. Taking off the leather bag, they open up the shells, and cut out with a knife pieces of flesh from inside."

There were coral reefs lying off the coasts of North Africa, near Sicily and Sardinia. Al-Idrisi, the 12th-century geographer, gives an account of coral gathering: "Coral is a plant which has grown like trees and subsequently petrified deep in the sea between two very high mountains. It is fished with a many-looped hemp tackle; this is moved from high up in the ship; the threads catch the coral branches as they meet them, and the fishermen then draw up the tackle and pick out from it the very considerable quantity of coral."

A 16th-century Arabic manuscript shows a furnace for making artificial rubies and sapphires. The Arabic text on the manuscript describes how it works.

Coral, along with pearl, was then used to decorate weapons and make prayer beads and jewelry. Today, like all jewelry, coral is worn in many styles, from long strands of beads to carved cameos and pins, but prices for this marine beauty can be as much as $50,000 for a 50-millimeter-diameter (2 inch) bead, as coral reefs are destroyed and coral as a jewel becomes scarcer.

13 CURRENCY

In the past money was alive, because camels, cattle, or sheep were used to "pay" for goods. In the time of Ibn Battuta, the 14th-century Muslim traveler, the Maldives used cowrie shells as currency, as they were highly treasured, and these reached distant regions such as Mali in West Africa. Today, we use credit cards, notes, and coins, but this is a small quantity compared to the amount of "invisible," intangible money shooting around the world as the financial markets make their electronic transfers. One day our coins and notes may be as useful as Ibn Battuta's cowrie shells would be today.

Dar al-Islam, or the Muslim world, spread its wings, even under separate rulers or sultans, using gold and silver coins as its international currency. If we are globetrotting today, we either take traveler's checks or risk having our purses full of different currencies. But in the 14th century, travelers in the Muslim world could scour every market nook and cranny and use dinars or dirhams, from capital cities to the smallest village.

Step outside the Muslim world and transactions were a different story. Again Ibn Battuta can tell us a lot about the world nearly 700 years ago as he had a surprising financial experience in China: "The people of China . . . buy and sell with pieces of paper the size of the palm of the hand, which are stamped with the sultan's stamp . . . If anyone goes to the bazaar with a silver *dirham* or a *dinar* . . . it is not accepted and he is disregarded."

In the seventh and eighth centuries, money was mostly made of gold and silver, and Muslims made their coins according to the Quran, which says: "When you measure, give an exact measure and weigh with an accurate scale" (Sura 17:35). It was the caliph's responsibility to ensure the purity and weight of the coins. The standard was established by the Sharia law as seven mithqals of gold to ten dirhams of silver. Any coins that did not measure up, foreign currency, and old coins were brought to the mint along with gold and silver bullion to be refined and struck into new currency. At the mint the bullion was first examined to determine its purity before being heated and made according to the established alloy standards.

An artistic re-creation shows 14th-century merchants conducting business.

CLOCKWISE FROM TOP LEFT: *showing the front and back of each coin: Two early Umayyad coins, 691-692—the lower coin shows the column placed on three steps topped with a sphere, replacing the Byzantine cross. A silver Nasrid dirham of Muhammed I, Granada. Early Fatimid coins, Al-Mahdiya, 949. A gold Nasrid dirham of Muhammed XII, Granada. A gold dinar of Caliph Abd al-Malik from the Umayyad dynasty, 696-697.*

Both dinars and dirhams were used by different Muslim rulers. The first caliph to make his own coins was Umayyad Caliph Abd al-Malik ibn Marwan, who ruled from 685 to 705. These dinars were the first gold coins with an Arabic inscription, as previously money had been silver Sassanian coins, and gold and copper Byzantine coins. By making his own coins in 691 or 692, Caliph Abd al-Malik could now keep his rule independent from Byzantium and unify all Muslims with one currency.

This new coin was copied from the Byzantine currency, the solidus. It was similar in both size and weight, and on the face were three standing figures, like the Byzantine coin, which had the figures of Heracles, Heraclius Constantine, and Heraclonas. A big difference was the Arabic testimony of Islam surrounding the design on the reverse: "There is no deity but Allah, The One, Without Equal. And Muhammad is the Apostle of Allah."

The Byzantine emperor was furious with this development, as new money meant competition and he refused to accept it, responding with a new coin. This angered Caliph Abd al-Malik, who made another coin with an upright figure of the caliph, wearing an Arab headdress and holding a sword, again with the testimony of Islam on the reverse, where the coin was also dated. Only eight of these early Arab-Byzantine dinars, dated according to the new Islamic calendar, have survived.

The coin throwing continued, and true to form the Byzantine emperor replied with yet another, and at this point in 697 the caliph had had enough, and introduced the first Islamic coin without any figures. On both sides of this new dinar were verses from the Quran, which made each piece an individual message of the faith. He then issued

FABLED COINS

There are two fabled Islamic coins, the One Thousand Muhurs and the One Hundred Muhurs. The first weighed in at 12 kilograms (26.5 pounds) of pure gold and the second was a baby in comparison, being a mere 1,094 grams of pure gold. Their estimated value today is about $10 million and $4 million, respectively.

The coins were originally minted for the Mogul emperors Jahangir, son of Akbar the Great, in 1613, and his son Shah Jahan, best known for building the Taj Mahal in 1639. The coins were presented to the highest dignitaries.

The One Thousand Muhurs was huge at 20 centimeters (8 inches) in diameter, and over the centuries four or five were mentioned as having been reserved for the ambassadors of the powerful rulers of Persia. Only one comparable coin is recorded from a plaster cast in the British Museum, a Two Hundred Muhurs, last reported in India in 1820 and since lost. None of the legendary giant gold Muhurs are known to have survived, and it is assumed they were melted down for their bullion value. But we know they did exist because travelers mentioned seeing gigantic coins in the treasury of Shah Jahan.

> *"Fatimid coins were of such high quality and so abundant that they became the most widespread Mediterranean world."*
>
> WIJDAN ALI, SCIENCE HISTORIAN

a decree making it the only currency to be used throughout Umayyad lands. All remaining Byzantine and Arab-Byzantine pieces had to be handed to the treasury, to be melted down and restruck. Those who did not comply faced the death penalty.

The new gold dinars weighed a bit less than the solidus, and the state controlled the accuracy of their weight along with the purity of the gold used. Umayyad gold coins were generally struck in Damascus, while silver and copper coins were minted elsewhere.

After this first coin, more of different values were struck, and after conquering North Africa and Spain, the Umayyads established new mints, each producing coins with the name of their city and date of minting.

The dinar continued to be the main currency used until 762, when Caliph Al-Mansur built Baghdad and the gold mint moved to the new capital. The names of persons responsible for the coins began to appear on silver coins called dirhams. But these had a short life because the next caliph, Harun al-Rashid, abandoned them when he came to power in 786. He minted dinars with the names of governors of Egypt instead, using the two active mints we know about, one in Baghdad and the other in Fustat, the seat of the governor of Egypt.

The Fatimids, who ruled between 909 and 1171, used dinars with Kufic scripts, and these became the most widespread trade coins of the Mediterranean world because of their high quality and because there were so many of them. When the Crusaders captured Palestine, they copied these coins instead of striking their own, and these ranged from excellent copies of the original to bad imitations.

It was from Andalusia that gold dinars traveled around Europe from around 711. Then, under the Nasrid rule in Granada from 1238 to 1492, the dinar became the dirham. These coins were heavy, carefully struck, and bore long legends with passages from the Quran and the rulers' family trees. None of the Nasrid coins showed a date, but they are identifiable by their motto, "None victorious save God." At the same time, in the Christian kingdoms of the north, Arab and French currencies were the only ones used for nearly 400 years.

After the 13th century, the Muslim Caliphate went from being ruled by one caliph to many small dynasties, each producing their own coins. Like currencies today, they carried the names of various governors from the semi-independent states. These were all minted independently, but still acknowledged the nominal leadership of the caliph.

Like today, coins were not the only ways of paying. Checks were around centuries ago as well. "Check" comes from the Arabic *sakk,* a written vow to honor payment for merchandise when its destination is reached. In the time of Harun al-Rashid in the ninth century, under a highly developed banking system, a Muslim businessman could cash a check in Canton, China, drawn on his bank account in Baghdad. The use of sakk was born out of the need to avoid having to transport coin as legal tender due to the dangers and difficulties this represented. Bankers took to the use of bills of exchange, letters of credit, and promissory notes, often drawn up to be, in effect, checks. In promoting the concept of the bill of exchange, sakk, or check, Muslims made the financing of commerce and intercontinental trade possible.

KING OFFA'S MYSTERY

Twelve hundred years ago, while the caliphs ruled the Muslim world, King Offa reigned in England. He introduced silver coins to his kingdom—and also minted a gold coin, the gold Mancus, worth 30 silver coins. The extraordinary thing about the Mancus was that it copied the Muslim gold dinar of the Abbasid Caliph Al-Mansur dated 157 A.H. or 774 C.E., carrying on one side the Arabic inscription "There is no Deity but Allah, The One, Without Equal, and Mohammad is the Apostle of Allah."

A significant difference from the original dinar is that King Offa stamped his name on it with the inscription of OFFA REX. Scholars have puzzled over why an English king would have made a replica Arab coin. Some say he had converted to Islam, but the more likely explanation is that it was produced for trade, or for pilgrims to use as they traveled through Arab lands. The coin most certainly would not have been made by an Arab craftsman because there is no understanding of the Arabic text: "OFFA REX" is upside down in relation to the Arabic Kufic script, and the word "year" is misspelled in Arabic. The coin was probably copied by Anglo-Saxon craftsmen.

Discoveries like this have helped us to redraw the international economic and trade relations of 1,200 years ago. King Offa's coin is evidence of how far Islamic currency had traveled by the eighth century. Archaeologists have found thousands of Muslim coins in modern-day Germany, Finland, and Scandinavia, showing that this currency was transported and traded from Muslim countries across Europe.

King Offa was not the only non-Muslim ruler to make an Arabic coin. An 11th-century Spanish Catholic prince, Alfonso VIII, ordered the minting of a decorative coin on which the inscriptions were written in Arabic and on which he referred to himself as the "Ameer of the Catholics" and the pope in Rome as the "Imam of the Church of Christ."

King Offa of Mercia in England made a copy of the gold dinar coin of the Abbasid Caliph Al-Mansur dated 157 A.H. (774 C.E.). It is a near-identical replica, including the profession of faith in Arabic on one side (right) and the name of King Offa on the other side (left).

Chapter Five

"Medicine is a science, from which one learns the states of the human body . . . in order to preserve good health when it exists, and restore it when it is lacking."

ELEVENTH-CENTURY IBN SINA FROM HIS BOOK *CANON*

HOSPITAL

HOSPITAL DEVELOPMENT • INSTRUMENTS OF PERFECTION SURGERY • BLOOD CIRCULATION • IBN SINA'S BONE FRACTURES NOTEBOOK OF THE OCULIST • INOCULATION • HERBAL MEDICINE PHARMACY • MEDICAL KNOWLEDGE

Medical care a thousand years ago was free for all and the treatments highly sophisticated. The hospitals of medieval Islam were hospitals in the modern sense of the word. In them was the best available medical knowledge, dispensed for free to all who came. It could even be said that they were a forerunner to the United Kingdom's National Health Service, and they flourished as rulers of Islam competed to see who could construct the most magnificent ones. Some hospitals were huge, others were surrounded with gardens and orchards, and most offered advanced social welfare to patients including treatment by music.

The facilities they used were custom-designed and the surgical instruments were outstanding. Forceps are just one of the instruments still used today, designed by Muslim surgeons more than a thousand years ago. Cutting-edge treatments, such as cataract operations, regular vaccinations, internal stitching, and bone setting, were also part of standard practice, as was a rigorous medical education in a teaching hospital.

OPPOSITE: *Fifteenth-century miniatures in Serefeddin Sabuncuoglu's* Cerrahiyyet'ul Haniyye *illustrate the treatment of patients and show various surgical procedures.*

01 Hospital Development

The idea behind Muslim hospitals a thousand years ago was to provide a range of facilities from treatments to convalescence, asylum, and retirement homes. They looked after all kinds of people, rich and poor, because Muslims are honor-bound to provide treatment for the sick, whoever they may be.

From the earliest Muslim times, these hospitals were funded by charitable religious endowments, called *waqf*, though money from the state coffers was also used for the maintenance of some hospitals. It was partly due to this funding they became strongholds of scientific medicine and an integral part of city life in less than two centuries.

Before the Muslims, the Greeks had temples of healing. In these, health care was based more on the idea of a miraculous cure rather than on scientific analysis and practice. A Greek Byzantine charitable institution, the *xenodocheion* (literally travelers' hostel or inn), came closest to being a hospital where care was given to the lepers, invalids, and the poor.

Islamic hospitals began in eighth-century Baghdad, and in some ways these resembled Byzantine travelers' hostels as they also looked after lepers, the invalid, and the destitute. But the first organized hospital was built in Cairo between 872 and 874. The Ahmad ibn Tulun Hospital treated and gave medicine to all patients free of charge. With two bathhouses, one for men and one for women, a rich library, and a psychiatric wing, it was an incredibly advanced institution. Patients deposited their street clothes and their valuables with the hospital authorities for safekeeping before donning special ward clothes and being assigned to their beds.

Other important hospitals included a large Baghdadi Hospital, built in 982, with a staff of 24 physicians. Twelfth-century Damascus had an even larger hospital, the Nuri Hospital. Here, medical instruction was given and druggists, barbers, and orthopedists, as well as oculists and physicians, were, according to manuals composed in the 13th century, examined by "market inspectors" on the basis of some set texts.

Ibn Tulun Mosque, Cairo, Egypt, was the first organized hospital that provided free treatments and medicines for patients.

AL-QAYRAWAN HOSPITAL

The Al-Qayrawan hospital in Qayrawan, Tunisia

The ninth-century Al-Qayrawan hospital was a state-of-the-art institute, with well-organized halls including waiting rooms for visitors, female nurses from Sudan, a mosque for patients to pray and study, regular physicians and teams of Fuqaha al-Badan, a group of imams who practiced medicine and whose medical services included bloodletting, bone setting, and cauterization. It also had a special ward for lepers called Dar al-judhama, built near the Al-Qayrawan hospital, at a time when elsewhere leprosy was deemed an untreatable sign of evil. It was financed by the state treasury, and by other people who gave generously to boost hospital income so that the best care could be provided.

In all, Cairo had three immense hospitals; the most famous was the Al-Mansuri Hospital. When the 13th-century Mamluk ruler of Egypt, Al-Mansur Qalawun, was still a prince, he fell ill with renal colic during a military expedition in Syria. The treatment he received in the Nuri Hospital of Damascus was so good that he vowed to found a similar institution as soon as he ascended to the throne. True to his word, he built the Al-Mansuri Hospital of Cairo and said, "I hereby devote these *waqfs* for the benefit of my equals and my inferiors, for the soldier and the prince, the large and the small, the free and the slave, for men and women."

The 1284 Al-Mansuri was built with four entrances, each having a fountain in the center. The king made sure it was properly staffed with physicians and fully equipped for the care of the sick. He appointed male and female attendants to serve male and female patients who were housed in separate wards. Beds had mattresses and specialized areas were maintained. Running water was provided in all areas of the hospital. In one part of the building the physician-in-chief was given a room for teaching and lecturing. There were no restrictions to the number of patients who could be treated, and the in-house dispensary provided medicines for patients to take home.

> *"[The hospital's] duty is to give care to the ill, poor, men and women until they recover. It is at the service of the powerful and the weak, the poor and the rich, of the subject and the prince, of the citizen and the brigand, without demand for any form of payment, but only for the sake of God, the provider."*
>
> THE CONSTITUTION ESTABLISHING THE AL-MANSURI HOSPITAL, CAIRO

From these early institutions, hospitals spread all over the Muslim world, reaching Andalusia in Spain, Sicily, and North Africa. These were all admired by Europeans, who later developed similar systems such as the Hospitaliers, fighters of the hospital, established by the French to treat their countrymen. Muslim physicians participated in establishing scores of southern European hospitals such as the famous Salerno hospital in southern Italy.

Hospitals of the Muslim world were managed efficiently. For example, Ibn Jubayr, a 12th-century traveler, praised the way in which the Al-Nuri Hospital (probably the earliest of its kind) managed the welfare of patients. He said, "The new one [the Nuri Hospital] is the most frequented and largest of the two [hospitals in Damascus], and its daily budget is about 15 dinar. It has an overseer in whose hands is the maintenance of registers giving the names of the patients and the expenditures for the required medicaments, foodstuffs, and similar things. The physicians come early in the morning to examine the ill and to order the preparation of beneficial drugs and foods as are suitable for each patient."

"He who studies medicine without books sails an uncharted sea, but he who studies medicine without patients does not go to sea at all."

SIR WILLIAM OSLER,
CANADIAN PHYSICIAN (1849–1919)

While traveling in the Near East, Ibn Jubayr also noted one or more hospitals in every city in the majority of the places he passed through, which prompted him to say that hospitals were one of "the finest proofs of the glory of Islam."

These hospitals were also forward-thinking, tackling ailments not only of the body. A ninth-century Baghdad hospital, where Al-Razi worked, had a ward exclusively for the mentally ill.

The present-day Sultan Qalawun is now a funerary complex. Earlier the site had housed the Al-Mansuri Hospital in Cairo, Egypt.

University hospitals were the foundation of training for new medical students, just as they often are today. Eight hundred years ago, these teaching hospitals provided firsthand practical and theoretical lessons for students.

Teaching was done both in groups and on a one-to-one basis just like today. Lectures were held in a large hall at the hospital and the subject matter was usually a reading from a medical manuscript by the so-called Reading out Physician. After the reading, the chief physician or surgeon asked and answered questions of the students.

Many students studied texts with well-known physicians, and since paper was plentiful in the Muslim world, manuscripts that have written on them "for personal own use" have been preserved. In Europe these same texts were scarce and seldom owned by the student.

Bedside teaching, another part of medical training with groups of students following the attending physician or surgeon on his ward rounds, was seen as very important. More advanced students observed the doctor taking the history of and examining patients and also making prescriptions for them in the outpatient department of the hospital.

One of these medical schools was in the Al-Nuri hospital in Damascus. Under the direction of the physician Abu al-Majid al-Bahili, the 12th-century ruler Nur al-Din ibn Zangi founded the hospital. It was named after him and he equipped it with supplies of food and medication, while also donating a large number of medical books, which were housed in a special hall.

It was a place for a medical career to blossom. Early in the 13th century, a physician named Al-Dakhwar served in the Nuri hospital at a low salary, then, as he increased in fame, his income from private practice brought him much wealth and he started a medical school in the city. This career route is familiar to many physicians today.

Many renowned physicians taught at the medical school, and physicians and practitioners sometimes assembled before the sultan, Nur al-Din, to discuss medical subjects. At other times they listened to the three-hour lectures that Abu al-Majid, the director of the hospital, gave his pupils. Among the well-known Muslim physicists who graduated from the medical school were Ibn Abi Usaybi'ah, a 13th-century medical historian, and Ibn al-Nafis, whose discovery of the lesser circulation of the blood, also in the 13th century, marked a new step in better understanding human physiology.

LEFT: *A miniature illustration shows an Ottoman chief physician.* RIGHT: *The entrance to the* Nur al-Din Bimaristan *or Hospital, in Damascus, Syria, makes an impressive statement. The building now houses the museum of Arab medicine and science.*

02 Instruments of Perfection

If someone showed you a tray of surgical instruments from a thousand years ago, could you tell the instruments apart from modern tools? If you are thinking that the thousand-year-old ones would be rough and crude, read on to find out more.

If we journeyed back to tenth-century southern Spain we could look over the shoulder of a cutting-edge surgeon called Abul Qasim Khalaf ibn al-Abbas al-Zahrawi, a man known in the West as Abulcasis. He would have already written *Al-Tasrif,* his medical encyclopedia, in which he included a treatise called *On Surgery,* which introduced a staggering collection of more than 200 surgical tools. Apart from some sketches of instruments carved in ancient Egyptian tombs, this may have been the first treatise in the history of medicine to illustrate surgical instruments. In fact, their design was so accurate that they have had only a few changes in a millennium, and it was these illustrations that laid the foundations for surgery in Europe.

Al-Zahrawi illustrated each instrument using clear hand-drawn sketches and also provided detailed information on how and when it was used. For example, in cauterization he states: "According to the opinion of the early [physicians] cauterization using gold is better than when using iron. In our opinion the use of iron is quicker and more correct."

He wrote about the scraper [*majrad*] tool and its use when treating a fistula in the nose: "Doctors give the name 'fistula' to what laymen call 'a quill.' When you have treated it with cautery or with caustic according to the instructions given previously, and it is not healed, there is no clear method of treatment except to cut down on the tumour at its ripening and let out all the humidity or pus therein, till you reach the bone. When the bone is reached and you see necrosis or

A 1964 Syrian commemorative stamp shows an artist's impression of tenth-century Spanish Muslim surgeon Al-Zahrawi.

blackness, scrape it with an instrument like this picture. It is called 'rough-head' and is made of Indian iron. Its head is round like a button but is engraved with markings finely engraved, like those of a file or a rasp. Place it on the site of the diseased bone and spin it between your fingers, pressing down a little with your hand, till you are sure all the diseased bone has been scraped away. Do this several times. Then let the place be dressed with stanching and styptic remedies. And if the place heals and flesh is generated there and the flow of sanies [pus from a wound] is stayed and there is no return after leaving for

"Al-Zahrawi remains a leading scholar who transformed surgery into an independent science based on the knowledge of anatomy. His illustration and drawing of the tools is an innovation that keeps his contribution alive, reflected in its continuous influence on the works of those who came after him."

L. LECLERC, 19TH-CENTURY FRENCH MEDICAL HISTORIAN

forty days, and there is no swelling, and nothing emerges, you may know it is perfectly healed."

The case of urethral stones was the subject of many pages of study. Al-Zahrawi devised an instrument *al-mish'ab* (the drill) for crushing these. He said, "Take a steel rod with a triangular sharp end . . . tie a thread proximal to the stone lest it slips back. Introduce it gently till it reaches the stone, turn it round to perforate it . . . urine comes out immediately, press on the stone from outside and crush it by your finger, it breaks and comes out with urine. If you do not succeed then do cutting."

Commenting on this, Geoffrey Lewis and Martin Spink, recent translators of Al-Zahrawi's book, described the originality of the instrument: "This device of Abulcasis does seem to have been in a manner a true lithotripter [a stone-crushing machine used to shatter urinary stones] many centuries earlier than the modern era and completely lost sight of and not even mentioned by the great middle-era surgeons Franco and Paré nor by Frère Cômethe, doyen of genito-urinary surgery."

Ibn Zuhr, a 12th-century Seville physician, improved on this device by fixing a diamond at the end of the steel rod. As well as drills, Al-Zahrawi also manufactured a knife to perform cystolithotomy.

Other instruments discussed by Al-Zahrawi include cauterization tools of various shapes and sizes; scalpels, very sharp knives that are used for making a variety of incisions; hooks, usually with a sharp or blunt half-circular end that are still used and named in the same way (blunt hooks were inserted in the veins to clear blood clots; sharp hooks were used to hold and lift small pieces of tissue so that they could be extracted and to retract the edges of wounds); forceps, metal instruments with two handles used in medical operations for picking up, pulling, and holding tissue (crushing forceps used two jaws for crushing and removing urinary bladder stones; delivery forceps had a semicircular end designed to pull the fetus from its mother, an instrument still used today).

So, rather than unsophisticated tools, surgical instruments from Muslim civilization bear much resemblance to those we still use today.

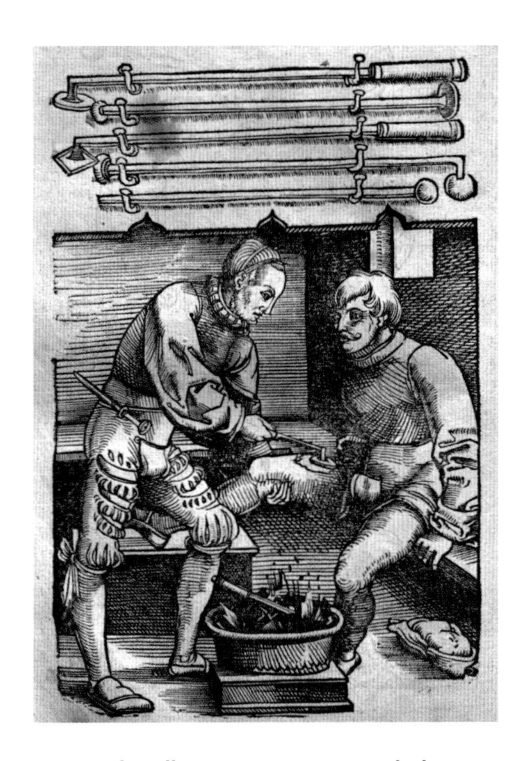

A 1532 woodcut illustrates cauterization, which appears in a work by Al-Zahrawi, also known as Abulcasis, in the Latin translation by Gerard of Cremona

SURGICAL PRECISION

Life-Saving Instruments We Still Use Today

LEGACY: **Developed more than 200 instruments, many used in modern-day medicine**

LOCATION: **Córdoba, Spain**

DATE: **10th-11th century**

KEY FIGURE: **Al-Zahrawi, surgeon**

Scalpels and knives, saws and scrapers, drills and forceps. The tools of surgery have changed surprisingly little over the last thousand years. Records left by Al-Zahrawi, a surgeon and scholar who lived in Spain at the height of Muslim civilization, show that he developed and used more than 200 medical instruments, many of which we still know today.

Born in Córdoba in 936, Al-Zahrawi, known as Abulcasis in the West, was a physician who carried out hundreds of operations and treatments during his lifetime, developing new techniques and treatments with which to improve his patients' prospects.

Al-Zahrawi wrote eyewitness accounts of his work and that of his peers in a 30-volume compendium called *Al-Tasrif*, which covered a huge range of medical situations, dental and surgical techniques. His writings also covered drug remedies based on mineral substances, herbs, and animal products.

The book was the first to illustrate surgical instruments, showing sketches of their form and describing how and when each one should be used. Although surgery was still dangerous and painful, suitable tools would have helped to treat patients suffering from bone diseases, tumors, bladder stones, and wounds, as well as assisting in childbirth.

Al-Zahrawi's sketches show scalpels, sharp and blunt hooks, scrapers, and saws. While many look very familiar today, others were further developed. In the 12th century, a doctor from Seville called Ibn Zuhr improved one tool by adding a diamond tip.

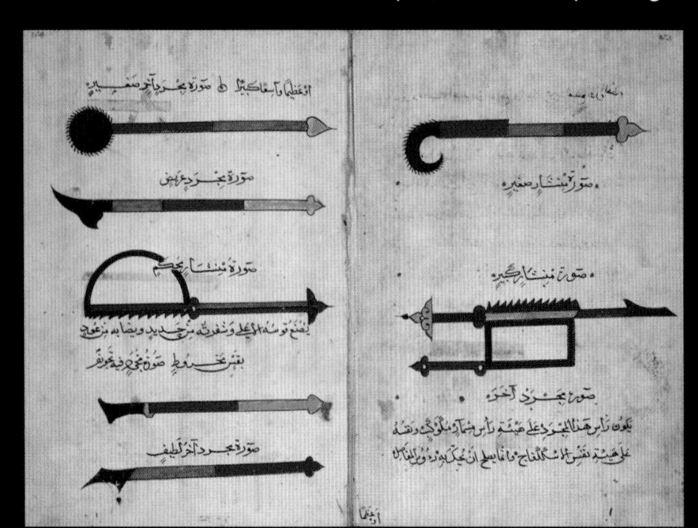

■ *A manuscript describing surgical instruments of Al-Zahrawi shows saws, or* Minshar, *of various shapes and scrapers, or* Mijrad, *used in orthopedic surgery.*

In *Al Tasrif,* Al-Zahrawi described correcting dislocated shoulders, setting broken bones in plaster casts, replacing missing teeth with replicas, and trying to treat cancers. His wealth of personal experience seemed to stand him in good stead with his patients.

In a typically sensitive gesture, he designed a knife with a concealed blade intended to calm nervous patients. This

An artist's rendering shows Al-Zahrawi performing surgery. Al-Zahrawi shared all he knew about medicine in his 30-volume encyclopedia Al-Tasrif. *Of the 200 medical tools he sketched and described, many have changed little in a thousand years.*

combination of bedside manner and innovative approach earned him the role of court physician to the ruler of Muslim Spain.

One of Al-Zahrawi's most memorable innovations was the systematic use of catgut for stitching a patient internally after surgery—as surgeons still do today. This idea, along with many others that Al-Zahrawi proposed, still bears a strong resemblance to the way modern medicine works.

In fact, through hospitals, medicines, and surgery, the modern world of health and medical treatment has countless links with the past. In early Muslim societies, a health care system developed that offered pioneering surgery, hospital care, and an increasing variety of drugs and medicines developed from ancient knowledge and new research.

Al-Zahrawi's innovations and observations spread into Europe when *Al-Tasrif* was translated into Latin by Gerard of Cremona, influencing medical practice right up to the Renaissance. Along with other medical writings from Muslim civilization, it was used as a manual for surgery in medical schools for centuries.

03 SURGERY

Modern surgery is a highly sophisticated culmination of centuries of innovation by dedicated people bent on saving lives. This life-saving ethic was beating in the heart of Muslim southern Spain a thousand years ago, where the Muslims performed three types of surgery: vascular, general, and orthopedic.

One of the most famous Muslim surgeons at this time lived in Córdoba at the height of the Islamic period. Al-Zahrawi, or Abulcasis, observed, thought, practiced, and responded to each of his patients with skill and ingenuity. So much so that he was recognized in his day as an eminent surgeon and was court physician to the ruler of Al-Andalus, Al-Mansur.

He revolutionized surgery by introducing new procedures, more than 200 surgical instruments, and giving detailed accounts of the full dental, pharmaceutical, and surgical disciplines of his time. His book, *Al-Tasrif*, also established the rules of practical medicine by emphasizing the do's and don'ts in almost every medical situation encountered.

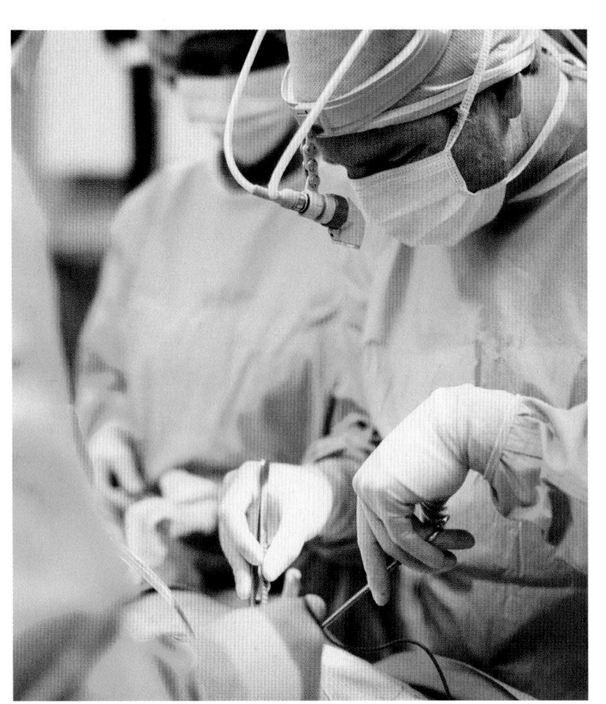

Al-Zahrawi has a list of firsts to his name and reading his curriculum vitae is impressive. New procedures he introduced included catgut for internal stitching, which is still used from the simplest to the most complicated surgery today. Catgut seems to be the only natural substance capable of dissolving and being accepted by the body.

Although Al-Zahrawi was the first to use catgut in surgery, it was Al-Razi who was the first to use animal (sheep) gut for sutures. Al-Zahrawi also used twisted fibers from strings of musical instruments for surgical purposes.

Responding to each case with ingenuity, he revolutionized medical procedures in many ways like using bone replacement for lost teeth; describing how to connect sound teeth to those that were loose by gold or silver wire; introducing a surgical treatment for sagging breasts; being the first to use cotton to control bleeding; performing a tracheotomy; regularly using plaster casts; and for stones in the urethra he introduced the technique of using a fine drill inserted through the urinary passage.

TOP: *Al-Zahrawi introduced catgut for internal stitching in the tenth century, and it is still widely used today.* BOTTOM: *Modern surgical tools used in today's operating rooms differ little from those invented by Al-Zahrawi.*

Physician Serefeddin Sabuncuoglu from Amasya, Turkey, showed various surgical procedures in his 15th-century Cerrahiyyet'ul Haniyye.

He also detailed how to remove a urinary bladder stone after crushing it with an instrument he designed. He discussed simple surgery like nose polyp removal and complicated procedures like the removal of a dead fetus using special forceps he devised. He mentions cauterizing or burning the skin to relieve pain and how to correct shoulder dislocation.

With all his innovations he kept his patients in mind, and in order not to frighten them in his surgical operations he invented a concealed knife to open abscesses. In the case of tonsillectomies, he held the tongue with a tongue depressor, and removed the swollen tonsil holding it with a hook and snipping it off with a scissor-like instrument. This had transverse blades, which cut the gland and held it for removal from the throat so the patient did not choke.

Al-Zahrawi displayed a sensible and humane reluctance to undertake the riskiest and most painful operations, as he was aware of the discomfort they inflicted on patients. This was a decisive breakthrough in the relationship between the surgeon and the patient.

For the first time in medical writing, Al-Zahrawi devoted a chapter, Chapter 61, of his surgery book to the technique of removing urinary bladder stones in women. *On Surgery* was only one of the 30 books to make up *Al-Tasrif*, so this makes you appreciate the amount of work he did.

He also described in detail his own technique of removing urinary bladder stones in males, adding several refinements to the technique described in the text *Sushruta Samhita* in Hindu medicine. Both Al-Razi and Al-Zahrawi stressed that the inner incision should be smaller than

A thousand years ago, patients were treated in hospitals for a wide range of health issues—just as they are today.

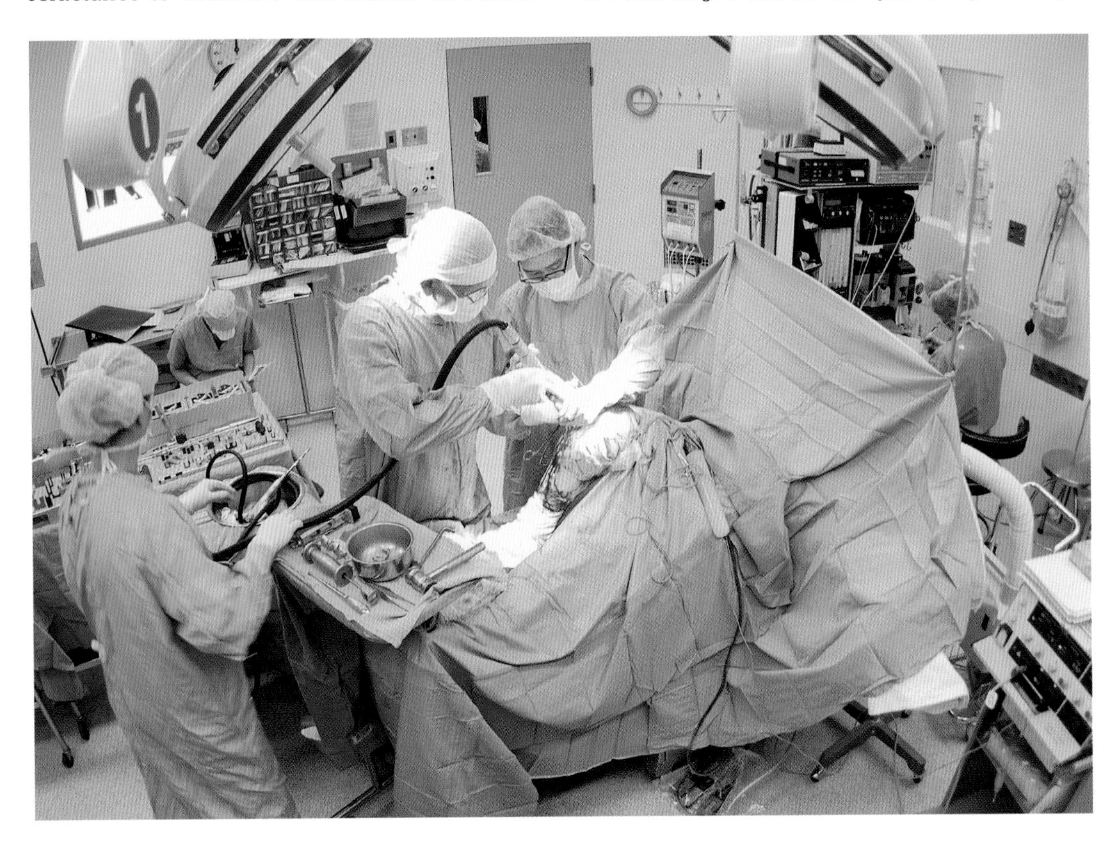

the external one to prevent leakage of urine. The stones should not be pulled out but extracted by forceps, and big ones should be broken and delivered out bit by bit. This demonstrates their care to avoid damage to the tissues, excessive bleeding, and formation of any urinary fistula. Al-Zahrawi also said every piece should be removed because even if one is left it will increase in size. This advice is still stressed nowadays.

"Surgeons must be very careful when they take the knife! Underneath their fine incisions stirs the Culprit—Life!"

EMILY DICKINSON, AMERICAN POET

In gynecology, his work, along with that of other Muslim surgeons, was pioneering. He gave instructions for training midwives on how to perform unusual deliveries and remove the afterbirth. He also designed and introduced vaginal specula.

There were many other medical doctors and surgeons in the Muslim world who carried out groundbreaking work, including Ibn Sina in the 11th century, who was from present-day Uzbekistan. He wrote *Canon,* which addressed the breadth of medicine, and you can read more about him in Ibn Sina's Bone Fractures.

In the opinion of Ibn Sina, cancer, *al-saratan* in Arabic, was a cold tumor that did not get inflamed, and was painless at first. Certain forms became painful and often incurable when they reached an advanced level. He said cancer grew out of the center just like the legs of a crab, from which it took its name. The internal cancers appeared without the patient's awareness, and despite their pain, patients could live quite long with them. The only forms of cancer upon which the surgeon could intervene were the "limited cancers." Here, the incision had to be perfect, so all of the tumor would be extracted. However, surgery was not always conclusive and definite, for the cancer could often reappear. Ibn Sina, in fact, advised against the amputation of the female breast, for it favored the spread of the disease. He suggested that oxide of copper or lead, although unable to cure the cancer, could be efficient in stopping the spread of the disease.

Ibn Sina, like Al-Zahrawi, spoke on many topics. On the retention of urine due to a bladder stone he explained: "If the patient lies on his back and his buttocks are raised and he was shaken, the stone moves away from the passageway . . . urine streams out, it may also be easy to push away the stone by a finger in the rectum . . . If that does not work, use a catheter to push the stone back . . . If it was difficult to be passed do not push hard." This is similar to how modern urologists handle an obstructing posterior urethral stone. They push it back either by a catheter or endoscopically.

According to Ibn al-Quff, a 13th century Syrian physician, surgical treatment of large bladder stones was easier than that of small ones because the large ones either stopped in the urethra or were in the cavity of the bladder, and here they could be more easily felt.

From all this evidence, we can see that a thousand years ago ailing people were treated in hospitals and looked after incredibly well. Unlike today, we do not have survival rates or statistics of success, but we do have copious notes from the great surgeons of the time. These notes of practices and research changed surgery irreversibly, for the betterment of all, even those of us in the 21st century.

04 Blood Circulation

The ancient Greeks thought that the liver was the origin of the blood, believing food reached the liver from the intestines through the veins. In the liver, blood would be filled with "natural spirit" before it continued the journey to the heart.

Then Galen, a Greek physician and scholar in the second century C.E., made further observations. He said that the blood reaching the right side of the heart went through invisible pores in the cardiac septum to the left side of the heart. Here it mixed with air to create spirit and was then distributed to the body.

For centuries, this explanation was accepted as the truth until William Harvey performed groundbreaking research into the circulation of the blood and the function of the heart in 17th-century Europe. Harvey argued that the heart was at the center of the circulatory system, and he was known as the person who discovered how our blood travels around our bodies.

But in 1924 a very important manuscript was found and made known to the world by an Egyptian physician, Dr. Muhyi al-Deen al-Tattawi. This discovery revealed a much earlier first description of pulmonary circulation.

This manuscript, *Commentary on the Anatomy of the Canon of Avicenna,* was written by Ibn al-Nafis, a Muslim scholar born in Damascus, Syria, in 1210 and educated at the famous Nuri Hospital. When he "graduated," he was invited to Cairo by the sultan of Egypt to work as the principal of the Nasiri Hospital, founded by Saladin.

As well as having a busy professional career as a physician and legal authority, Ibn al-Nafis wrote a number of books on a variety of subjects including *Commentary on the Anatomy of the Canon of Avicenna.*

Avicenna (also known as Ibn Sina) was a polymath who excelled in philosophy, law, and medicine. Ibn al-Nafis's treatise, a commentary on Avicenna's monumental *Canon,* was groundbreaking its own way. In it, Ibn al-Nafis accurately described the pulmonary circulation, explaining the role of the heart and lungs, and emphasizing that blood was purified in the lungs, where it was refined on contact with the air inhaled from the atmosphere outside the body.

On how the blood's pulmonary circulatory system worked, Ibn al-Nafis explained that the system was based on the movement of blood from one chamber of the heart to the lungs and then back to a different chamber of the heart. According to him, blood enhanced with vital *pneuma* (air from the lungs) flowed through the arteries to all parts of the body. His innovation was to say that the venous blood from the right ventricle of the heart (to be enhanced with air from the lungs) had to pass through the lungs before entering the left ventricle, at which point it could enter the arteries as arterial blood.

In his own words he said: "The blood from the right chamber of the heart must arrive at the left chamber, but there is no direct pathway between them. The thick septum of the heart is not perforated and does not have visible pores as some people thought or invisible pores as Galen

TOP: *A title page from a book translating the work of Greek physician Galen. It was only through Arabic that the work of Greek scholars such as Galen could be found.*

> *"The thick septum of the heart is not perforated and does not have visible pores as some people thought, or invisible pores as Galen thought."*
>
> IBN AL-NAFIS, MUSLIM SCHOLAR

thought. The blood from the right chamber must flow through the pulmonary artery to the lungs, spread through its substance, be mingled with air, pass through the pulmonary vein to reach the left chamber of the heart."

In modern language, this is translated as follows: Blood that has waste in it comes into the right atrium through the large vein called the vena cava. Filled with this waste-rich blood, the right atrium then contracts, pushing the blood through a one-way valve into the right ventricle. In turn the right ventricle fills and contracts, sending the blood into the pulmonary artery, which connects with the lungs. There, in the capillaries, the exchange of carbon dioxide and oxygen takes place. The blood is now oxygen-rich as it enters the pulmonary veins, returning to the heart via the left atrium. The left atrium fills and contracts, pushing oxygen-rich blood through a one-way valve into the left ventricle. The left ventricle contracts, forcing the blood into the aorta, from which its journey throughout the body begins.

These important observations were not known in Europe until 300 years later, when Andrea Alpago of Belluno translated some of Ibn al-Nafis's writings into Latin in 1547. Following this, a number of attempts were made to explain the phenomenon, including by Michael Servetus in his book *Christianismi Restitutio* in 1553 and Realdus Colombo in his book *De re Anatomica* in 1559. Finally it was Sir William Harvey, in 1628, who was credited with for the discovery of circulation, while Ibn al-Nafis remained the pioneer of the "lesser," or pulmonary, circulation.

It was only in 1957, 700 years after his death, that Ibn al-Nafis was credited with the discovery.

THE BLOOD CIRCULATION SYSTEM

In the 13th century Ibn al-Nafis explained the pulmonary blood circulation system, i.e., the system of oxygenation of oxygen-poor blood by the lungs. The right ventricle of the heart pumps deoxygenated blood to the lungs through the pulmonary arteries where it is oxygenated and then returned to the left atrium of the heart through the pulmonary veins. In the 17th century William Harvey discovered the full blood circulatory system in which the blood returns to the heart from the body extremities (the blue arrows to the heart on the diagram).

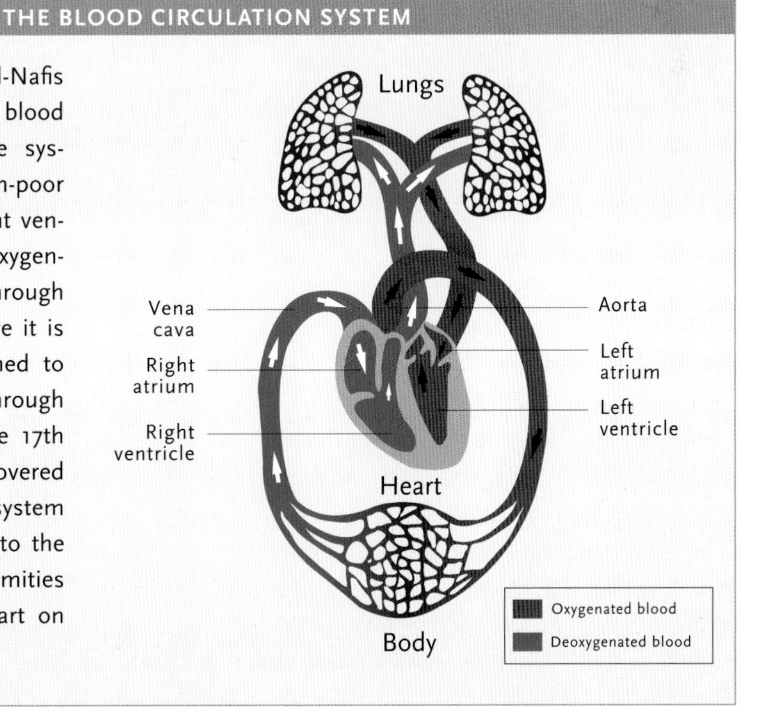

05 Ibn Sina's Bone Fractures

Ibn Sina, known as Avicenna in the West, was so highly regarded that he was compared to Galen, the ancient Greek physician, and was known as the Galen of Islam. Because of his great fame, many nations competed to celebrate his anniversary, Turkey being the first in 1937, 900 years after his death.

To appreciate his contribution in developing the philosophical and medical sciences, all members of UNESCO celebrated a thousand years after his birth in 1980.

He was born in Afshana, now in Uzbekistan, and left age 21, spending the rest of his life in various Persian towns, becoming a renowned philosopher and medical doctor. Through his life he composed 276 works, all written in Arabic, except for a few small books written in his mother tongue, Persian. Unfortunately, most of these works have been lost, but there are still 68 books or treatises available in Eastern and Western libraries.

He wrote on all branches of science, but he was most interested in philosophy and medicine, so some recent historians called him a philosopher rather than a physician, but others say he was the "Prince of the Physicians" during the Middle Ages.

The majority of his work was in medicine. Forty-three works were in this area; 24 in philosophy; 26 in physics; 31 in theology; 23 in psychology; 15 in mathematics; 22 in logic; and 5 in Quran interpretation. He also wrote on asceticism, love, and music, and he also wrote some stories.

Al-Qanun fi al-Tibb or *Code of Laws in Medicine* was his most important work, and is known in English as the *Canon*. It was written in Arabic, and has been described as the most famous medical textbook ever written, because it is a unique reference full of all known medical knowledge gathered from many civilizations up to his time.

"Medicine was absent until Hippocrates created it, dead until Galen revived it, dispersed until Rhazes [al-Razi] collected it, and deficient until Avicenna [Ibn Sina] completed it."

DE POURE, EUROPEAN PHYSICIAN

By the 12th century the essentials of the *Canon* were condensed to make the ideas more readily accessible, and commentaries were written to clarify the contents. The most popular short version was called *The Concise Book of Medicine*, written in Syria by Ibn al-Nafis, who died in 1288.

The *Canon* was made up of five books. The first was concerned with general medical principles; the second with materia medica; the third with diseases occurring in a particular part of the body; the fourth with diseases not specific to one bodily part, such as fevers and also traumatic injuries such as fractures and dislocations of bones and joints. The final book contained a formula giving recipes for compound remedies.

AVICENNAE
ARABVM
MEDICORVM PRINCIPIS.
Canon Medicinæ.
QVO VNIVERSA MEdendi scientia pulcherrima, & breui methodo planissimè explicatur.
Eiusdem
De { Viribus cordis. Remouendis nocumentis in regimine sanitatis. Syrupo acetoso.
CANTICA.
VENETIIS, Industria ac sumptibus IVNTARVM.

The cover of the Latin edition of Ibn Sina's Canon of Medicine

The fourth book had two treatises; one was called "Fractures as a Whole," and the second was "Fractures of Every Bone Separately."

"Fractures as a Whole" described the causes, types, forms, methods of treatment, and complications of fractures, talking about fractures in general, while "Fractures of Every Bone Separately" looked at the special characteristics of fractures of each bone. Ibn Sina, by using this form of explanation, was very close to following the format of modern medical textbooks.

He drew attention to the necessity of not splinting the fracture immediately, advising postponing it beyond the fifth day. Today, this is called the Theory of Delayed Splintage, and Professor George Perkins (1892-1979), who was based at St. Thomas' Hospital in London, is considered the pioneer of this theory.

> *"Anyone who wants to be a good doctor must be an Avicennist."*
>
> OLD EUROPEAN COMMON SAYING

Ibn Sina talked about what is now called "Bennett's fracture 1882" a thousand years before Bennett.

The arrangement, comprehensiveness, and methods of explanation of the *Canon* were similar to the layout of modern medical textbooks with regard to classification, causes of diseases, epidemiology, symptoms and signs, and treatment and prognosis. This made the *Canon* the most widely used medical book in both Muslim and European countries; it was known to Europeans in the 12th century from the Latin translations of Gerard of Cremona. It remained in use in medical schools at Louvain and Montpellier until the 17th century, and according to the journal of UNESCO it was still in use at Brussels University until 1909, well into the age of "modern medicine."

The school of Salerno is depicted in this 14th- to 15th-century Latin translation of the manuscript Canon of Medicine *by Ibn Sina.*

THE DOCTOR'S CODE

A Book of Medical Knowledge from Many Civilizations

LEGACY: Influenced medical practice until the 19th century

LOCATION: Persia

DATE: 10th–11th century

KEY FIGURE: Ibn Sina, known as Avicenna, doctor and polymath

The 11th-century scholar Ibn Sina wrote and taught widely on medicine, philosophy, and natural sciences. Known in the West as Avicenna, his most influential book was *Al-Qanun fi al-Tibb*—translated as *Code of Laws in Medicine*, but known most commonly as the *Canon*.

In the *Canon*, Ibn Sina collected together medical knowledge from across civilizations creating a master reference work. Made up of five volumes, the book covered medical principles, medicines, diseases of various body parts, general disease, and traumas.

Ibn Sina described in detail the causes, types, and complications of fractures, and the various ways to treat them. He advised against splinting a fractured limb right away, but recommended waiting five days—a procedure now universally adopted. In his writings, he included detailed instructions for understanding traumatic injuries to every bone—and even described a thumb injury now known as Bennett's fracture hundreds of years before the scholar after whom it is named.

One hundred and forty-two properties of herbal remedies were included in Ibn Sina's *Canon*. With historical roots in Egypt, Mesopotamia, China, and India, herbs had been important to health in ancient Greek and Roman societies. In early Muslim civilization, an increase in travel and trade made new plants, trees, seeds, and spices available, along with the possibilities of new herbal medicines.

The *Canon* was certainly comprehensive—but when Ibn Sina wrote it, nearly a thousand years ago, he could not have foreseen how long its wisdom would last. Gerard of Cremona translated the *Canon* into Latin in the 12th century, and soon the medical communities of Europe were all using the book, like the doctors of Muslim civilization before them. By the 13th century, concise Latin versions of the *Canon* had been published, along with commentaries to clarify its contents. The *Canon* was still consulted by some doctors until the early 1800s.

Ibn Sina wrote prolifically all his life—and on many topics. Along with various medical books, he wrote 26 books on physics, 31 on theology, 23 on psychology, 15 on mathematics, 22 on logic, and several influential works of philosophy. He also found time to write about love and music.

Along with other scholars of Muslim civilizations, Ibn Sina questioned superstitious beliefs and sought to develop a rational understanding of the Earth's systems. He discussed sources of water and the formation of clouds in his writing, including in *The Book of Healing*, a comprehensive chapter on mineralogy and meteorology, examining how mountains form, the concept of geological time, and what causes earthquakes.

بسم الله الرحمن الرحيم

-ntury manuscript shows the opening page of the first book of the* Canon of Medicine, *-s originally called* Al-Qanun fi al-Tibb, *or* The Code of Laws in Medicine. *Its author, Ibn Sina, Avicenna in the West, influenced medical treatment and thought across Muslim and European from the 11th century until as late as the 19th.*

06 Notebook of the Oculist

Nearly every medical book by Muslims a thousand years ago covered some aspect of eye diseases. Their studies were limited only because animal eyes were used instead of human eyes, because the dissection of the human body was considered disrespectful. However, that did not stop the oldest pictures of the anatomy of the eye from being constructed.

Muslim eye surgeons or ophthalmologists of the tenth to the thirteenth centuries were performing operations, dissecting, discovering, and writing about their findings in textbooks and monographs. According to Professor Julius Hirschberg, an eminent 20th-century German professor of medicine, 30 ophthalmology textbooks were produced, and 14 of them still exist today.

Modern terms were used like conjunctiva, cornea, uvea, and retina. Operations on diseases of the lids like trachoma, a hardening of the inside of the lid, were also common practice. The treatment of glaucoma, an increase in the intraocular pressure of the eye, under the name of "headache of the pupil," was popular, but the greatest single contribution in ophthalmology by the Muslims was in the treatment of cataracts.

The term for cataract in Arabic is *al-ma' al-nazil fil'ayn*, meaning "water descends into the eye," which refers to the water accumulating in the lens, making it soggy and cloudy.

To restore vision, Al-Mawsili, from tenth-century Iraq, designed a hollow needle and inserted it through the limbus, where the cornea joins the conjunctiva, to remove the cataract by suction. This type of cataract operation, among others, is still carried out today with some added modern techniques, such as freezing the lens before suction.

Scholars in Muslim civilization took eye diseases very seriously, just as physicians do today.

From his study and practice he then wrote the *Book of Choices in the Treatment of Eye Diseases*, which discussed 48 diseases. This manuscript (No. 894) can be found in the Escorial Library in Madrid, Spain.

Until the 20th century, Al-Mawsili's work was only available in Arabic and a 13th-century Hebrew translation. The German version was made as recently as 1905 by Professor Hirschberg, who wrote that Al-Mawsili was "The most clever eye surgeon of the whole Arabian Literature."

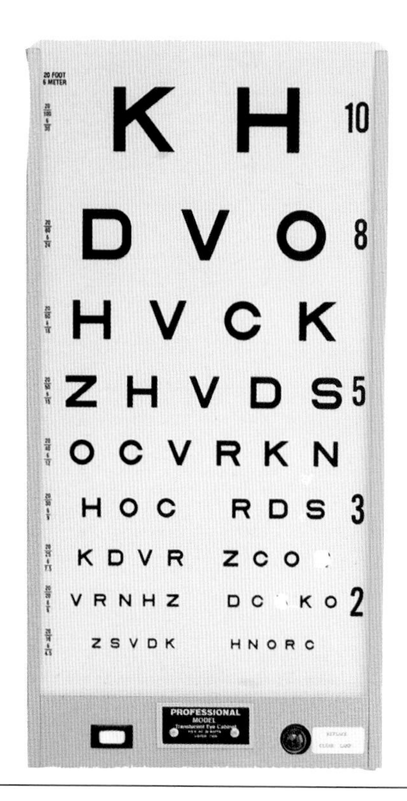

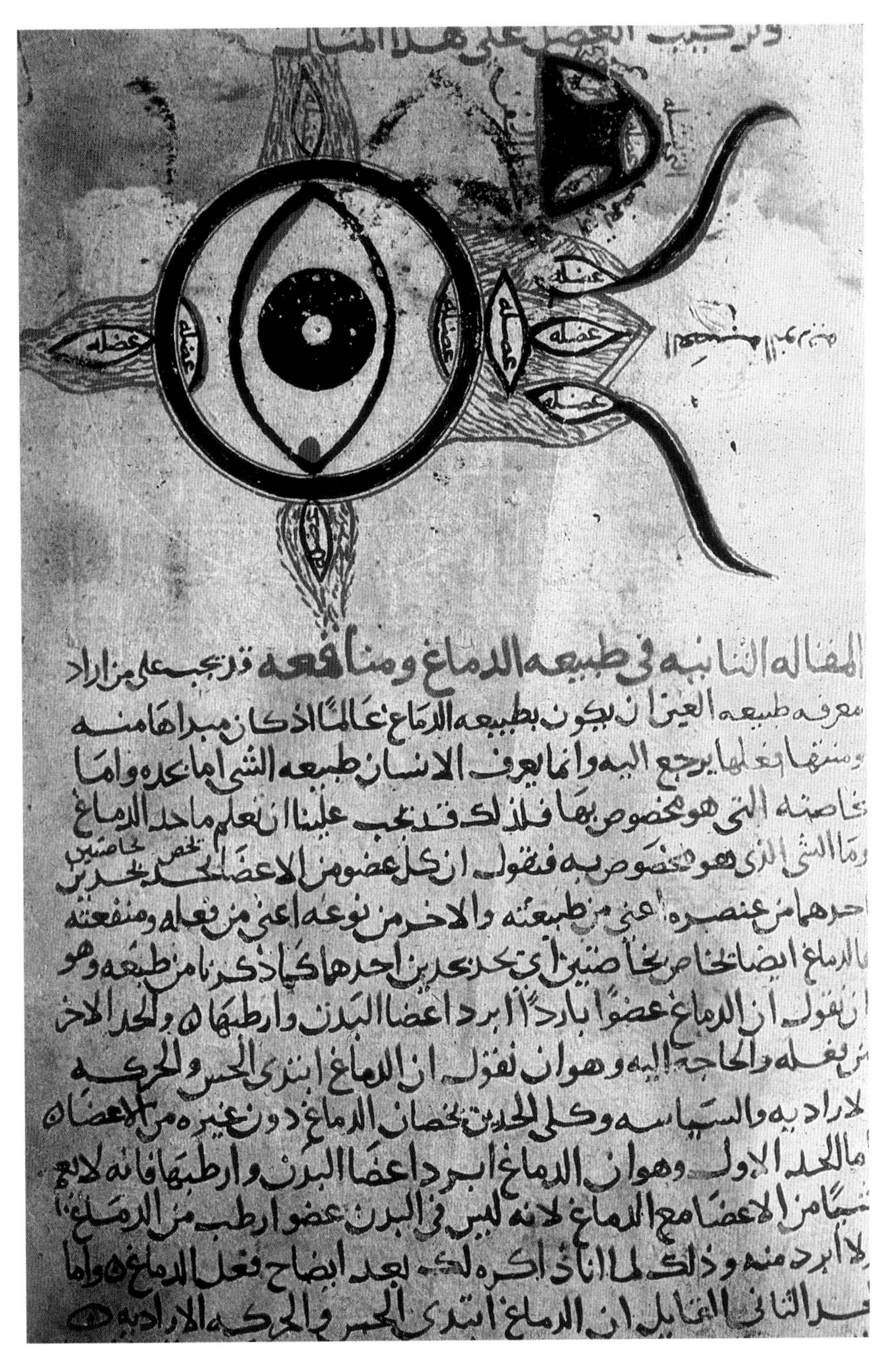

The anatomy of the eye is illustrated in a 12th-century manuscript that refers to the treatise on ophthalmology by Hunayn ibn Ishaq, a ninth-century Christian from Baghdad. During Muslim civilization, Muslim and non-Muslim scholars worked side by side.

A contemporary of Al-Mawsili and the most famous of all the oculists of Islam was Ali ibn Isa, also from tenth-century Baghdad, Iraq. He wrote the *Notebook of the Oculist* and this was the most complete textbook on eye diseases, which was translated into Latin and printed in Venice in 1497. Again Professor Hirschberg and his fellow eye surgeon J. Lippert translated it into German in 1904, and the English version, by American oculist and academic Casey Wood, appeared in 1936.

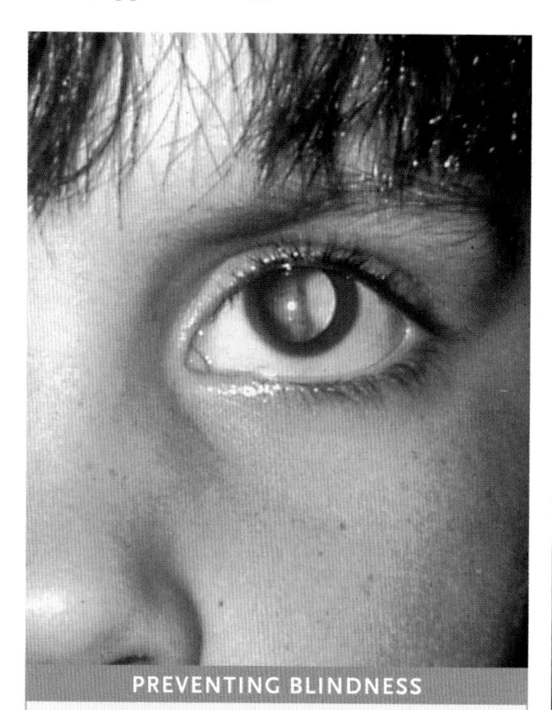

PREVENTING BLINDNESS

"Muslim physicians have been in the forefront of the effort to prevent blindness since 1000 C.E., when al-Razi became the first doctor to describe the reflex action of the pupil. At about the same time, . . . al-Mawsili invented the technique of suction-removal of cataracts by the use of a hollow needle."

OPTOMETRY TODAY,
A PUBLICATION OF THE ASSOCIATION OF OPTOMETRISTS, ENGLAND, MARCH 28, 1987

"During this total darkness in medieval Europe, they [the Muslims] lighted and fed the lamps of our science [ophthalmology]—from the Guadalquivir [in Spain] to the Nile [in Egypt] and to the river Oxus [in Russia]. They were the only masters of ophthalmology in medieval Europe."

PROFESSOR JULIUS HIRSCHBERG
CONCLUDING HIS ADDRESS TO THE
AMERICAN MEDICAL ASSOCIATION, JULY 1905

Ibn Isa's book *Notebook of the Oculist* was the authoritative textbook on ophthalmology for centuries, describing 130 eye diseases, including several forms of trachoma and ophthalmia.

It is also the oldest Muslim work on ophthalmology that is complete and survives in the original state. Dr. Cyril Elgood, a 20th-century British medical historian, wrote: "The first part is devoted to anatomy, the second to the external diseases of the eye, and the third part to internal diseases of the eye which are not visible upon inspection . . . The nearest approach that Ali makes to the modern conception of eye disease as a manifestation of general disease is when he urges the practitioner to realize that defective vision may be due to a disease of the stomach or brain just as much as to an incipient cataract."

Ibn Isa was not the only eye surgeon to urge that diseases of the eye could be signs of other ailments. Abu Ruh Muhammad ibn Mansur ibn Abdullah, known as Al-Jurjani, from Persia around 1088, wrote a book called *The Light of the Eyes*. One chapter dealt with diseases that lay hidden, but whose signs were clear in the eyes and vision, like third nerve paralysis, blood disorders, and toxicity.

An oculist who has been immortalized in a bust in Córdoba, southern Spain, is Muhammad ibn Qassum ibn Aslam al-Ghafiqi. He lived and practiced in Córdoba, writing a book called *The Right Guide in Ophthalmic Drug*. The book is not just confined to the eye but gives details of the head and diseases of the brain. Reporter Rageh Omaar said in the BBC's *An Islamic History of Europe* that Al-Ghafiqi's treatment of the eye disease trachoma was carried out until World War I. His bust in the municipal hospital of Córdoba was erected in 1965 to commemorate the 800th anniversary of his death.

In the United Kingdom today, cataracts are the most common cause of blindness in people older than 50, but there's good news from the Royal College of Ophthalmologists: "Cataract surgery has excellent outcomes and makes an enormous difference to patients' lives." Hundreds of thousands of cataract operations take place every year in the United Kingdom, making it the most commonly performed elective operation in the country. Who would have thought that Al-Mawsili's work in the tenth century would have laid the foundations for an incredibly popular 21st-century surgery.

By the 13th century, eye surgeons were already investigating the eye's structure and developing new ways to tackle disease.

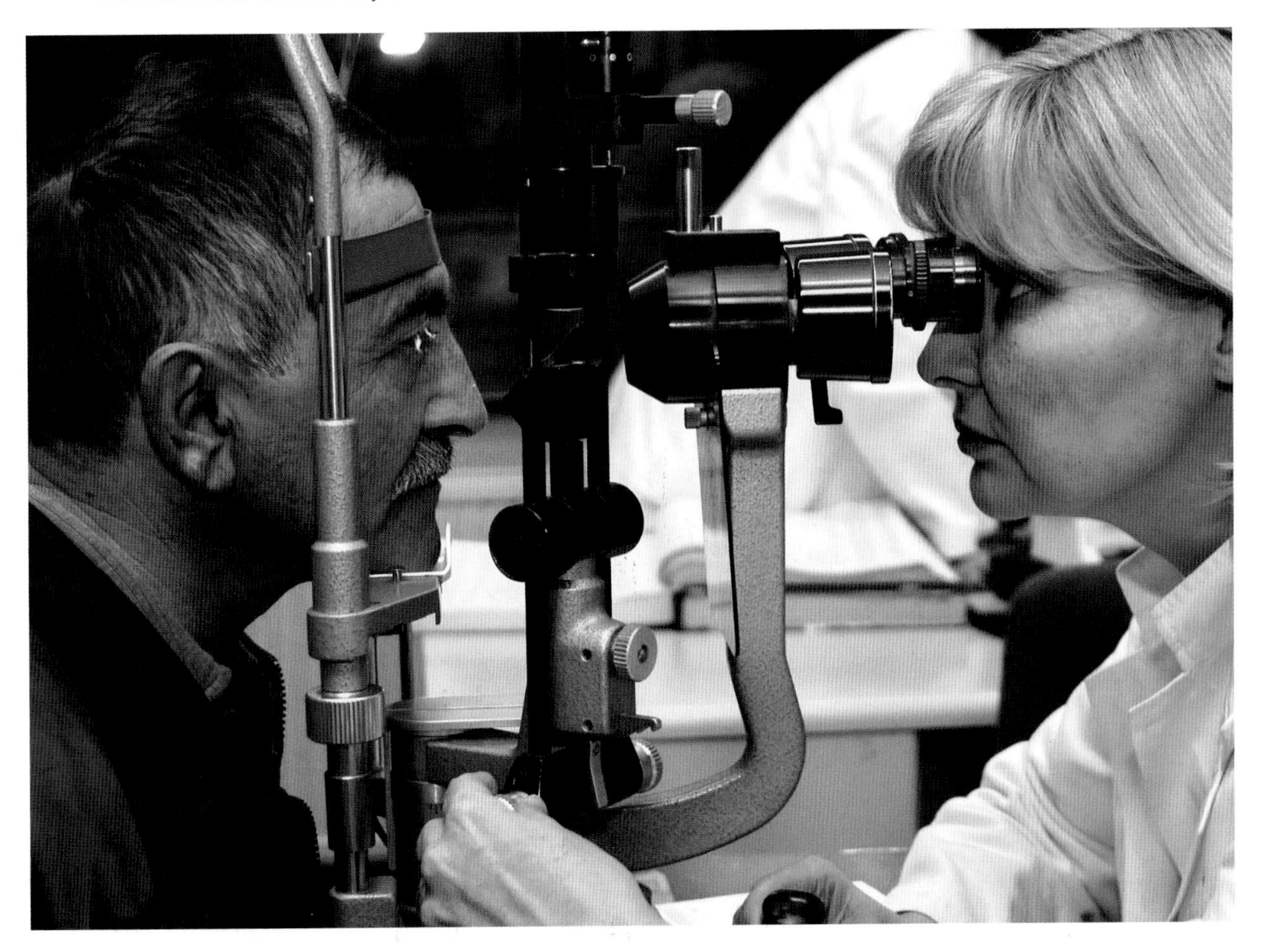

07 INOCULATION

Vaccination today can be a controversial issue, and it was rejected when it was first brought to England from Turkey nearly 300 years ago. The Anatolian Ottoman Turks knew about methods of vaccination. They called vaccination *Ashi,* or engrafting, and they had inherited it from older Turkic tribes.

Vaccination is a process where a person is given a weakened or inactive dose of a disease-causing organism. This stimulates the immune system to produce antibodies to this specific disease. Today, the development of new vaccines takes eight to twelve years, and any new vaccine has to be rigorously tested before it can be accepted as safe.

The Turks had discovered that if they inoculated their children with cowpox taken from the breasts of cattle, they would not develop smallpox. This kind of vaccination and other forms of variolation were introduced into England by Lady Montagu, a famous English letter writer and wife of the English ambassador at Istanbul

TOP: *Lady Mary Wortley Montagu, 1689-1762, introduced the smallpox vaccination from Turkey into England.*
BOTTOM: *An 1802 caricature titled the "Cow-Pock" by James Gillray shows Dr. Jenner vaccinating patients at the St. Pancras Smallpox and Inoculation Hospital.*

between 1716 and 1718. She came across the Turkish methods of vaccination and became greatly interested in smallpox inoculation after consenting to have her son inoculated by the embassy surgeon, Charles Maitland.

While in Istanbul, Lady Montagu sent a series of letters to England in which she described the process in detail. On her return to England she continued to spread the Turkish tradition of vaccination and had many of her relatives inoculated. She encountered fierce opposition to the introduction of inoculation, not only from the church authorities, who used to oppose any intervention, but also from many physicians. Through her tenacity though, inoculation became increasingly widespread and achieved great success.

> *"For more than two hundred years, vaccines have made an unparalleled contribution to public health . . . Considering the list of killer diseases that once held terror and are now under control, including polio, measles, diphtheria, pertussis, rubella, mumps, tetanus, and Haemophilus influenzae type b (Hib), one might expect vaccination to have achieved miracle status."*
>
> RICHARD GALLAGHER, EDITOR OF THE INTERNATIONAL MAGAZINE AND WEBSITE *THE SCIENTIST*

The breakthrough came when a scientific description of the vaccination process was submitted to the Royal Society in 1724 by Dr. Emmanuel Timoni, who had been the Montagus' family physician in Istanbul. This was further augmented by Cassem Aga, the ambassador of Tripoli, who provided a firsthand account of inoculation and its safety record in Tripoli, Tunis, and Algiers, which gave valuable reassurance about the long safety record of the practice in Muslim countries, and for which he was elected fellow of the Royal Society in 1729. Inoculation was then adopted both in England and in France, nearly half a century before Edward Jenner, to whom the discovery is attributed.

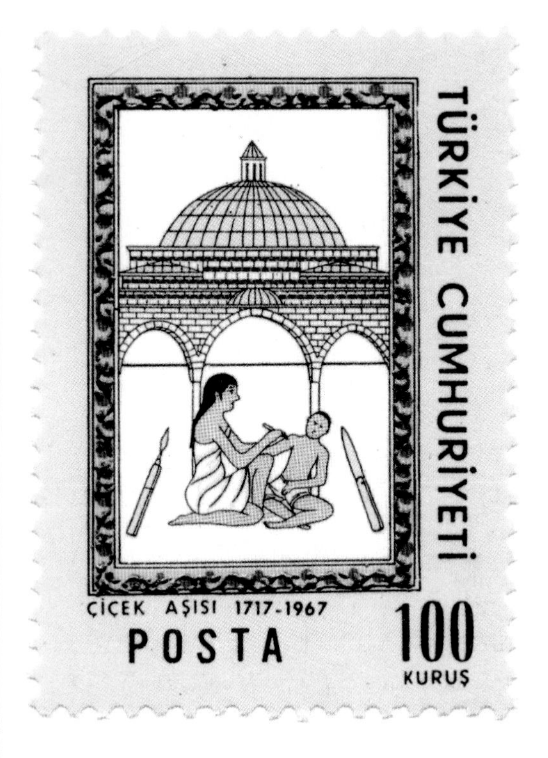

A stamp issued by the Turkish Postal Authority in 1967 depicts the 250th anniversary of the first smallpox vaccination.

It is currently believed that in 1796 Jenner "heard" that cowpox provided immunity to smallpox when he saw the case of James Phipps, an eight-year-old boy, who was infected with cowpox from a cut on the hand of a milkmaid, Sarah Nelmes.

In 1967, Turkey commemorated the 250th anniversary of the first smallpox vaccination. The stamp shows a child being inoculated. In the background is an Islamic dome and in the foreground a surgeon's scalpel.

08 Herbal Medicine

A thousand years ago gardens were also scientific "field" laboratories, looked after by eminent scientists who wrote manuals on the medical properties of plants. Herbal medicine was not seen as an alternative medicine but was very much a part of medical practice, with many hospitals keeping gardens full of herbs for use in medicines, and new drugs were discovered and administered.

This kind of herbal discovery has been made since the dawn of civilization. There are records from Egypt, Mesopotamia, China, and India that reflect a tradition that existed before we discovered writing. In the West, the first "herbal" (a book listing and explaining the properties of herbs) was Greek and written in the third century B.C.E. by Diocles of Carystus, followed by Crateuas in the first century C.E. The only work that has survived, *De Materia Medica*, was written in 65 C.E. by Dioscorides. He remains the only known authority among the Greek and Roman herbalists.

"And the leaves of the tree were for the healing and the restoration of the nations."

THE BIBLE, REVELATION 22:2

An illustrated page from a 15th-century Arabic botanical treatise

As the Muslim lands grew, merchants and travelers came across exotic plants, trees, seeds, and spices previously unknown to them. They collected and brought back a huge number of samples of raw ingredients, along with knowledge and information about their use, combing the world and its harshest of environments, going as far afield as the steppes of Asia and the Pyrenees. The discovery and wide use of paper also meant that on-the-spot detailed recording of their journeys and observations could be made.

With this vast amount of data and material, coupled with their scientific medical knowledge, many new traditional and herbal medicines became available. All these discoveries meant that a huge amount of information was built up and spilled out of colossal encyclopedic works.

CLOCKWISE FROM TOP LEFT: *A vine from a 15th-century Arabic botanical treatise. An illustration from Dioscorides'* De Materia Medica *depicts the physician handing his student a mandrake root, which was regarded as a highly effective medicine. A botanical species from a treatise by Ibn al-Baytar of Málaga. The tapping of a balsam tree as shown in a 15th-century Persian manuscript.*

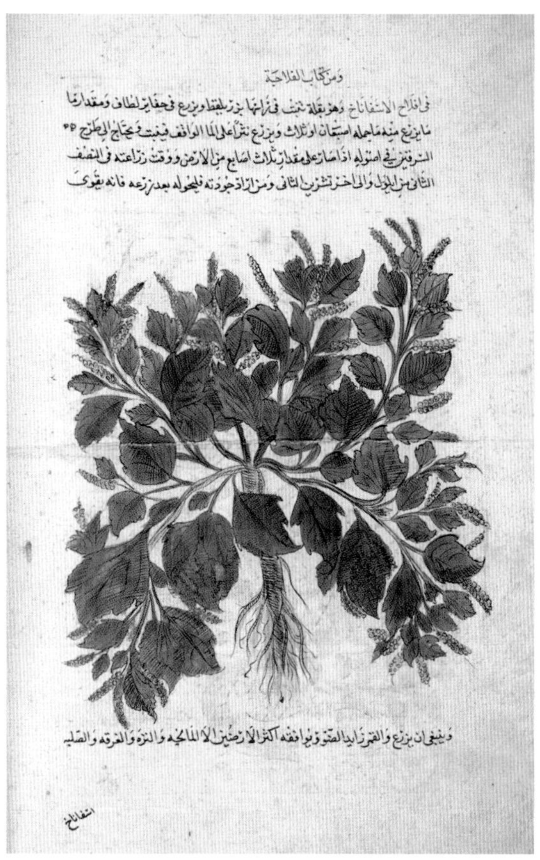

The 13th-century treatise by Ibn al-Baytar of Málaga depicts different botanical species—the example at left is from the manuscript Al-Kafi *and at right from the manuscript* Al-Filaha. *The treatise gives the physiology of plants and descriptions of their sowing environment as well as their maintenance.*

Ibn Samajun, who died in 1002, wrote *Collection of Simples, Medicinal Plants and Resulting Medicines*. This was a classification of plants and their medical properties based on the work of his predecessors. Also in the 11th century, Ibn Sina in his *Canon* listed 142 properties of herbal remedies.

Botany, the scientific study of plants, and the use of plants in medicine went hand in hand. While men like Abu Hanifa al-Dinawari, called "the father of modern botany," were compiling vast lists of plants in books like his *A Treatise on Plants*, others, like Al-Razi, a tenth-century medical scholar, used colchicum as a drug for the treatment of gout.

As botany became an academic science, chemistry was advancing at an incredible rate, and both these developments helped to propel herbal medicine into the mainstream. Coupled with the appearance of improved water-raising machines and new irrigation techniques in the tenth century, experimental gardens sprouted and herbs were cultivated.

Al-Andalus, or Muslim Spain, was a springboard for herbal development. In 11th-century Toledo, Spain, and later in Seville, the first royal botanical gardens of Europe made their appearance. Initially they were pleasure gardens, but they also functioned as trial grounds for the acclimatization of plants brought from the Middle East.

You can read about Ibn al-Baytar of Malaga in the Pharmacy section, but the basis for his work, *Dictionary of Simple Remedies and Food*, an enormous pharmacological encyclopedia, reflects his botanical skills; in this he studied 3,000 different plants and their medical properties.

One of the best herbal medicine books was produced by Al-Ghafiqi, who died in 1165. This was called *The Book of Simple Drugs*. It was exceptionally accurate and was republished by Max Meyerhof in Egypt in 1932.

In the tenth century, Ibn Juljul wrote a commentary on Dioscorides's 900-year-old *De Materia Medica* and translated it into Arabic, adding many new substances such as tamarind, camphor, sandalwood, and cardamom. He also identified many new plants and their properties along with their medicinal values for treating various diseases.

A very simple but major breakthrough that Muslims made in herbal medicine was watching how the herb affected the patient. Now this seems quite an obvious thing to do, but they were the only ones using and relying on scientific methods of experimentation and observation at that time.

Elsewhere in medieval Europe, books on herbs were rare and known only among a small number of scholars, and until the end of the 15th century many Europeans were using the Arabic texts and Arabic versions of Greek texts translated into Latin. So between 1500 and 1600 there were about 78 editions of Dioscorides, the Greek scholar.

"And in it, their drink is mixed with ginger."

QURAN (76:17) MENTIONS GINGER AS ONE OF THE DRINKS OF PARADISE. TODAY, GINGER IS USED TO RELIEVE NAUSEA AND VOMITING.

The success of the European scholars was measured by what they borrowed from Muslim botanists and how they made Dioscorides more prominent, but things were not going well. The once great Salerno school was in decline because of a lack of ability in Latin, Greek, and Arabic, and they did not fully understand the Greek texts as most of the time they were translations of translations.

European herbalists were frustrated by ignorance, malpractice, faults in earlier bad Greek translations, and also from not being able to identify ingredients correctly because they were described in local dialects. All this led Sir Thomas Elyot, a 16th-century English diplomat and scholar, to inform his readers that he derived no understanding from the ancients and that they gave "no little profyte concernynge myne owne helthe."

Fortunately, herbal medicine has done away with using mother's blood, which was sometimes added in certain medieval European recipes. Today, in the United Kingdom, one Briton in five uses complementary medicine and, according to a recent survey, one in ten uses herbalism or homeopathy. Around £130 million is spent on oils, potions, and pills every year in Britain, and the complementary and alternative medicine industry is estimated to be worth £1.6 billion annually.

For Muslims today herbal medicine is regaining its importance as many herbal physicians have started to emerge, although in villages and rural areas herbal medicine has persisted through the centuries as an integrated part of tradition.

09 PHARMACY

On nearly every business street and in most supermarkets, a pharmacy or chemist can be found. But they are not modern-day concepts, as pharmacies were springing up in Baghdad, Iraq, more than a thousand years ago.

At the beginning of the ninth century pharmacists were independent professionals running their own pharmacies. These family-run businesses operating in the markets were periodically (especially in the 12th and 13th centuries) inspected by a government-appointed official, *Al-Muhtasib*, and his aides. They checked the accuracy of weights and measures, as well as the purity of the drugs used, removing impostors and charlatans with the threat of humiliating corporal punishment.

Hospitals of a thousand years ago also had their own dispensaries, producing drugs and other pharmaceutical preparations.

So the practical side of pharmacology was well developed and supported by scholars like Sabur ibn Sahl in the ninth century, who was the first physician to describe a large variety of drugs and remedies for ailments; Al-Razi, who promoted chemical compounds in medicine; Ibn Sina, who described 700 preparations, their properties, actions, and their indications; and Al-Kindi, who determined and applied the correct drug dosage, which formed the basis of medical formulary.

Other influential scholars included Al-Zahrawi of Spain, who in the tenth century pioneered the preparation of medicines by sublimation and distillation, which meant a whole range of new drugs could now be produced. As he had already used catgut for internal stitching, he also administered drugs by storing them in catgut parcels, which were ready for swallowing. So when you take a drug capsule today remember that its forerunner is more than a thousand years old.

Al-Zahrawi's work *Al-Tasrif* was translated into Latin as *Liber Servitoris* and told the reader how to prepare simple and more complex compound drugs. He also gave methods of preparing substances such as litharge or lead monoxide, white lead, lead sulfide (burnt lead), burnt copper, cadmium, marcasite iron sulfide, yellow arsenic, and lime, and numerous vitriols and salts.

Abu al-Mansur Muwaffaq broke new ground when he wrote *The Foundations of the True Properties of Remedies* in the tenth century. This described arsenious oxide and he knew about silicic acid. One use of this today is in pills that help form a protective membrane in easily irritated stomachs. He made a clear distinction

TOP: *This 12th-century Persian pharmacy jar was used by apothecaries to store dried herbs, minerals, and other medicines. The glazed surface of pottery drug jars could be easily cleaned.* BOTTOM: *An Arabic edition of Dioscorides's* De Materia Medica *illustrates a pharmacy with chemists preparing medications.*

An Arabic manuscript with pharmacological tables ascribed to 13th-century Ibn al-Baytar shows descriptions of symptoms, the location of the ailment, applications of the medicine, and what dosage should be used.

between sodium carbonate and potassium carbonate, and drew attention to the poisonous nature of copper compounds, especially copper vitriol, and also lead compounds.

In the 11th century, Al-Biruni wrote one of the most valuable works in the field called *The Book of Pharmacology*, giving detailed knowledge of the properties of drugs, and outlining the role of the pharmacy and the functions and duties of the pharmacist.

A primary aim of the pharmacists was that their work had to be expertly organized, making it of maximum practical value to the apothecary and medical practitioner. This meant that they listed drugs alphabetically in tables for easy referencing and quick usage, and medical encyclopedias were available as full works or sections on medical specialities.

These early drug treatises passed into Europe with all this vital pharmaceutical information, influencing 13th-century European pharmacists like Johannes de Sancto Amando and Pietro d'Abano, a professor in Padua, Italy, from 1306 to 1316. Works that took this European journey included books by Ibn al-Wafid of Spain, who was published in Latin more than 50 times. His main work was called *The Book of Simple Drugs* and ran to 500 pages, taking 25 years to compile. The Latin translation, *De medicamentis simplicibus*, is only a fragment of all his work.

The 13th-century Málaga Muslim Ibn al-Baytar was a leading botanist as well as the author of the largest pharmacological encyclopedia that has survived to our time. *Dictionary of Simple Remedies and Food* describes more than 3,000 botanical samples listed in alphabetical order. A Latin version of the book was published in 1758, and its complete translation appeared in 1842.

European pharmacists were truly inspired by these works, so *Compendium aromatariorum*, written by well-known 15th-century physician Saladin of Ascolo, was divided into seven parts. It follows, exactly, the earlier Muslim categorization of subjects.

Physician Ludovico dal Pozzo Toscanelli worked at the Florentine College of Physicians, which produced a 17th-century edition of the *London Dispensatory*. This listed botanicals, minerals, simple and compound drugs for external and internal use, oils, pills and cataplasms, all revealing a possible Muslim connection.

An interest in Muslim pharmacy was recently revived by American historian Martin Levey. Before he died in 1977, he had translated Arabic texts and unearthed huge lists of therapeutic treatments, including compound drugs, pills, pastilles, powders, syrups, oils, lotions, and toothpastes.

10 Medical Knowledge

Muslim physicians of a thousand years ago would be happy to learn that a few decades, sometimes centuries, after their deaths their works were being translated into Latin, making them accessible to the whole of Europe.

Tunisia was a hotbed of medical knowledge because of a pioneering hospital called Al-Qayrawan, which was built in 830. As well as being a practicing hospital, Al-Qayrawan had medical scholars producing enormous medical tomes of knowledge, which were taken to Europe by people such as Constantine the African.

> *"The European medical system is Arabian not only in origin but also in its structure. The Arabs are the intellectual forebears of the Europeans."*
>
> DR. DONALD CAMPBELL,
> 20TH-CENTURY HISTORIAN OF ARABIAN MEDICINE

In the 11th century, this Tunisian Christian scholar translated medical encyclopedias so they were available to Latin-speaking Europeans. This revolutionized medical studies in Europe, while also creating a generation of prominent medical teachers. Constantine's best-known translation is of *The Royal Book* by tenth-century physician Ali ibn Abbas al-Majusi, known in Latin as the *Pantegni.* It was printed in Lyons, France, in 1515 and in Basel, Switzerland, in 1536.

The Guide for the Traveler Going to Distant Countries or *Traveler's Provision* was a medieval bestseller written by physician Ibn al-Jazzar, who practiced and studied at Al-Qayrawan hospital.

DE CONSERVANDA
BONA VALETVDINE,
Liber Scholæ Salernitanæ.

DE ANIMI PATHEMATIS, ET
remediis quibuſdam generalibus.
CAPVT I.
1 Anglorum Regi ſcribit ſchola tota Salerni.
2 Si vis incolumem, ſi vis te reddere ſanum,
3 Curas tolle graues, iraſci crede profanum.

The cover of a 16th-century illustrated work shows Constantine the African lecturing at the school of Salerno. Constantine was an 11th-century Tunisian Christian who translated medical encyclopedias.

There he died in 955, more than 80 years old, leaving 24,000 dinars and 25 quintars (one quintar is 45 kilograms) of books on medicine and other subjects. His legacy also included a treatise on women's diseases and their treatment. Such writings earned him immense fame and made him very influential in medieval western Europe.

Constantine translated *Traveler's Provision* into Latin as *Viaticum peregrinantis* and Synesios translated it into Greek and Hebrew as *Zedat ha-derachim,* which propelled it to international bestseller and most read status. *Traveler's Provision* was a systematic and comprehensive medical work accepted into the so-called *Articella* or *Ars medicinae*, a compendium of medical textbooks widely used in medical schools and universities at Salerno, Montpellier, Bologna, Paris, and Oxford. It contained remarkable descriptions of smallpox and measles.

The translated Arabic works soon became popular in all centers of learning, including Salerno, which with its medical school was a major center of learning in Europe.

Other translated medical works that had a major impact on Europe included those by Ibn Sina, known as the "Prince of Physicians" in the West. His 11th-century *Canon* was another enormous medical encyclopedia, which remained the supreme authority in the world for around six centuries. His scientific, philosophical, and theological views left their mark upon many important figures such as Albertus Magnus, St. Thomas, Duns Scotus, and Roger Bacon.

The first known alphabetical classification of medical terms, listing the names of illnesses, medicines, physiological processes, and treatments, was called *Kitab al-Ma'a* or *The Book of Water*. Written by Al-Azdi, also known as Ibn al-Thahabi, it was called *Kitab al-Ma'a* because the word *al-Ma'a*, water, appears as the first entry.

Al-Razi's 20-volume *Comprehensive Book* covered every known branch of medicine. Translated into Latin as *Liber Continens,* it was probably the most highly respected and frequently used medical textbook in the Western world for several centuries. It was one of the nine books that composed the whole library of the medical faculty at the University of Paris in 1395.

Then there was the work of Al-Zahrawi, an outstanding physician in Córdoba, southern Spain, around the year 1000. The surgical part of his 30-volume medical work *Al-Tasrif* was translated into Latin by Gerard of Cremona, with various editions published in Venice in 1497, in Basel in 1541, and in Oxford in 1778. The book became a manual of surgery for most European medical schools, such as Salerno and Montpellier, playing a central part in the medical curriculum for centuries.

Lastly, we look at the work of Ibn al-Nafis, a Syrian physician who died in 1288. He has left us *The Complete Book on Medicine,* which was compiled in 80 volumes. Manuscripts of portions of this huge work are now available in collections in Damascus, Aleppo, Baghdad, and Oxford, as well as Palo Alto in California, which has a large fragment in Ibn al-Nafis's own handwriting.

Much medical knowledge also came through direct contact with Muslim physicians as they treated some Crusaders. Even Richard the Lionheart was treated by the personal physician of Saladin.

First published in 1858, Gray's Anatomy *is a leading medical encyclopedia today. It follows on the tradition begun by Muslims, whose treatises gained equal popularity at universities.*

چشمه
باب همایون

Chapter Six

"Allah hath promised . . . gardens under which rivers flow, to dwell therein, and beautiful mansions in gardens of everlasting bliss."

QURAN (9:72)

TOWN

TOWN PLANNING • ARCHITECTURE • ARCHES
VAULTS • THE DOME • THE SPIRE • INFLUENTIAL IDEAS
CASTLES AND KEEPS • PUBLIC BATHS • THE TENT
FROM KIOSK TO CONSERVATORY • GARDENS • FOUNTAINS

Life in cities like ninth- and tenth-century Córdoba in Spain and Baghdad in Iraq was a pleasurable experience. This was high civilization with free education and health care, plus public amenities such as baths, bookshops, and libraries lined the paved streets, lit at night. Rubbish was collected on a regular basis by a donkey cart and some sewage systems were underground.

Neighborhoods were peaceful, with houses off main thoroughfares, connected by narrow, winding, and shade-giving streets, all within earshot of the local mosque. Business and trade were kept to the main streets and public squares. Gardens, both public and private, were an imitation of Paradise with attention and care to details. Huge water-raising machines could be seen pumping water from rivers into the fields and to the cities. The fountains of the Alhambra Palace in Granada, Spain, still use the 650-year-old water systems devised by Muslim engineers.

Advances in architecture saw huge mosques and crevice-spanning bridges. Domes and minarets dominated the skyline and were so impressive that Crusaders took these designs, and sometimes the Muslim architects, back with them to Europe to improve upon the Roman designs built there.

OPPOSITE: *A 16th-century manuscript from the* Hünernâme *depicts daily activities of people in Istanbul's Topkapi Palace.*

01 Town Planning

Just as traditional European towns have certain features such as market squares, churches, and parks, Muslim towns were also designed according to the local population's needs, based on four main criteria: weather and landscape, religious and cultural beliefs, Sharia (Islamic law), and social and ethnic groupings.

Many of these cities were in fiercely hot climes, so a lot of shade was needed. To provide this, towns were planned with narrow covered streets, inner courtyards, terraces, and gardens.

Religion was vital to cultural life, so the mosque, like a church, had a central position. Around the mosque threaded narrow, winding, quiet streets that led away from the public places into cul-de-sacs and private life.

Social and legal issues were handled by the religious authorities, who lived in central places close to the main mosque, the main public institution. The law, for example, set the height of the wall above the height of a camel rider so a passerby could not see into a property.

How and where people lived was based on families and groups of people from the same families and tribes with similar ethnic origins and cultural views. Separate quarters, called *Ahyaa,* developed for each group, so there were quarters for Arabs, Moors, Jews, Christians, and other groups such as Andalusians, Turks, and Berbers in the cities of the Maghreb, North Africa. Some North African cities were divided into quarters for

An aerial view shows the Andalusian village of Zuheros in Córdoba, Spain.

Muslims, Christians, and Jews, but this was often voluntary and not exclusive. Within these quarters they had kinship solidarity, defense, social order, and similar religious practices.

These separate quarters did not prevent the society from being socially cohesive, as the general trend reflected the Quranic verse: "O mankind, indeed We have created you from male and female and made you peoples and tribes that you may know and appreciate one another. Indeed, the most noble of you in the sight of Allah is the most righteous of you. Indeed, Allah is Knowing and Acquainted" (Quran 49:13).

Near the main mosque would be a central area for social gatherings, then a souk, or market, and then a citadel near an outer defensive wall surrounding residential quarters, all joined by an intricate street network to the outer wall.

The souk was split into areas for spices, gold, fish, perfume, and much more, with items such as candles and incense sold close to the mosque. There would also be booksellers and binders nearby, too.

The citadel, like a Western castle, was the palace of the governor, within its own walls. It was a district on its own, with its own mosque, guards, offices, and residence. It was usually in a high part of the town near the outer wall.

Neighborhoods clustered around mosques and could not be farther than the reach of the muezzin's call to prayer. Densely packed, each area had its own mosque, school, bakery, and shops. They even had their own gates, which were usually closed at night.

All this was surrounded by a well-defended wall with a number of gates, and outside the wall were Muslim, Christian, and Jewish cemeteries. Just beyond the main gate were private gardens and fields and also the weekly market with its many animal stalls.

The most elaborate city of its day, the New York City of the ninth century, was Córdoba.

A painting illustrates a narrow Córdoba street, a typical feature of old Muslim town planning.

The "physical sides [of Córdoba] reveal an ingenious and inventive Muslim culture. They were clearly driven to improve on the past, to modernize the city and make it a better place to live in, not just for the rulers but for everyone . . . There were dozens of libraries, free schools, and houses had running water, and what's more, the streets were paved and they were lit, the kind of amenities London and Paris would not have for a further seven hundred years," said reporter Rageh Omaar in the BBC's *An Islamic History of Europe*.

The streetlights were oil burners and lanterns, lit at sunset, and each city district employed people to maintain them. Litter was also collected on the back of donkeys, which took it outside the city walls to special dumps. The streets were drained by a system of great sewers and cleaned daily.

THE GOOD LIFE

Gracious Living in the Towns of Muslim Civilization

LEGACY: **Old streets still remain intact in some ancient towns, providing glimpses of life 1,000 years ago**

LOCATION: **From Córdoba to Damascus and Baghdad**

DATE: **Eighth to sixteenth century**

If you have ever wished that you could go back in time to see a town as it would have looked a thousand years ago, you are in luck. The Spanish cities of Córdoba and Seville still retain areas of their old towns in which you can glimpse how life was lived centuries ago under Muslim rule.

Towns planned in Muslim civilization centered around the mosque, with its crucial role in religious and civil life. Nearby would be the market, where traders sold food, spices, candles, and perfumes. Business districts would also incorporate bookshops, libraries, and health centers.

Indeed, early hospitals had a wide role in Muslim society. As well as offering medical treatment and convalescence, they acted as asylums for those suffering mental illness; they were also used as retirement homes, particularly for the poor. The first such hospitals in Muslim civilization were in Damascus and Baghdad, followed by the Ahmad ibn Tulun Hospital in Cairo, built between 872 and 874.

Bathhouses, or *hammams*, were a feature of every town, often in elegant buildings with sumptuous tiled walls, fountains, and decorative pools. A visit to the *hammam*, with its steam rooms, heated baths, and cold plunge pools, was part of every Muslim's daily routine. Men and women would bathe at different times of day, and visiting the bathhouse would be a social experience full of opportunities to exchange news and catch up with friends.

Away from this bustling center, along narrow streets, you would find residential zones. Houses had inner courtyards with gardens and terraces, kept private with walls high enough to stop a camel rider from peering over. Compared to other European cities of their day, these towns were advanced, with paved roads, litter collection, and even covered sewers.

Córdoba was one of the world's most advanced cities in the tenth century, and even had oil lamps to light its streets after dark. Farther east in Muslim territory, Cairo had multistory buildings and roof gardens.

The Al-Azhar Mosque, Cairo, Egypt, was founded in 972, and is pictured here in 1831. The mosque played a central role in the everyday life of Muslims. It was located in the heart of the city, with homes and businesses branching out from it in different directions.

■ *A 16th-century manuscript shows the town plan of Diyabakir in southeast Turkey.*

Gardens symbolized an earthly paradise for many people during Muslim civilization. Intended to promote contemplation, gardens and gardening spread across the Muslim world from Spain to India, from the eighth century onward. As well as providing food for the kitchen, these green spaces with their distinctive features still influence garden design today.

Geometrical flowerbeds, shallow canals, and fountains emerged as new features of gardens in the ninth century. You can still see such gardens surrounding the Alhambra in Granada and the Taj Mahal in India—along with similar features in European formal gardens designed centuries later.

Fountains and garden water features became very popular in palaces and mosques across the Muslim world. Although water was scarce in many hot countries of the Muslim world, as far back as the ninth century ingenious engineers were applying their skills to creating fabulous fountains. Their water displays incorporated jets that changed shape, or later, even mechanisms for telling the time.

02 ARCHITECTURE

Many European buildings today have distinct characteristics and features such as domes and rose windows on cathedrals, the arches of train stations, and vaults in churches. It may surprise you to learn that many of these were developed and perfected in architectural terms by Muslims, and flowed into Europe a thousand years ago via southern Spain and Sicily. Building designs and ideas were also taken home by scholars, Crusaders, and pilgrims visiting Jerusalem as they traveled overland through Muslim countries and cities like Córdoba, Cairo, and Damascus. These designs improved upon the Roman designs, producing a spectacular mix.

For Muslims, architecture had to get across a number of ideas, like Allah's or God's infinite power, which was shown in repeated geometric patterns and arabesque designs. Human and animal forms were rare in decorations because Allah's work was matchless. So instead, highly stylized foliage and flower motifs were used. Calligraphy added a final touch of beauty to the building by quoting from the Quran, while large domes, towers, and courtyards gave a feeling of space and majestic power.

The decoration of these buildings concentrated on visual aesthetics, because although Islam opposes unnecessary spending, it does not oppose having a comfortable life or enjoying it, as long as people live within the boundary of God's law and guidance. This means Muslims do not have to live miserably. The Muslim wisdom "Strive for your earthly life as you live forever and strive for your hereafter as you will die tomorrow" sums up the Muslim attitude to architecture, too; if you are going to make something, make it modestly and beautifully.

Rose windows are a good example of this. Looking at the facades of most European cathedrals and churches you cannot help noticing their imposing beauty and how they decorate the walls above the main entrance. Historians have connected the origin of these huge circular windows to Islam, and the six-lobed rosettes and octagon window on the outer wall of the Umayyad Palace of Khirbat al-Mafjar. This was built in Jordan between 740 and 750.

The Crusaders saw this and introduced it into their European churches, first in Romanesque

The circular window at Khirbat al-Mafjar, Jordan, from 740 to 750, is thought to have influenced the design of the stained-glass rose window in Durham cathedral in Durham, United Kingdom.

architecture (11th to 12th centuries) in places like Durham Cathedral, and later in Gothic architecture. The rose window had a function of letting light in, while supposedly symbolizing the eye of the Lord. Others, though, claim the idea is from the Roman oculus, a circular window in the dome of the Pantheon in Rome, but this was more like a circular opening pierced in the roof.

This example is just a taste of what you will discover in the following sections about the varied world of Muslim architecture and how it influenced building styles over the centuries.

Muslim architecture often has environmentally friendly features. To reduce smoke pollution from the thousands of candles and oil lamps, Sinan designed the interior space of the Suleymaniye Mosque in Istanbul so that the soot was channeled by air circulation into a filter room before being discharged into the city. The collected soot was conveyed into a water fountain, where it was mixed and stirred to produce high-quality ink that was used in calligraphy. This ink also repelled bugs and bookworms, which prolonged the life of the manuscripts.

The Suleymaniye Mosque (1550–1557), designed by architect Sinan, crowns one of Istanbul's seven hills. The buildings include a madrasa, hospital, dining hall, caravansary, hammam, hospices, and shops. It was also an environmentally friendly design, containing a filter room that cleaned the air of soot from candles and oil lamps before exhausting it into the atmosphere.

MASTER ARCHITECT SINAN

The Selimiye Mosque in Edirne possesses the highest, most earthquake-defying minarets in all of Turkey. It is the work of master architect Sinan, who was the architect for the Ottoman Empire. He designed and built a staggering 477 buildings during his long career in the service of three sultans in Turkey during the 16th century, acknowledging the importance of harmony between architecture and landscape, a concept that did not surface in Europe until the 16th century. His Turkish designs revolutionized the dome, allowing for greater height and size—an outstanding advance in civil engineering, which later became his trademark.

The Selimiye Mosque in Edirne, Turkey

03 Arches

Arches are essential in architecture because they span large spaces while also bearing huge loads. Strong and flexible, they have been made bigger and wider, and today we see them in buildings from shopping centers to bridges. They are so common nowadays that it is easy to forget how advanced arches were for their time thousands of years ago.

In the simplest arch the thrust comes from the weight of the masonry on top of the arch, and sideways from the cumulative wedge action of the *voussoirs*, or the arch bricks. This gives the arch "elasticity" and it can be compared to a hanging load chain—"the arch stands as the load chain hangs." This silent dynamism of the arch was known in the Muslim world through the saying "the arch never sleeps."

Like others before them, Muslims were the masters of the arch. They loved this motif as much as they loved palm trees, imitating the curve of their graceful branches in their constructions. The spherical nature of the universe was an inspiration for its development, too.

Knowledge of geometry and the laws of statics meant that various types of arches were created. What Muslims did structurally was to reduce the thrust of the arch to a few points, the top and sides. These could then be easily reinforced, leaving other areas free from support, so lighter walls and vaults could be built, saving materials in building.

LEFT: *A portion of the west elevation of the Great Mosque at Córdoba after the fourth enlargement (961-976) shows the following in brickwork: a flat arch (lintel) immediately above the doorway, a semicircular horseshoe relieving arch above it, blind cross arches above the panel to the doorway, and a five-lobed (or cinqfoil) arch above the window.* RIGHT: *The clock tower of Big Ben of the Palace of Westminster in London shows the adoption of a series of arches of the five-lobed form.*

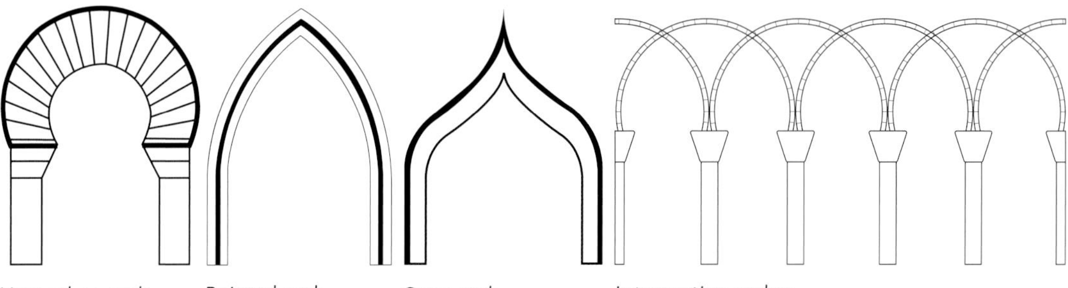

The Egyptians and the Greeks used lintels, while the Romans, and later the Byzantines, built semicircular arches, possibly because of the abundance of strong marble. The Romans used an odd number of arch bricks with a capstone or keystone being the topmost stone in the arch. This shape was simple to build but not very strong. The sides would bulge outward, so they had to be supported by masonry pushing them back in.

All these predecessors of the arch were inherited by Muslims, who had grand plans for their mosques and palaces. For these, they needed strong arches spanning great distances, which looked good as well. So they developed new forms like the horseshoe, multi-foil, pointed, and ogee arch, all crucial for architectural advancement.

The Horseshoe Arch

The horseshoe arch was based on the semicircular arch, but it was extended slightly beyond the semicircle. It was not so strong but looked impressive, and was the first Muslim arch adaptation, used in the Umayyad Great Mosque of Damascus, which was built between 706 and 715. In Islam, the horseshoe is a symbol of sainthood and holiness, and not luck as in other cultures. Structurally, the horseshoe arch gave more height than the classical semicircular arch.

The first time the horseshoe arch appeared on European soil was in the Great Mosque of Córdoba, whose building started in 756 and lasted 40 years. The arch then traveled north with the Mozarabs, the Christian Spanish living in Andalusia. They were artists, scholars, builders, and architects, moving between the southern and northern Christian parts of Spain.

These arch designs could be found in great illustrated manuscripts, the architect's master plans, drawn by the Mozarabs. One was called *Beatus of Lebana* and its author, named Magins, worked at the monastery of St. Miguel de Esacalda, near León. This was a large religious building in the Moorish style with horseshoe arches, and was built by monks arriving from Córdoba in 913.

The horseshoe arch is known in Britain as the Moorish arch. It was popular in Victorian times, and used in large buildings like the railway station

Intersecting arches at Bab Mardum Mosque, now called Church of Cristo de la Luz, were built between 998 and 1000 in Toledo, Spain.

entrances in Liverpool and Manchester. These were designed by John Foster in 1830, and the arches of these two buildings are like those in the Gate of Cairo. Today, you can see the horseshoe arch in the front gate of Cheetham Hill Synagogue in Manchester (1870).

Intersecting Arches

Muslims were so confident of their mastery of the arch that they carried out some spectacular experiments with forms and techniques of its construction. One of these was the introduction of intersecting arches, which provided an additional structural bonus. It meant they could build bigger and higher, and add a second arch arcade on top of a first, lower level. This can be seen best in the Great Mosque of Córdoba.

The Pointed Arch

The main advantage of the pointed arch was that it concentrated the thrust of the vault on a narrow vertical area that could be supported by a flying buttress, a major feature of European Gothic architecture. This meant that architects could lighten the walls and buttresses, which had previously been massive to support semicircular arches. Other advantages included a reduction of the lateral thrust on the foundations, allowing for level crowns in the arches of the vault, making it suitable for any ground plan.

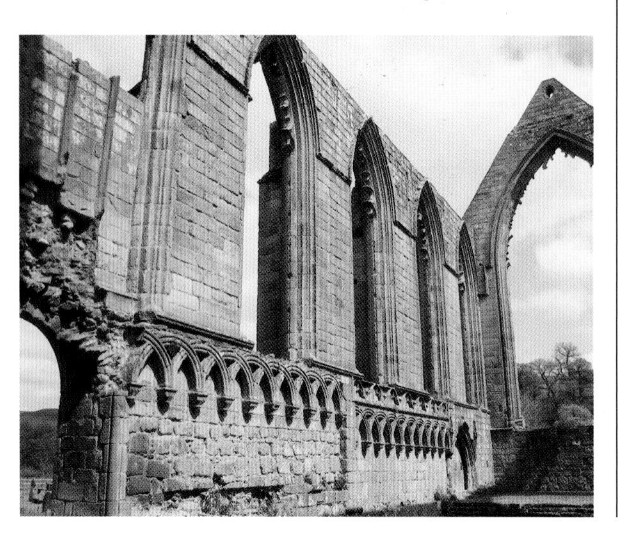

Many people think that the pointed arch, on which Gothic architecture is based, was an invention of European architects trying to overcome problems in Romanesque vaulting, but it came to Europe from Cairo via Sicily with Amalfitan merchants. They were trading with Egypt in 1000, and it was here that the beautiful Ibn Tulun Mosque of Cairo displayed its mighty pointed arches. In Europe, it was first used in the porch of the Abbey of Monte Cassino in 1071, which Amalfitan merchants generously financed.

At this time in the late 11th century, Monte Cassino became the retiring place for Tunisian Christian scholar Constantine the African, whom you can read about in the Translating Knowledge section in the School chapter. A physician, translator, and distinguished scholar in mathematics, science, and theology, he also had a great deal of experience in Muslim building techniques, gained from the Muslim Fatimid of North Africa. Constantine would have undoubtedly given his opinion during the building process in Monte Cassino.

The pointed arch was passed north when St. Hugh, the abbot of Cluny in southern France, visited Monte Cassino in 1083. Five years later, work on the third church of Cluny started and it eventually had 150 pointed arches in its aisles. This was destroyed in 1810. But the journey of the arch did not stop there, as the next person in its travel chain was Abbot Suger, who visited Cluny between 1135 and 1144. He and his engineers went on to build the church of St. Denis, the first Gothic building.

The adoption of pointed arches and other Muslim motifs in Cluny and Monte Cassino, the two most influential churches in Europe, encouraged the rest of Christian Europe to take them on. Like any new fashion it rapidly spread across

The pointed arch, found in buildings like 12th-century Bolton Abbey, United Kingdom, may have come to Europe from Cairo's Ibn Tulun Mosque, built in 876.

much of France, especially in the south, then to Germany in the mid-12th century, and eventually to the rest of Europe. In Britain there were many buildings that had these arches, almost all of them religious buildings.

The Multi-foil Arch

It was in Samarra in Iraq that the first multi-foil arch was designed before passing into the rest of the Muslim lands, including Spain and Sicily, and then to Europe. Its first appearance was in the windows of Al-Mutawakkil Mosque, built between 848 and 849 in Samarra. These windows were on the enclosure and spanned by cinqfoil arches.

The multi-foil arch reached North Africa and Andalusia, where it became very popular, decorating most Moorish buildings, especially Córdoba Mosque. From the tenth century, Europeans fell in love with it and adopted it in their buildings, plans, and arts. Its most popular use was in the trefoil form, which suited the concept of the Trinity in Christianity. Like many of these arches, those seen in the Córdoba Mosque were the main inspiration.

LEFT: *An Indo-Islamic version of the four-pointed arch can be seen at the 16th-century tomb of Humayun outside Delhi.* CENTER: *The interior of the Great Mosque of Córdoba displays superimposed arches on the left and multifoil arches to the right. These were built in succession between the eighth and tenth century.* RIGHT: *A typical ogee arch, known as a Gothic arch in Europe.*

Ogee Arch

After the semicircular arch, one of the most important arches was the ogee arch, otherwise known as the Gothic arch in Europe. This is an elegant arch, a stylized development of the pointed arch. The arch curve is constructed in the form of two "S" shapes facing each other and was used mostly for decoration, sometimes with a stone knot at the top. The new arch design was developed in Muslim India, and later reached Europe in the 14th century, becoming particularly popular in late Gothic 16th-century architecture in Venice, England, and France.

You can see it everywhere in England, because almost all churches and cathedrals have a full ogee arch, or use it in the form of an ogee molding. They are used in decorative screens, entrances, and later Gothic styles. They are used in decorative screens, entrances, and later Gothic styles.

04 VAULTS

An architectural vault is a stone arch that makes a ceiling or canopy, making it possible to have a roof over a large space made of bricks, stone blocks, or a mixture of mortar and debris. Until metal girders and trusses were introduced in the 19th century, the only alternatives to stone vaults were long wooden rafters or stone lintels. These were much simpler materials to use but were not as sophisticated and were more expensive, and the building was limited by the length of the wood.

Vaults, like arches, were used by the Romans, but Muslims refined them so they could build bigger and higher. They made vaults that were as strong, but finer, with thinner curtain walls, so more light was let in. Until the 11th century, most of Europe used thick Roman vaults, which needed robust (as thick as 2 meters, or 6.6 feet) and short walls to carry them, but when Europeans saw the Muslim vaults of Córdoba, they began to imitate their design and techniques. Some of these became typical of the Romanesque period (tenth to twelve centuries) in Europe, and they were first seen in great cathedrals, like Durham in England.

Ribbed tunnel vaults at 12th-century Ste.-Marie-Madeleine

Rib Vaulting

The Great Mosque of Córdoba, called the Mesquita, was the springboard for much of European architecture. Its vast hall of polychrome, horseshoe, and intersecting arches, ribbed vaults, and domes all made their way north, and it is worth noting that ribbed vaults do not appear in churches that existed then, such as those in the León region (western Spain), because they were built before the Great Mosque of Córdoba.

A ribbed vault was a ceiling or canopy of stone that was strengthened by single semicircular arches added beneath the vault to provide extra support. These added arches looked like ribs, and they supported the crown. This meant a large amount of the thrust of the vault was concentrated on these ribs, relieving the pressure on the walls, enabling the builder to make them thinner and higher.

Instead of using the old rubble mix or the large massive pieces of stone used by the Romans, Muslim architects introduced small stones or bricks between the ribs, arranging them like the building of a wall in the early stages of construction.

The earliest form of rib vaulting was traced to the eighth-century Abbasid Palace of Ukhaydar in Iraq. This architecturally rich desert palace contains eight transverse arches and ribbed vaults. This system of ribs is also found in many of the tunnel vaults of the Ribat of Susa in Tunisia, built in 821-822, and these greatly influenced the cross vaults of the nave of St. Philibert at Tournus, built at the end of the 11th century, of Ste.-Marie-Madeleine at Vézelay (1104-1132), and of Fontenay Abbey (1139-1147). The idea of building vaults like this came from contact with North Africa, especially the town of Susa in Tunisia.

The cistern of Ramla in Palestine is made of pointed arches standing on cruciform piers of masonry, which were covered with six barrel vaults reinforced with walls. It was built by Harun al-Rashid in 789. A similar vault was built in Susa, Tunisia, in the two main mosques of Banu Fatata (834-841) and the Great Mosque (850-851). This idea appears in the Notre Dame d'Orcival cathedral, built in the 12th century in Puy-de-Dôme in Auvergne, France.

Gothic Rib Vaulting

As you have already read, ribbed vaults were known to the Muslims more than 150 years before they appeared in Christian cathedrals and churches. A second type of rib, which became known in Europe as the Gothic rib, was more complex, and it first appeared in the great mosques of Muslim Toledo and Córdoba.

The ribs of Córdoba inspired European architects and their patrons to adopt them in the Romanesque and Gothic movements, and really the history of Gothic architecture is also the history of the rib and flying buttresses.

Bab Mardum Mosque in Toledo has a unique form of rib vaulting that later developed into the quadripartite vault—a vault with supporting ribs in the form of diagonal and intersecting arches, which is accepted as the origin of the Gothic style.

Rib vaulting in the maqsura dome of the Great Mosque of Córdoba was added in the tenth century.

This Toledo mosque was built by Muslim architects Musa ibn Ali and Sa'da between 998 and 1000. It was in the shape of a square made up of nine small compartments, and covered with nine different ribbed cupolas or domes. Each dome is a little vault supported by intersecting arches that look like ribs thrown in the most fantastic way across one another.

French art historian Elie Lambert said, "The Arab architects . . . knew and employed in their vaults, since the end of the tenth century, not only the same principle of the rib, but also the system of crossed arches, a system which became later known in France as the quadripartite vaulting."

Similar vaulting was used in another mosque that was later transformed into a house named Las Tornerías in 980. It also had nine ribbed domes combining a variety of ribs that dominated the central vault, making it an impressive-looking house because it also used polychrome horseshoe and trefoil arches.

Similar ribbed domes can be seen in a large number of Spanish buildings, especially those

> *"The Arab architects . . . knew and employed in their vaults, since the end of the tenth century, not only the same principle of the rib, but also the system of crossed arches . . . known in France as the quadripartite vaulting."*
>
> ELIE LAMBERT, FRENCH ART HISTORIAN

LEFT AND RIGHT: *These church roofs and cloisters demonstrate Gothic rib vaulting.*

LEFT: *A honeycomb dome, or* muqarnas, *can be seen at Alhambra Palace in Granada, Spain.* RIGHT: *A muqarnas vault is found at the entrance at Lotfollah Mosque in Isfahan, Iran.*

built by the Mozarabs. They can also be seen in churches built along the route of the pilgrimage to Santiago de Compostela, where these ribs decorate the domes of buildings of the Almazán church in Castille, Torres del Río in Navarre, and in the Pyrenees in St. Croix d'Oloron, and the hospital of St. Blaise. Ribs are also found at the Templar church at Segovia and the 12th-century chapter house at Salamanca.

The traveling of the ribs was due to the improvement of relations between Mozarabs and Muslims at the time of 'Abd al-Rahman III, as well as the great cultural and artistic achievements of his reign. In this time of peace and tolerance, art flourished. In less calm times, the capture of Toledan mosques, including Bab Mardum, must have given European artists and architects valuable lessons. The French, in particular, benefited because they were closely connected to Toledo after it was taken by the Spanish Christians.

Muqarnas

The last vault we will visit is the stalactite vault, or *muqarnas*. They are three-dimensional forms made from geometrical shapes and carved into vaults, domes, niches, arches, and wall corners. Developed in tenth-century Persia, the idea was later spread by the Seljuks, a Turkish dynasty that ruled across Persia, Anatolia, and Turkey between 1038 and 1327. By the late 11th century, the muqarnas became a common architectural feature all over the Muslim world.

One of the best examples of a muqarnas is the honeycomb of the Alhambra Palace in Granada, designed more than 700 years ago. The honeycomb vault of the Hall of the Abencerrajes was organized in an eight-pointed star made of a large number of interlocked small squinches of lozenge shapes, projecting from the walls in cells very much like the honeycomb. These symbolized the honey juice, which the good believer is promised in Paradise. It was also designed with 16 windows, two for each side of the star, which let in an enormous amount of light. This all helped to re-create in the Alhambra Palace a vision of the promised Paradise and its eternality, which would reward those who strove to reach it.

05 THE DOME

The dome is effectively a three-dimensional arch, and in Islamic architecture it had two main symbolic meanings: to represent the vault of heaven and the divine dominance engulfing the emotional and physical being of the faithful. It also had a functional use, which was to emphasize particular areas, such as the nave, or the *mihrab*, while also lighting the inside of the building.

The development of domes had to overcome the problem of how to make a square bay from a domed, arched shape. The Byzantines, Romans, and Persians managed this a considerable time before Muslims by using pendentives, triangular segments of a sphere placed at the corners to establish the continuous circular or elliptical base needed for the dome. These pendentives took the weight of the dome, concentrating it at the four corners where it could be supported by the piers beneath.

Muslims used pendentives for a while, but later developed squinches that threw arches at the corners, creating small niches. The use of these culminated in impressive stalactite squinches, or vaults, known as *muqarnas* that decorated the inside of the domes.

Semicircular Dome

The most common form of the dome is the semicircular form, which is the oldest and most widespread. Early mosque domes were small and built on the crossing before the mihrab, as in the mosques of Qayrawan (670-675), and the Umayyad mosques in Damascus (705-707) and Córdoba (756-796). Over the centuries domes grew in size and number, and were later used in the center and sometimes covering the entire roof of mausoleums. Under the Ottomans, the size of domes grew to cover entire sanctuaries, surrounded by smaller domes like those in Suleymaniye Mosque.

Traditionally, domes had been made using a mixture of mortar, small stones, and debris, but this required a lot of wood, and the masons had

TOP: *St. Paul's Cathedral in London shows Sir Christopher Wren's adoption of the Islam-inspired architecture of the duality of dome and towers.* BOTTOM: *Semicircular domes at the Sultan Ahmed Mosque, or Blue Mosque, in Istanbul, Turkey.*

"The physics and philosophy of the Arabians spread themselves in Europe, and with these their architecture."

SIR CHRISTOPHER WREN (1632–1723), BRITISH ARCHITECT

to wait for it to dry. To make the building process easier, the wooden centering was replaced with brick coursing and the use of four squinches made of radiating semicircles to produce a circular base for the dome.

The Muslims also used ribs, which enabled them to construct the dome in a similar way to ribbed vaulting.

The Bulbous Dome

The bulbous dome, or onion-shaped dome, was favored particularly by the Mughals, who spread it to Persia, the Indian subcontinent, and South Asia. So familiar today in Moscow, Russia, bulbous domes first appeared in Europe in Venice where they were used to decorate the lanterns of the domes of St. Mark's Cathedral. The domes correspond to the ogee arch or Gothic arch as a new architectural fashion after its widespread use in the Muslim world, especially Asia and Persia in the 14th century. The bulbous cupolas fit aesthetically perfectly with this form of arch.

The bulbous dome was gradually introduced to eastern Europe, first in wooden architecture before stone, and this probably came from the Mosque of the Dome of the Rock in Jerusalem, as well as from Syria where illustrations in Umayyad mosaics have been found showing the early development of these domes.

The Duality of Dome and Minaret

The duality of the dome and minaret created an aesthetic appeal that was imitated by many Western architects, including Sir Christopher Wren.

The imperial palace of Tsarskoe Selo outside St. Petersburg in Russia dates from 1717.

In his greatest ever project, St. Paul's Cathedral in London, the Muslim influence can be seen in the structure of the domes, in the aisles, as well as in the use of the combination of rounded dome between two towers. Sir Christopher Wren studied the architecture in Ottoman and Moorish mosques, becoming convinced of the Muslim roots of Gothic architecture, establishing the Saracenic Theory.

Wren explained: "This we now call the Gothic manner of architecture (so the Italians called what was not after the Roman style), though the Goths were rather destroyers than builders: I think it should with more reason be called the Saracen style . . . This manner was introduced into Europe through Spain; learning flourished among the Arabian all the time that their dominion was in full power . . . the physics and philosophy of the Arabians spread themselves in Europe, and with these their architecture: many churches were built after the Saracenic mode."

POWERFUL DOMES

New Building Methods That Exalted the Heavens

LEGACY: **The influence of Muslim architecture spread around the world**

LOCATION: **Istanbul, Turkey**

DATE: **15th century**

KEY FIGURE: **Sinan, architect**

From the Dome of the Rock and the Church of the Holy Sepulchre in Jerusalem, to St. Peter's Basilica in Rome, Hagia Sophia in Istanbul, and the Taj Mahal in Agra, some of the world's most memorable buildings are roofed with domes. The Byzantines and Persians were the first to build great domes. But as Muslim architects adopted and improved upon them, their popularity and diversity increased.

Domes appealed to many Muslim building designers. To them they symbolized the vault of heaven, and the overarching power of God. In mosques, a dome could emphasize a significant area of the building such as the *mihrab,* a niche in one of the walls indicating the direction of Mecca. Domes appear on churches, palaces, and public buildings as well as mosques, taking different shapes reflecting the local culture.

Born in 1489, Sinan started out as a humble stonemason and carpenter, learning his father's trade. But he went on to become chief architect in the Ottoman State, a coveted role he held during the reign of three sultans. He designed and built more than 470 buildings in his lifetime, developing techniques to construct taller and wider domed roofs than had ever been seen before.

Sinan designed and built impressive schools, mosques, and public buildings, approaching his work with an eye for harmony between architecture and the landscape. His work appeared in Damascus, Mecca, Bosnia, and elsewhere, but perhaps his most impressive building is his last, the Selimiye Mosque in Edirne, Turkey.

How much did the architecture of Muslim civilization influence other cultures? In the domes, arches, and towers of buildings across the world, we see strong echoes of ideas that developed and grew in Muslim lands. Travel and trade increased the interchange between nations, spreading new architectural ideas through fashion or because the ideas offered practical ways for constructing larger or stronger buildings.

An illustration depicts Russian architecture, showing bulbous domes at an orthodox church.

■ *Byzantines and Persians were the first to build great domes, but Muslim architects adopted and improved upon their ideas. The Dome of the Rock in Jerusalem dates from 691.*

But ideas also spread via other routes. The Normans, as they conquered areas of Europe, encountered Muslim architecture in Sicily, resulting in the emergence of a distinctive Gothic style. Because of high-profile marriages such as that of Edward II and Eleanor of Castile in 1254, English architects at court came to know and adopt Muslim architectural ideas into royal chapels and palaces.

Onion-shaped domes are well known today in Russian Orthodox churches and were particularly popular throughout the Mughal Empire in India. In the mid-17th century, Sir Christopher Wren drew consciously on Muslim influences when he designed St. Paul's Cathedral in London, with its combination of dome and towers.

Domes form part of the fairy-tale appearance of the Sea Cathedral in Kronstadt, near St. Petersburg, built in 1913. The Kremlin in Moscow is also topped with domes. Brighton's Royal Pavilion, designed by John Nash in 1815 for the Prince Regent, features the bulbous onion domes that had been popular throughout the Mughal Empire in India.

06 The Spire

Minaret comes from the Arabic word *manarah,* which means "lighthouse," but not in the meaning of sea lighthouse as some writers thought. It has rather a symbolic significance referring to the light of Islam, which radiates from the mosque and its minaret.

By the eighth century, in the Great Mosque of Damascus the minaret had become an essential feature of Muslim religious architecture. Minarets have two main parts; the lower part has a strong blind base with little or no decoration at all, and the higher part is very graceful and richly decorated. This sectioning of the tower is seen in many English towers such as Sir Christopher Wren's St. Mary le Bow Tower.

The earliest surviving Muslim tower is the Qal'at of Benu Hammad, which was built in 1007 in eastern Algeria. With its huge size expressing the power of Benu Hammad, the tower was used as a watchtower as well as a minaret. It was richly decorated, with openings providing light and reducing the weight of the structure. Various types of arches were used on the frames of these windows, including trefoil, cinqfoil, semicircular, and polylobed arches.

It was features like these that rapidly influenced the character of the Romanesque and Gothic towers of the West. Good examples of this are the Church of St. Abbondio, Como, Italy (1063-1095), Church of St. Etienne, Abbaye aux Hommes at

Qal'at of Benu Hammad is the earliest surviving Muslim tower of its kind, built in 1007 in Algeria. Features it displayed—rich decoration and arch designs of the upper sections—were later seen in the Romanesque and Gothic towers in Europe.

Caen, France (1066-1160), and St. Edmund at Bury in England (1120). In all cases, the influence of Qal'at Benu Hammad is unquestionable, and the European trade links with North Africa must have been responsible for its transfer.

In Europe the tower first appeared in the tenth-century Romanesque period, but became associated with Gothic architecture. Some believed the tower came from the minaret, as it began appearing in European castles and gatehouses when the Crusaders came home. This view is possibly due to the increased number of towers appearing all over Europe after the Crusades, but it is obviously incorrect as towers as well as church towers appeared much earlier.

Spires were never used until the minaret was built, and in England there was no spire before 1200, the first being that of St. Paul's Cathedral in London, finished in 1221. (The spire was destroyed by lightning in 1561 and the church by the Great Fire of London in 1666, rebuilt by Wren in 1710.) The minarets of Al-Jeyushi Mosque in Cairo, built in 1085, were particularly influential in Italy and England. Square-shaped minarets continued to influence European towers, as seen in Palazzo Vecchio at Piazza della Signoria (1299-1314) in Italy. Piazza Ducale in Italy is particularly striking when it is compared to the Umayyad Mosque in Damascus. The Italian tower has the same gradual progress of the square-shaped tower and the same bulbous dome at the top end. The arcade of the cloister, which the tower emerges from, shows a similar visual and structural combination to the one used in the Umayyad Mosque.

This graceful, circular form of minaret was also imitated in Germany in buildings like the Holy Apostles Church in Cologne (1190), in Amiens Cathedral (1009-1239), and in Worms Cathedral (11th to 13th centuries) in Rhineland. The Cologne tower has particularly breathtaking proportions as it soars into the air.

TOP: *The church of St. Abbondio, in Como, Italy, dates from the 11th century.* BOTTOM: *The minaret at the Umayyad Great Mosque in Damascus was built in the early eighth century.*

07 Influential Ideas

A lot of Muslim architecture reached Europe through captured artists, and a development of the Norman style appeared at the same time as the first countercampaign against the Muslims in Spain and in the Holy Land. One artist who was taken prisoner was Lalys. His new master was Richard de Grandville of England, who had Lalys design the abbey of Neath in South Wales in 1129. Lalys then became the architect of Henry I.

The Normans brought a lot to English architecture after their 1066 invasion. They also occupied Sicily, where they made contact with Muslims. It was here that they became great builders instead of destroyers. In fact, as Rageh Omaar from the BBC's *An Islamic History of Europe* said: "Architecturally little remains in Sicily from the Muslim time and the buildings that look Islamic are not. They were built in the 11th century by the Norman conquerors, who were fascinated by Arabic culture. The 12th-century Norman king Roger II of Sicily was particularly passionate about Muslim architecture. He was also fluent in Arabic." It was these architecturally Islamized Normans who later played a leading part in building Europe. Gothic-style architecture also developed under these Norman kings.

"The 12th-century Norman king Roger II of Sicily was particularly passionate about Muslim architecture. He was also fluent in Arabic."

RAGEH OMAAR FROM THE BBC'S *AN ISLAMIC HISTORY OF EUROPE*

The Palatine Chapel, in Palermo, Sicily, was designed and decorated by Muslim artists in the reign of Norman King Roger II.

Edward I sent ambassadorial exchange missions to Persia to make allies of the Mongols, who had taken the region and were enemies of the Muslims. This mission was led by Geoffrey Langley in 1292 and lasted a year. It included Robertus Sculptor, who is thought to have brought back with him a number of ideas, such as the ogee arch, which were then introduced to English architecture at the end of the 14th century.

Later, Edward I had good contacts with Persia, and his crusading experience plus his marriage to Eleanor of Castille provided further contact with Muslim Spain. These contacts are commemorated in English folklore by Morris dancing, first known as Morisco. The Muslim contacts also led to Tudor architecture, such as the star polygon plan at Windsor, in the tower of Henry VII and in

The Taj Mahal, in Agra, India, was built in 1630.

the windows of his chapel, and the turrets of Wolsey's great gate at Oxford, now called Tom Tower.

Others to take back ideas were pilgrims and artists visiting Egypt, like Simon Simeon and Hugh the Illuminator. Both were Irishmen who visited the Holy Land in 1323, and who passed through Egypt and saw the Mausoleum of Mustapha Pasha (1269-1273) in Cairo. This had Muslim perpendicular decoration that became a common feature of Gothic architecture in the United Kingdom.

The chapels of the Knights Templar Order, founded by nine French knights in Jerusalem in 1118 after the first Crusade, were built with a centralizing form, which was derived from the city's Dome of the Rock Mosque. This form of church later spread west and can be seen in the circular Temple Church of 1185 in London. The rotunda, which is late Norman, and the Gothic choir, built in 1240, have a number of common features, and they are both subject to the same geometric system. Some Western scholars insist that this system came to Europe and France from the Greeks, especially Plato and Vitruvius, but others wonder at this perfect timing, questioning why the French did not rediscover Plato earlier or later, but coincided with the time when these features were very evident in Islamic structures visible to Crusaders, other travelers, and traders.

Another famous building is the Taj Mahal, in India, built by the Mughal sultan Shah Jahan in memory of his wife, Mumtaz Mahal, who died while giving birth to their 14th child. The mosque is called the "teardrop on eternity" and was finished in 1648, after using precious and semiprecious stones as inlay and huge amounts of white marble that nearly bankrupted the Mughal Empire. The Taj Mahal is completely symmetric—except for the tomb of the sultan, which is off center in the crypt room below the main floor.

More amazing Islamic architecture includes the Cathedral Mosque in Córdoba, Spain, and the Alhambra Palace in Granada. All of these still fascinate people today, and the Taj Mahal surpasses the Alhambra for the most visitors with three million a year, while the Alhambra draws 2.2 million or 7,700 people a day.

08 Castles and Keeps

Cities today are no longer designed with a potential siege mentality, but look around the world and the fortifications of the past are now accessible to us as tourist sites, such as the Tower of London.

Even though the European Crusaders had superiority in ammunition and manpower when they went to Jerusalem, the Muslims were able to sustain attacks, and for a considerable time. The impressiveness of their military structures and castles was not lost on the Europeans, who took these architectural ideas home with them. The invincible designs of the castles of Syria and Jerusalem were imitated in Western lands with key features like round towers, arrow slits, barbicans, machicolations, parapets, and battlements soon appearing.

Before the Crusaders lost vital battles to Saladin in the 12th century, most Christian military towers had square keeps. Saladin's round towers impressed upon the Crusaders the need to leave out projecting angles, because they encouraged flanking fire. The first recorded European example to abandon the square and adopt the round tower was Saone, which was built in 1120.

European castles, like this one in Bavaria, Germany, imitated the designs of castles in Syria and Jerusalem.

The Loopholes

The loopholes, or arrow slits in fortified walls, were first used around 200 B.C.E. by Archimedes to protect Syracuse. These long and narrow slits meant a bowman could shoot at the enemy, but be protected from returning fire. They were also used in the fortifications of Rome, and were improved and popularized by Muslims in the Palace of Ukhaydar, an eighth-century Iraqi palace, and the ninth-century Sussa Ribat in Tunisia. The first recorded use of them in England was in London in 1130.

The Barbican

The barbican is a walled passage added to the entrance of a castle in front of the main defensive wall. This delayed the enemy's entrance into the castle, and also gave the defenders more opportunity to hold up the attackers by forcing them into a small space. The enemy could then be attacked from above and from the sides. The word "barbican" is taken from the Arabic *bab al-baqarah*, meaning "gate with holes."

The returning Crusaders often brought Muslim masons with them, and they built these features into the defenses of European castles in the 12th century. There were also peaceful periods in the Crusades when the architects and builders with the Crusaders could watch and learn how local Muslims designed and built their fortifications.

Christian masons also had to earn their living, especially in times of peace, and some of them were hired by Muslims to help in repair or in new constructions. The story of Eudes de Montreuil demonstrates such an encounter as he accompanied St. Louis on the Crusade between 1248 and 1254, and worked at Jaffa and then in Cyprus.

Bonding Columns

Muslims also used bonding columns inside masonry to strengthen the walls. They had taken and developed this technique from Roman architect and engineer Marcus Vitruvius Pollio. The walls of the harbor of Acre were built in this way. It was the emir of Egypt, Ahmad ibn Tulun, who in 883 instructed that a harbor be built with the strongest form to repel the waves and enemy attacks alike. So timber beams were inserted into the masonry of the wall, as steel is today, to bind its two faces together. After the Crusaders' occupation of Acre in 1103, they learned this construction technique and introduced it into their military architecture, such as that in Caesarea in 1218.

Machicolations

Machicolations were an important feature in Muslim defenses. These were holes or gaps in the overhang of a parapet. Through them defenders could fire arrows and drop stones or oil on their attackers. They appeared first in Qasr al-Hayr near Rusafa in Syria in 729, and came to Europe in the 12th century, first at the Château Gaillard built by Richard the Lionheart, following his return from the Crusades. Then they appeared in Norwich in 1187 and in Winchester six years later. Like many of these defenses, the returning Crusaders learned the idea from the Muslim world.

Battlements

Battlements are a series of stone indentations and raised sections added to the tops of walls of buildings. Originally they gave cover to the defenders, but in modern times they are decorative. They became popular in Europe in the 12th century with returning Crusaders. There is a great likeness between the battlements of the 15th-century church at Cromer in Norfolk, the Palazzo Ca' d'Oro in Venice, and buildings in Cairo, such as the 13th-century Zayn al-Din Yusuf Mosque, and the tenth-century Al-Azhar Mosque respectively.

Although the Crusades were a bloody time, there were interspersed moments of peace, where ideas were talked about and swapped. The vast movement of people also meant the movement of ideas, which helped Eastern concepts migrate to the West.

TOP: *A drawing shows an early 16th-century model castle. Two of these models would be wheeled into a large field and would hold 60 fighting men, in full army dress, ready to practice military maneuvers.* BOTTOM: *The Crusaders were impressed by Saladin's round towers and built similar towers in Europe. This example is in Podzamcze, Poland.*

09 PUBLIC BATHS

Spas and health clubs have sprung up over the world today, letting all luxuriate in their steam and fine soaps, but this was not always the case. In the so-called Dark Ages of Europe, the Roman bath became particularly unfashionable.

After the disintegration of the Roman Empire, the Romans and most of their public amenities disappeared. For the Romans, the bath was in an elaborate building complex, complete with a medium heated room or *Tepidarium*, a hot steam room or *Caldarium*, and a room with a cold plunge pool or *Frigidarium*. In some of the larger baths there were other sections with changing rooms called *Apodyterium*, a reading room, and a sports area. But these treatment centers were for the rich and political elite only.

While these baths fell into disrepair as the Roman Empire gradually collapsed, on the other side of the Mediterranean the Arabs, who had been under Roman rule in countries like Syria,

Men relax inside Cagaloglu Hamami, a Turkish bath, or hamman, *in Istanbul that was built around 1690.*

inherited the tradition of using the bath. As the Romans left, the Arabs and then Muslims gave them special importance because of Islam's emphasis on cleanliness, hygiene, and good health. Reporter Rageh Omaar presenting the BBC's *An Islamic History of Europe* said that there were "thousands of *hammams* in a city of quarter of a million."

> *"Indeed, God loves those who turn to Him constantly, and He loves those who keep themselves pure and clean."*
>
> QURAN (2:222)

The bathhouse, or *hammam*, was a social place and it ranked high on the list of life's essentials. The Prophet Muhammad said, "Cleanliness is half the faith." Hammams then were elaborate affairs with elegant designs, decor, and ornamentation. Under Mamluk and Ottoman rule, they were especially sumptuous buildings in their rich design and luxurious decorations, furnished with beautiful fountains and decorative pools.

The hammam was, and still is, a unique social setting for some Muslim communities, playing an important role in the social activities of the community. As an intimate space of interaction for various social groups, it brought friends, neighbors, relatives, and workers together regularly to undertake the washing ritual in a relaxing atmosphere. Group bonds strengthened, friendships rekindled, and gossip exchanged. This therapeutic ritual was carried out by both men and women at separate times, with the women usually bathing in daylight and men in the evening and night.

The intrigue and sociability at the hammam did not just stop at scrubbing and gossip, as traditionally the setting played a significant role in matchmaking. In conservative communities such as those of North Africa, women who were looking for suitable brides for their sons would go to the hammam. Here they had the perfect opportunity to have a closer look at the bride-to-be and select the most physically fit. However, this tradition has gradually lost its popularity as arranged marriages in these societies are becoming increasingly rare and the role of the ladies' public bath has been reduced.

The exterior of the public bathhouse in Tbilisi, Georgia, is covered with dazzling and colorful tiles.

It is also customary in many parts of the Muslim world for the new bride to be taken with her friends to the hammam, where she is prepared, groomed, and adorned in stylized designs with *henna*, the herbal paste that leaves a reddish brown color on the hair, hands, and feet. The groom is also escorted there the night before he meets his bride.

The art of bathing in hammams is guided by many rules, such as men must always be covered in lower garments, and women are forbidden to enter if men are present. Quite a few books have been written about this; one is *Al-Hammam and Its Manners* from the ninth century by Abu Ishaq Ibrahim ibn Ishaq al-Harbi.

The sophistication of the bathing process in 14th-century Baghdad involved private chambers and three towels, causing Ibn Battuta to say, "I have never seen such an elaboration as all this in any city other than Baghdad."

As we have said, the bath was known to Europe in Roman times, but it fell out of use as Rome fell. In the 1529 work by Sir John Treffy, *Grete herbal,* we can read about bathing attitudes: "Many folke that hath bathed them in colde water have dyed."

Hundreds of years later, baths were rediscovered during the Crusades when the Crusaders encountered Muslim-style baths in Jerusalem and Syria. This rediscovery was brief, though, as numerous churches banned their use, partly because they belonged to "the culture of Muslims, the infidels," and partly because of the spread of adultery and bad sexual habits and diseases following their immoral use, because the manners of the Muslim hammam were not followed.

By the 17th century, hammams were rediscovered when Europeans travelers encountered Turkish baths. This was at the same time that it became fashionable to use Eastern baths and Levantine flowers. In England, in places such as London, Manchester, and Leeds, this was a real craze. The first Turkish bath, or *bagnio,* was opened as early as 1679 off of Newgate Street, now Bath Street, in London, and was built by Turkish merchants. Turkish baths were also built in Scotland, in Edinburgh, where the famous Drumsheugh Baths were designed by John Burnet in 1882. The elaborate nature of the bath was re-created in all its glory, as this contained a suite of Turkish baths with a dome supported on a brick and stone structure, with geometrical lattice windows in frames of horseshoe arches. Meanwhile, the facade was decorated with an elegant Moorish arcade with iron grilles in a geometric pattern.

So, it is believed that the hammam is the origin of most of the modern health and fitness clubs and retreat centers now found around the modern world. Saunas, however, are said to be of Scandinavian heritage. Sweating flushes out impurities and helps us lose fat. Steam and hot water increase blood circulation and raise the pulse and metabolic rate. The relaxation in the *Al-Barrani* (translated as "the Exterior"), the equivalent to the rest room or Roman *Apodyterium,* lets the body rest and benefit from previous exercise, while the social interaction and the friendly atmosphere benefit all.

The exterior of an old Turkish bath stands in the shadow of Sir Norman Foster's famous building known as the London Gherkin. The former bathhouse is now a pizzeria.

A 16th-century Turkish manuscript shows a public bath on wheels as part of a procession on craftsmanship that paraded in front of Sultan Murad III on the occasion of his son's circumcision.

10 The Tent

Tents these days conjure up images of rain-drenched campsites or beautiful wedding marquees. They have a practical and social function, are large or small, and remain true to their roots from the time when Muslims and Bedouin Arabs used them as a shelters and meeting places.

They could be elaborately decorated royal structures in a sultan's ceremonies, which were beautifully colored affairs with silk crowns and a raised section to add extra splendor and majesty. Inside were comfortable seats and canopies, colorful carpets, plus some of the sultan's favorite weapons and toiletries. The tent followed the sultan in his travels: for war, hunting trips, and other visits and ceremonies.

Europeans fell in love with the Ottoman tent the first time they set eyes on it. In the beginning it was reserved for royals and the rich, for grand parties and royal ceremonies. The French king Louis XIV was its greatest admirer and he had many ceremonial tents, *à la Turque*. These usually accompanied extravagant processions and royal parties with firework displays. His fashion statements caught on with the rest of the royal households of Europe who did not want to be left out of the latest craze and the tent dominated most of the 17th century.

Louis XIV had a real interest in the Islamic world, and he gathered knowledge about it through travelers such as François de la Boullaye-le-Gouz and Jean-Baptiste Tavernier. La Boullaye even arrived at the royal court wearing Persian dress. Louis also had in his service two renowned Arabic linguists, Laurent d'Arvieux and Antoine Galland.

In Vauxhall Gardens, London, one of these tents was built in 1744, and it held a dining area with 14 tables. The two most famous Turkish tents in England were built around 1750, in the gardens

A 16th-century miniature from the Hunernâme, *by Mehmed Bursevî, shows the ascension to the throne of Sultan Selim I. The various uniforms classify ranks in the government. A senior officer is leaning to kiss the hem of the sultan's cloak. Kissing the hem is an Ottoman tradition to demonstrate loyalty and obedience.*

of Painshill, Surrey, owned by the Honorable Charles Hamilton, and in Stourhead, Wiltshire, owned by Henry Colt Hoare. John Parnell made a watercolor illustration of the tent at Painshill after he visited it in 1763.

The site of the tent at Stourhead was originally intended for a mosque with minarets, but the idea changed into a tent that was dismantled in the 1790s. A third Turkish tent was built at Delgany, Wicklow, Ireland, by David La Touche in the late 18th century, but tents never really caught on there because of the weather.

European imitation of Turkish tents also took on a lot of the Islamic architectural styles, and in the 18th century architect John Nash produced a "total exotic exterior effect" of a Royal Pavilion, which greatly pleased his royal patrons. He used the Eastern scenery described by 18th-century landscape painter Thomas Daniell. Daniell was also the author of *Oriental Scenery,* and was hired as a consultant to help design a British residence with such features as a bulbous dome with corner *chattris* and overhanging eaves, cusped arches, and pinnacles. It was Daniell who inspired Nash, who was commissioned by George IV to remodel an unfinished structure at the Royal Pavilion in Brighton. So he combined bulbous domes with concave-shaped roofs, imitating the Turkish caliph's tents that covered the banqueting and music rooms of the building. He also used minaret-like structures to disguise the chimneys.

This style of building still exerts a strong influence, and one still survives at Canterbury Park in Hampshire. The roof of the Rotunda in Vauxhall Gardens was a tent with blue and yellow alternating stripes, supported by 20 pillars. English writer Nathaniel Whittock in 1827 described it as a Persian Pavilion.

Other famous people to enjoy and own tents included France's Empress Josephine, who had a Muslim tent room at Malmaison, and King George IV often dined there. The Marquess of Hertford, nicknamed "the Caliph," had a tent room made for him by Decimus Burton at St. Dunstan's House. This burned down in 1930 and was rebuilt in a different design.

This Ottoman manuscript is a memoir of the military movement during an expedition of Suleyman the Magnificent against Hungary. The tents of different colors possibly refer to different regiments pitched around the River Ibri of Mitrovica, located in the Leposavic municipality of Kosovo. The writing insets give a snapshot of the camp on 23rd Safar 950 A.H. (or May 24, 1543), and reveal that the regiments moved six miles in two days.

11 FROM KIOSK TO CONSERVATORY

What we now think of as a garden summerhouse and the bandstand in the local park or town square came from what was called a Turkish kiosk, or *Koshk*. This was a domed hall with open and arched sides, attached to the main mosque under the Seljuks. Gradually it evolved into the summerhouses used by Ottoman sultans.

The most famous of these kiosks were the Cinili Koshk and Baghdad Koshk. The Cinili Koshk was built at the Topkapi Palace, Istanbul, in 1473, by Muhammad al-Fatih and had two stories topped with a dome, with open sides overlooking the gardens of the palace. The Baghdad Koshk was also built at the Topkapi Palace in 1638-1639, by Sultan Murad IV. This also had a dome, and the view it gave of the palace's gardens and park, as well as the architecture of the city of Istanbul, was amazing.

Lady Wortley Montagu, wife of the English ambassador to Constantinople, wrote a letter on April 1, 1717, to Anne Thistlethwayte mentioning a "*chiosk*," describing it as "raised by nine or ten steps and enclosed with gilded lattices," but it was European monarchs who brought it to Europe. The king of Poland particularly liked it, as did the father-in-law of Louis XV, Stanislas of Lorraine, who built kiosks for himself based on his memories of his captivity in Turkey. These kiosks were used as garden pavilions for serving coffee and beverages, but later were converted into the bandstands and tourist information stands decorating many European gardens, parks, and high streets.

All good designs evolve, and in this case the kiosk evolved into what we now call conservatories, glass rooms built in gardens or on the sides of many European houses. The earliest conservatories were those made by Humphrey Repton for the Royal Pavilion at Brighton. They were sumptuous affairs, with corridors connecting the pavilion to the stables, and with a passage of flowers covering the glass. They joined the orangery, a greenhouse, an aviary, an enclosure for pheasants, and hothouses. The pheasantry area was particularly Muslim in concept, as it was an adaptation of the kiosks on the roof of the palace in the Fort of Allahabad in India.

European glasshouses, or conservatories, evolved from kiosks built at the Topkapi Palace in Istanbul.

A 16th-century miniature shows Sultan Murad III and his sons sitting in a kiosk from the Shahinshahnane-i-Murad *by Mirza Ali ibn Hacemkulî III.*

12 GARDENS

Sunny days are spent cutting the grass while hoping it does not rain too much. Insects are dealt with, moles are moved on, and birds are made to feel welcome. Lawns, with their herbaceous borders, dominate many gardens in Europe, especially in the United Kingdom. Back in the Middle Ages though, gardens in Europe were limited to the courts of nobles or monasteries, and their main use was for herbs, vegetables, and some fruits for self-sustenance.

For Muslims, gardens have always been a source of wonder and enchantment because plants, trees, animals, insects, and all of nature are a blessed gift of Allah and a sign of His Greatness. Islam permits us to use, enjoy, and change nature, but only in ethical ways, so Islamic gardens were designed to be sympathetic to nature, and gardens to this day enjoy an elevated status in a Muslim's mind.

Gardens such as Eden were repeatedly described in the Quran as places of great beauty and serenity, and as ideal places for contemplation and reflection. These heavenly paradises were re-created and spread across the Muslim world, from Spain to India, mainly from the eighth century onward. About a hundred years later, the Abbasids innovated designs of their own. From that point on, gardens with geometrical flower beds, shallow canals, and fountains were built everywhere in Islamic Persia, Spain, Sicily, and India to provide peaceful seclusion from the outside world. Just a look at the Alhambra in Granada, Spain, or the Taj Mahal in India shows this.

A 16th-century miniature shows Suleyman the Magnificent. The Sarai gardens in Turkey of Suleyman the Magnificent cultivated tulips.

Gardens were not only for meditation; many had a practical function, and Arab rulers collected plants. Kitchen gardens not only supplied food, but also gave rise to a type of Arabic poetry known as the *rawdiya*, the garden poem, which conjured up the image of the Garden of Paradise.

It was in Toledo in 11th-century Muslim Spain, and later in Seville, that the first royal botanical gardens of Europe made their appearance. They were pleasure gardens, and also trial grounds for the acclimatization of plants brought from the Middle East. In the rest of

The Generalife gardens at the Alhambra, Granada, Spain, show the geometrical planting and water features typical of Islamic gardens.

Europe these gardens appeared about five centuries later in the university towns of Italy. Today, the influence of the Muslim garden can be widely seen in Europe, from the Stibbert Garden in Florence to the Royal Pavilion in Brighton, England.

It was not just the concept of gardens that spread with the Muslims, because they also brought flowers from the East that you can now buy down at the local garden center. Such travelers include the carnation, tulip, and iris.

Some people believe the word "tulip" comes from *Dulband,* which means turban, as people used to wear the flower on their turban. Others say the word "tulip" is an anglicized version of *dulab*, which is Persian for tulip. From Persia, the tulip reached Constantinople through an ambassadorial gift exchange, where it was largely planted in the Sarai gardens, especially in the Topkapi Palace in Istanbul.

The tulip's journey into Europe has been like a well-thought-out invasion of perfume and color. It started in 1554 with Count Ogier de Busbecq, the Habsburg (Austrian-Hungarian) ambassador to Suleyman the Magnificent, when he took one with him. About ten years later, it reached its now famous "home" in Holland. The Duke of Sermoneta, Francisco Caetani, was a tulip collector and had 15,147 in his Italian garden in the 1640s. The Huguenots, France's persecuted Protestants, took the tulip

"Early Muslims everywhere made earthly gardens that gave glimpses of the heavenly garden to come. Long indeed would be the list of early Islamic cities which could boast huge expanses of gardens. To give only a few examples, Basra is described by the early geographers as a veritable Venice, with mile after mile of canals criss-crossing the gardens and orchards; Nusaybin, a city in Mesopotamia, was said to have 40,000 gardens of fruit trees, and Damascus 110,000."

ANDREW M. WATSON, *AGRICULTURAL INNOVATION IN THE EARLY ISLAMIC WORLD,* 1983

A 17th-century manuscript shows Sultan Babur holding a plan and watching his gardeners measure flower beds.

Tulips became hugely popular in Europe after they arrived from countries in the Muslim world.

with them into different countries as they fled. Finally, in the 1680s an Englishman called Sir George Wheler brought it to Britain from the Serail gardens of Constantinople.

The carnation and iris were less well traveled as flowers but popular in decorating Persian and Turkish ceramics. With its fan shape, the carnation was a successful combination with the tulip in Iznik pottery. This design was also copied in European decoration and appeared in a number of Lambeth chargers, ceramics produced at Lambeth, England, dating from 1660 to 1700.

The iris was used in horizontal and circular forms by Persian potters, particularly under the Safavid dynasties in the 16th and 17th centuries. The iris went on, like the carnation, to influence European designs such as the Bristol delftware ceramics.

The British love gardening and still cultivate these flowers, and flower shows are booming. One of the biggest of shows is Chelsea, and if these figures are anything to go by, gardening and gardens are far from fading: Each show costs about £3 million and over show week 60,000 pieces of cake, 110,000 cups of tea and coffee, and more than 28,000 rounds of sandwiches are sold as the keen gardeners sustain their appetites for all things green.

13 FOUNTAINS

Fountains soothe the two senses of sight and sound at once. They provide a calming atmosphere and screen out urban noises like traffic, road drills, and barking dogs in today's ever noisier world. They also provide privacy, with quietly spoken words not reaching others in the vicinity, and are a bath for birds.

Water features are an integral part of gardens today, just as they were a thousand years ago in the Islamic world. Then they were a display of ultimate wealth, as water was scarce, and a water display was regarded as a thing of wonder. Fountains became cornerstones of Islamic art and architecture, and one of the best examples is the fountain in the Lion Gardens of the Alhambra, Spain, which is more than 650 years old. It was commissioned by Sultan Mohammed V for the Court of Lions, and built between 1354 and 1359.

The fountain has a round basin, encircled by 12 lions carved from marble that originally would have been richly painted, mostly in gold. The lions represent the 12 signs of the zodiac and the 12 months. Water was carried to them by aqueducts from the surrounding mountains, and it flowed from their mouths via an elaborately timed system of channels in the floor.

Each hour one lion would produce water from its mouth, giving the impression of 12 months elapsing as though they were 12 hours. The sense of timelessness created was highly significant, because the magnificent palace was considered as a paradise on Earth, and time in paradise is nonexistent as the dwellers there live in eternal happiness.

At the edge of this great fountain is a poem written by Ibn Zamrak. This praises the beauty of the fountains and the power of the lions, but it also describes their ingenious hydraulic systems and how they actually worked, which had baffled all those who saw them. To this day the system has remained exactly the same. It is just gravity and water pressure.

The Banu Musa Brothers' Fountains

Muslim engineers spent a lot of time and effort inventing various ways of representing water and controling the way it flowed, because water is connected with Paradise. Some of the most ingenious

TOP: *Water features and fountains are a soothing sight in gardens across the world.* BOTTOM: *The Lion Fountain at the Alhambra, Spain, is more than 650 years old. It is believed that the 12 lions form a water clock. Water used to spurt out of the lions in a consecutive manner such that the water emerging from the first lion indicated one o'clock, and so on for each hour.*

THE THREE BASIC STYLES OF FOUNTAIN BY THE BANU MUSA BROTHERS

The balance was a pipe that carried the water from the main reservoir and had two positions: horizontal and raised. When horizontal, water went from the reservoir to the left tank, which fed pipes that went through to the bud making a spear-shaped fountain. As this was happening, small containers attached to the arm of the balance slowly filled with water. These eventually tipped the balance arm to its raised position. When raised, water from the main reservoir was channeled into the tank on the right, feeding the shield-shaped bud. The small containers on the side slowly emptied, until the balance returned to its horizontal position and the process repeated over and over as long as there was water in the main reservoir.

Lily

Shield

Spear

An illustration shows the three main designs of fountain by the Banu Musa brothers.

people to do this were the Banu Musa brothers in the early ninth century.

These brothers, Jafar Muhammad, Ahmed, and Al-Hasan, wrote the *Book of Ingenious Devices,* which included fountains that continuously changed their shape. For the ninth century, and even today, these fountains produced a sense of mysticism and amazement because of their splendor and variety of watery shapes.

> *"Surely the God-fearing shall be among gardens and fountains."*
>
> QURAN (51:15)

The Banu Musa brothers' fountain designs were full of fine technology, like worm gearing, valves, balance arms, and water and wind turbines. All this showed their competence as designers, and as craftsmen. The three basic shapes were shield, spear, and lily, and all three could emerge from the same fountain—indeed, the most breathtaking fountains were those that could change shape, from a spear to a shield and back again, at certain intervals. But first a large vessel of water had to be placed high above the fountain and out of sight, to give it sufficient pressure to obtain the desired water shape.

Some fountains used worm gears and a clever hollow "navel" valve, so called since it is shaped like a human's navel. It was this valve that directed where the water would go to produce which spouting shape.

The use of the worm and wheel to transmit motion from the flowing water to the revolving pipe was a major leap forward in the invention of control systems engineering, which was essential for the invention of automatic machines during the industrial revolution.

Fountains today are carrying on this tradition of incorporating the latest fine technology, but now this involves light and music in time with jets of water. A millennium later, water, plus human ingenuity, is still amazing us.

Sicily
Greece
Smyrna
Adaliah
Tabriz
Caspian Sea
ASTRABAD
Urumiah
Reshd
Balfrush
TEHERAN
KHORASSAN
Senna
Tunis
Monastir
Candia
Cyprus
Aleppo
Palmyra
Hamadan
Great Salt Desert
G. of Cabes
Jerbah I.
MEDITERRANEAN SEA
Damascus
BAGDAD
PERSIA
Ispahan
TRIPOLI
Mesurata
Grenah
Derne
Abereton
Mouths of the Nile
Acre
SYRIA
Ruins of Babylon
Shuster
KERMAN
Bengazi
G. of Sidra
BARCA
Jerusalem
Syrian Desert
Wakisah
Euphrates
Basorah
Kerman
Ghadames
TRIPOLI
Muklar
Alexandria
Ghizeh
Isthmus of Suez
Suez
Soweih
FARS
Shiraz
Mizda
Sockna
Oasis of Siwah
Benisouef
EGYPT
Mt. Sinai
Mt. Horeb
ARABIA
Shakuk
Zeghan
Zella
Augela
Siout
R. Nile
Moila
LACHSA
PERSIAN GULF
Lar
LARISTAN
Gombe
El Kasar
Khaibar
El Katif
FEZZAN
Farafreh
Girgeh
Thebes (in Ruins)
Esneh
Medina
NEDJED
Zuela
West.n Oasis
Daraieh
Zabara
Sohar
MOURZOUK
Tibesti
Gr.t Oasis
Edfou
Assouan
Tegerhy
LIBYAN DESERT
1st Cataract
HEDJAZ
Yembo
L. Salome
El Yeman
OMAN
El Baad
Ebsamboul
Derr
MECCA
Tibesti Oasis
Route
2nd Cataract
Jidda
Arabian Desert
Tarajit
Salt Mines
3rd Cataract
4th Cat.
NUBIA
Nawarik
Leith
Seggedem
Koufodeh
Masser
New Dongola
DONGOLA
5th Cataract
Suakim
Bedr
SAHARA
Bilma
Borgoo
Kuka
Meroe
Berber
Hali
Hora
HADRAMAUT
L. Boushashem
Shendy
Jisan
Barths Route
R. Barkuu
6th Cataract
Masuah
Mareb
Dofar
Northern Limit of the Negro Races
R. Nile
Khartoum
YEMEN
Sana
C. Fartak
KANEM
Lari
L. Hadiba
WADAY
Cobbe
TIGRE
Axum
Mocha
GULF OF ADEN
BORNOU
R. Yeou
Lake Tchad
Ghazal R.
DAR SALEY
Wara
DARFUR
Bent Fezarah
KORDOFAN
El Obeid
SENNAR
Mt. Abba Yared 15200 ft.
AMHARA
Perim
Babelmandeb
Kouka
L. Fittre
Sennar
Gondar
Zeyla
C. Guardafui
Angornou
New Birnie
BEGHARMI
L. Dembea
Berbera
Somauli
MANDARA
Mora
NIGRITIA
Begharmi
Blue Nile
ABYSSINIA
SHOA
Ankobar
SOMAULIES
Ras Mabber
White Nile
Chadda R.
Youlo
R. Shary
FERTIT
G.t Lake
Hurrur
Hanilli
Atah R.
Tribes
R. Keilak
Sakka
Ras al Khyl
Pagan
R. Dome
Donga
Keke
Mien
Bonga
L. Abbo
Webbe R.
AJAN
Cameroon Mts.
Kurkur
R. Bako
Ras Awath
UNKNOWN
Uadir
Berize
Toorhu
Biafra
R. Cameroons
Andoma
Doko
Shivelli
Lhedil
Bimberi
Gatu
Gondokoro
Mts. of the Moon
BIAFRA
R. Asua
GALLA
Gananeh
INTERIOR
Koki
Magadoxo
Luta Nzigeh L. about 2200 ft.
Karpuma Fs.
TRIBES
Grredi
LOWER GUINEA
Ashira
Mundo
Kari
Ripon Fs.
L. Baringa
Galwen
Brava
Equinoctial Line
Jnba
Adjumba
ANZICO
Mt. Mfumbira 10000 ft.
Victoria Nyanza 3554 ft. above the Ocean
Mt. Kenia
Juba R.
Barkaro R.
Rusisi Lake
Mt. Booro
Port Durnford
INDIAN OCEAN
LOANGO
Konta bella
Rumanikas Palace
Speke
Formosa B.
Lumerezis
Mnanza
Mt. Kilimanjaro 20000 ft.
Melinda
Congo R.
Sundi
Ujiji
ZANGUEBAR
Mombas
Coolo
Lake Tanganyika 1800 ft. above the Sea Discovered by Burton 1859.
Mininga
SUAHELI
Pemba I.
Almirante Is.
Muroque
Kazeh
Sonho
St. Salvador
Grant
Zanzibar T. & I.
Zaire R.
Oha
Marora
CONGO
Mafia I.
Ambriz
Coave R.
Quiloa T. & I.
St. Paul de Loando
Matiamvo
CHANNEL
Coanza R.
ANGOLA
LONDA
Lindy
C. Delgado
Cazembe
L. Nyassa
Dr. Livingstone's Route
L. Dilolo
CAZEMBE
Livuma R.
Comora Is.
C. Amber
S. Felipe
Benguela
Bihe
Shinte
Mazavamba
Lupata Mts.
Port Luke
Nano
Luambesi R.
L. Shirwa
MOZAMBIQUE
Ibo
Manambatou
Caconda
Mocanda
Pomba B.
BENGUELA
BANYETI
East Cape
Negro
Honda
Leeambye R.
BATOKA
Zumbo
Mozambique
Tete
C. St. Andrews
Antongil Bay
Route of Dr. Livingstone
BANYAI
Mokaimba
Port Louis
MAKOLOLO
Sena
Quilimane
Tamatave
OYAMPO
R. Chobe
Mths. of the R. Zambese
Ondonga
R. Embarra
Linyanti
Victoria Falls
Mazaro
ANTANANARIVO
Andevorante
MADAGASCAR
C. Frio
OVAHERERO
Kamakama
MONOMOTAPA
Sterile Coast
L. NGAMI
MATEBELE
Tantamane
CIMBEBAS
Okavare
Batuana
Zouga
Sofala
Mourondava
St. Denis
Bourbon
Tunobis
Mahuku
Europa I.
Ambaram
Barmen
Nchokotsa
Limpopo R.
FALASA
Mananzari

CHAPTER SEVEN

"The Earth is spherical despite what is popularly believed . . . the proof is that the sun is always vertical to a particular spot on Earth."

IBN HAZM, A TENTH-CENTURY
MAN OF LETTERS FROM CÓRDOBA, SPAIN

WORLD

PLANET EARTH • EARTH SCIENCE • NATURAL PHENOMENA
GEOGRAPHY • MAPS • TRAVELERS AND EXPLORERS • NAVIGATION
NAVAL EXPLORATION • GLOBAL COMMUNICATION
WAR AND WEAPONRY • SOCIAL SCIENCE AND ECONOMICS

TODAY, EVERYONE CAN EXPERIENCE A DIFFERENT COUNTRY FOR THE PRICE OF AN AIR TICKET, but globe-trotting is not a modern concept. Even though they did not have planes, trains, or automobiles, medieval Muslims were inspired to travel as pilgrims on the annual hajj to Mecca, to gain knowledge, and for trade. They were renowned for discovering their world. From their experiences we have comprehensive travel books that show us the medieval world in detail.

Medieval Muslims were also making observations and calculations about their surroundings to make sense of their environment. Al-Biruni discussed the theory of the Earth rotating on its own axis 600 years before Galileo, while also explaining the ebb and flow of tides. Others noted why the sky was blue, causes of rainbows, and size of the Earth's circumference. Communications were improving and the desire of one sultan a thousand years ago to eat fresh cherries initiated the birth of pigeon post, or mail. The previous century saw Al-Kindi laying the foundations for code breaking, which led to passing secret messages in times of war. Open this chapter to sense the global vision of pioneers of a thousand years ago.

OPPOSITE: *Muslim explorers were renowned for their eyewitness accounts of the world in which they traveled.*

01 PLANET EARTH

There was a time when the idea of the world as a tilting, wobbling, land-and-sea-covered molten globe spinning on its own axis, while tracing an elliptical path around a fiery orb, would have been an absurd suggestion. Only through centuries of observation and experimentation by succeeding civilizations can we now be sure that this is really the case, and it is called planet Earth.

Ptolemy in 127-151 was among the earliest thinkers in these great debates. As a great astronomer and mathematician of antiquity, he estimated the change in longitude of the fixed stars to be about 1° per century, or 36 seconds annually, when he described the then supposed Earth-centered system of the universe. Today, this movement is known as "the precession of the equinoxes," and is understood as the Earth slowly wobbling on its rotation axis through its orbit, caused by the gravitational pulls of the sun and the moon on the Earth's equatorial bulge.

What we also know today is that over a cyclical period of 25,787 years, this wobble influences the time at which the Earth is closest to and farthest from the sun, and ultimately, it also affects the timing of the seasons. This also means the stars and constellations slowly drift westward.

Muslim astronomers obtained increasingly accurate figures about the precession of the equinoxes than Ptolemy had. The renowned tenth-century Baghdad astronomer Muhammad al-Battani said it was 1° in 66 years, or 54.55 seconds per annum, or 23,841 years for a complete rotation. Ibn Yunus, who died in 1009, said it was 1° in 70 years, or 51.43 seconds per annum, or a rotation in 25,175 years. This compares amazingly well with the present-day figure of about 50.27 seconds per annum, or about 25,787 years for a complete rotation.

It is the Earth's axis tilted to the plane of the elliptical orbit that is the main cause of the seasons; so, for example, when the Northern Hemisphere is tilting toward the sun, we are in summer. As the Muslims discussed the phenomenon of seasons, they were also studying and calculating the tilt of the Earth.

Discovering the exact degree of tilt became a matter for intense deliberation among astronomers and mathematicians in the centuries following Ptolemy. In the late tenth century, a Tajikistan mathematician and astronomer named Al-Khujandi built a huge observatory in Rayy, near Tehran, Iran, to observe a series of meridian transits of the sun. These let him calculate, with a high degree of precision, the tilt of the Earth's axis relative to the sun.

Today, we know this tilt is approximately 23°34', and Al-Khujandi measured it as being 23°32'19", so he was quite close. Using this information, he also compiled a list of latitudes and longitudes of major cities.

A century before this discovery, the enlightened ninth-century caliph Al-Ma'mun engaged a group of Muslim astronomers to measure the Earth's circumference. They did it by measuring the length of the terrestrial degree, which they found to be 56,666 Arabian miles or 111,812 kilometers (69,477 miles), which brought the circumference to 40,253.4 kilometers (25,012 miles). Today, we know the exact figure of the Earth's circumference is 40,068.0 kilometers (24,897 miles) at the Equator, and 40,000.6 kilometers (24,855 miles) through the Poles, so the astronomers were not far off either.

Al-Biruni, an 11th-century polymath, said with a touch of dry humor: "Here is another method for the determination of the circumference of the Earth. It does not require walking in deserts." He calculated the figure by using a highly complex geodesic equation and wrote it all up in his book *On the Determination of the Coordinates of Cities.* Len Berggren, a contemporary writer, said: "It doubtless gladdened al-Biruni's heart to show that a simple mathematical argument combined with a measurement could do as well as two teams of surveyors tramping about in the desert."

Al-Biruni's book also made a systematic and detailed study of the measurements of the Earth's surface. He measured latitudes and longitudes, and determined the antipodes and the roundness of the Earth. He was a man genuinely ahead of his time, and even discussed the theory of the Earth rotating about its own axis 600 years before Galileo.

Many educated Muslims, including Al-Biruni, at this time took it for granted that the Earth was round. Ibn Hazm, a tenth-century man of letters from Córdoba, said, "The Earth is spherical despite what is popularly believed . . . the proof is that the Sun is always vertical to a particular spot on Earth." This is another example of where Muslim scientists were carrying out groundbreaking research that was based on observation and experimentation rather than hearsay and myth.

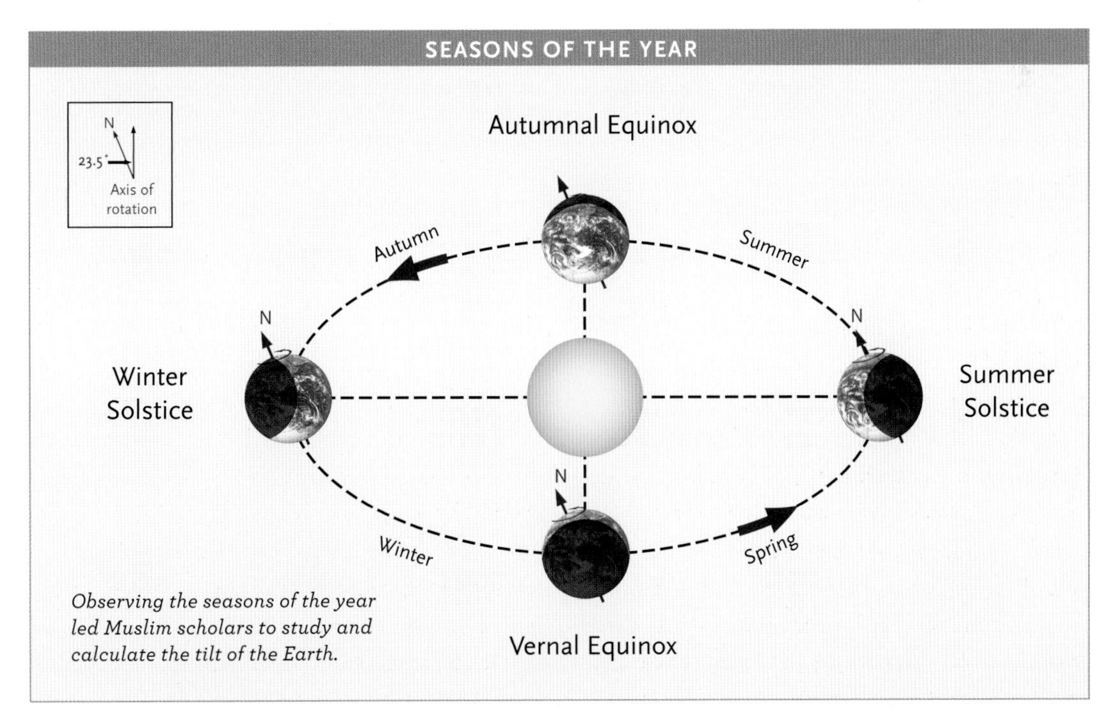

Observing the seasons of the year led Muslim scholars to study and calculate the tilt of the Earth.

02 EARTH SCIENCE

Muslim civilization was outstanding in its natural outlook toward the universe, humanity, and life. Muslim scientists thought and wondered about the origin of minerals, rocks, mountains, earthquakes, and water.

The ancient Egyptians, Mesopotamians, Indians, Greeks, and Romans knew of certain varieties of mineral, precious stones, and gems. Most of the lands of these people became part of the Islamic State or Caliphate. Once their writings on gems and minerals, like other subjects, had been translated into Arabic in the first 300 years of the Islamic world, Muslim scientists and explorers carried on the work and research.

The enormous area that the Islamic world covered meant that Muslims could study and develop earth sciences not only in the Mediterranean area, as the Greeks had done, but also in Europe, Asia, and Africa. Knowledge of minerals, plants, and animals was gathered from areas as far away as the Malay Islands and brought together in books such as 11th-century scholar Ibn Sina's *The Book of Cure*, which was essentially an encyclopedia of philosophy and natural sciences.

Ibn Sina, known as Avicenna in the West, was a true product of Muslim civilization at the height of its scientific growth, but he is better known today in medicine and philosophy than earth science. However, in his *Book of Cure* there is an important chapter on mineralogy and meteorology, where he presented a complete coverage of the knowledge of his day regarding what happens on the Earth. Through its Latin translation it became known in Renaissance Europe, where it was a source of inspiration to the founders of geological thought in Europe, men such as 15th-century Leonardo da Vinci, 17th-century Nicolas Steno, and 18th-century James Hutton.

Ibn Sina was not the only Muslim scholar who was pushing the boundaries of knowledge.

Early 11th-century scholar Al-Biruni spent most of his time studying in India, where he correctly identified the sedimentary nature of the Ganges River Basin.

Another big name in the field of earth science was Al-Biruni, who was a contemporary of Ibn Sina.

Born in Khwarizm in 973, Al-Biruni cannot be covered by only one label professionally, because he wrote prolifically in many areas, including mathematics, astronomy, medicine, philosophy, history, pharmacy, and earth science or mineralogy.

RECOGNIZING GEMS

Gems and minerals fascinated ancient peoples.

"I bought some raw pebbles brought from India. I heated some of them, they became more red. There were two very dark pieces, one was with reddish color, the other was less red. I put both pieces in a crucible and directed the flame at them for a period sufficient to melt 50 *mithqal* of gold. I took the pieces after they cooled. I noticed that the less red piece became purer with a rose red color. The other, deep red piece lost its color and became like Sarandib [now Sri Lanka] quartz. I then examined this latter piece and found that it was softer than the *yaqut* [ruby] . . . I concluded: When redness is lost with heating, the heated material is not *yaqut*. This conclusion cannot be reversed; i.e. if the heated material stays red it is not necessarily *yaqut*, because iron stays red after heating."

11TH-CENTURY SCIENTIST AL-BIRUNI INVESTIGATING RUBIES, FROM HIS BOOK *TREATISES ON HOW TO RECOGNIZE GEMS*

"We must clear our minds . . .
from all causes
that blind people
to the truth—old custom,
party spirit,
personal rivalry or passion,
the desire for influence."

AL-BIRUNI IN *VESTIGES OF THE PAST*

A great deal of his time was spent in India, where he learned the language and studied the people, religion, and places. This he wrote up in his vast book called *Chronicles of India*. As well as speaking Hindi, he also knew Greek, Sanskrit, and Syriac, although he wrote all his books in Persian and Arabic. His time in India meant he looked intently at its natural history and geology, and he correctly described the sedimentary nature of the Ganges Basin. His great mineralogical work was called *Treatises on How to Recognize Gems*, and it made him a leading scientist in this area.

Other scholars made their contribution to the science we now know as geology:

Yahya ibn Masawayh (died 857) wrote *Gems and Their Properties*. Al-Kindi (died about 873) wrote three monographs, the best of which was *Gems and the Likes* but which is now lost. Al-Hamdani, a tenth-century scholar, wrote three books on Arabia in which he described methods of exploration for gold, silver, and other minerals and gems along with their properties and locations. The tenth-century group of scholars known as *Ikhwan al-Safa'* (the Brothers of Purity) wrote an encyclopedic work that included a part on minerals, especially their classification.

Of the enormous number of works written on the subject of minerals, stones, and gems, much has been lost, but a few works have survived and are now in print.

03 Natural Phenomena

It is children who usually ask us the difficult questions: "Why is the sky blue?" "Where does the rainbow end?" "Why does the sea lap at the sand?" Today, we take much of the natural world around us for granted, but Muslim minds of the ninth century were thinking deeply about these questions out of a curiosity to understand their surroundings, and Allah's creation motivated them.

Before and at the time of Ibn Hazm, who was a tenth-century man of letters from Córdoba, astrologers believed that stars and planets had souls and minds and that they influenced people. Ibn Hazm took a more pragmatic view and said, "The stars are celestial bodies with no mind or soul. They neither know the future nor affect people. Their effect on people however can be through their physical characteristics, such as the effect of the sun's heat and rays on the planets and the effect of the moon on the tides of seas."

Another scholar of the 11th-century, Al-Biruni, explained that the increase and decrease in the height of the ebbs of tides occurred in cycles on the basis of changes in the phases of the moon. He gave a very vivid description of the tide at Somnath, a city in India, and traced it to the moon.

As they studied the heavens, some scholars, like Al-Kindi, commented on the blueness of the sky. He did this in a short treatise with a long title: *Treatise on the azure color which is seen in the air in the direction of the heavens and is thought to be the color of the heavens.* More simply, he was telling people why the sky was blue. Al-Kindi said that it was due to the "mixture of the darkness of the sky with the light of the atoms of dust and vapor in the air, illuminated by the light of the sun." His words, like the length of the title, explain it fully: "The dark air above us is visible by there being mingled with it from the light of the Earth and the light of the stars a color midway between darkness and light, which is the blue color. It is evident then that this color is not the color of the sky, but merely something which supervenes upon our sight when light and darkness encounter it. This is just like what supervenes upon our sight when we look from behind a transparent colored

The moon looks larger than normal against a cityscape in this photo illustration. Ibn al-Haytham studied and explained the visual effect of why the moon appears larger than it is, concluding it is an optical illusion.

terrestrial body at bright objects, as in the sunrise, for we see them with their own colors mingled with the colors of the transparent object, as we find when we look from behind a piece of glass, for we see what is beyond of a color between that of the glass and that of the object regarded."

Al-Kindi was on the right lines, for the sky is not really blue, in spite of the confused and impossible views that passed for knowledge, even in highly educated circles in his time. He could compete with these views because he was a widely read man and excelled in science, mathematics, and music, and was a physician in ninth-century Baghdad.

Ibn al-Haytham also went against the conventional wisdom of his day. It was a thousand years ago in Cairo that he was placed under house arrest because he could not regulate the flow of the Nile as the caliph had asked him to do. He knew that if the ancient Egyptians had not been able to do it, then neither would he. To save his skin and continue his studies, he pretended to be mad. The house arrest suited him because it meant that he could concentrate all his time on observing the rays of light that came through holes in his window shutters.

The time he had for observation and experimentation meant he could explain phenomena like rainbows, halo effects, and why the sun and moon seem to grow in size when they near the horizon. He said it was the effect of the atmosphere that increased the apparent size of sun or moon as they neared the horizon, adding that the increased size was a visual trick played by the brain. He showed that it was through atmospheric refraction that the light of the sun reaches us, even when the sun is as much as 19 degrees below the horizon, and on this basis he calculated the height of the atmosphere at ten miles.

Kamal al-Din al-Farisi, who died in around 1319, repeated and improved on Ibn al-Haytham's work by observing the path of the rays in the interior of a glass sphere. He hoped to determine the refraction of solar light through raindrops, and his findings enabled him to explain the formation of primary and secondary rainbows, which is essentially the splitting up of white light by a prism.

So next time a child asks you "Why?" maybe telling him or her about the work of these medieval Muslims would be a good starting point for a personal journey of discovery.

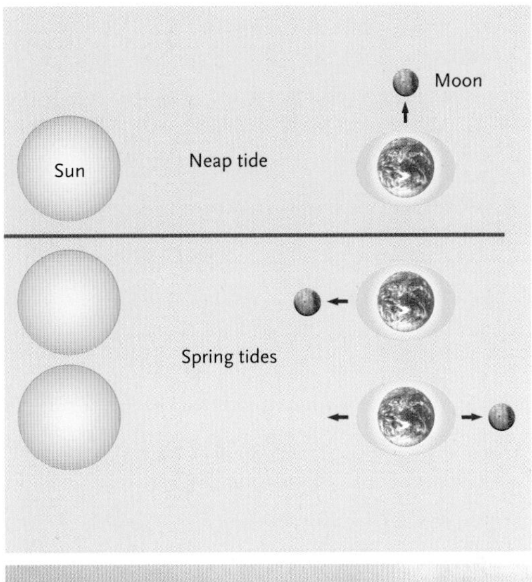

TOP: *The gravitational pull of the moon causes the rise and ebb of tides, as described by Al-Biruni in the early 11th century.* BOTTOM: *About the same time as Al-Biruni, Ibn al-Haytham was describing the phenomenon of rainbows.*

04 GEOGRAPHY

Travelers, explorers, and merchants: People living in Muslim civilization were outward-looking, observing and recording their surroundings near and far. Their interest in geography was partly due to the environment in which they lived. They had to move, along with their precious flocks and herds, in search of fresh and better pastures, so knowledge of their surroundings including that of plants and wild animals was vital. In these circumstances, the science of geography developed as a practical necessity.

The holy pilgrimage, or the hajj, was also a valuable source of material. Many pilgrims used word-of-mouth accounts of routes to Mecca and Medina, as they came from distant regions. These were later put in written form, so travel guides passed on to others, helping them on the long and difficult journey of their holy pilgrimage from all corners of the Muslim world.

The orientation of the mosques toward Mecca was another impetus to study geography, as was the need to know the direction of the Ka'bah in Mecca for daily prayers. Finally, wars and invasions and the political and administrative requirements of the expanding Muslim world created another dimension in the search for geographical knowledge.

With the development of more accurate astronomy and mathematics, giant steps were made in the progress of geographical study, as map plotting became one of its respected branches. Al-Khwarizmi, a ninth-century Persian scholar, was one of the earliest scientific descriptive geographers, and a highly talented mathematician. His famous book, *The Form of the Earth*, inspired a generation of writers in Baghdad and Muslim Spain, or Al-Andalus, to unearth, analyze, and record geographical data.

Another geographer named Suhrab, at the beginning of the tenth century, wrote a book describing various seas, islands, lakes, mountains, and rivers of the world. His notes on the Euphrates, Tigris, and Nile are very significant, while his account of the canals of Baghdad is the main basis for the reconstruction of the medieval plan of that city. This reconstruction was done

Geographical knowledge developed in Muslim societies through travel.

in 1895 by Guy Le Strange, who incorrectly read Suhrab's name as that of a well-known physician named Ibn Sarabiyun (or in Latin, Ibn Serapion). Le Strange also used work by al-Ya'qubi, who was from the ninth century, in his reconstruction. The two texts, Suhrab's account of the water system network and al-Ya'qubi's description of the highways coming from Baghdad, complemented each other very well.

Al-Muqaddasi was a tenth-century Muslim geographer. He traveled throughout the Muslim world, observing, corroborating, weighing and sifting evidence, taking notes and writing. The outcome of years of travel was *Best Divisions for Knowledge of the Regions*, completed in 985. It appealed to a variety of people while also being an entertaining read. Like many before and after him, his reasons for scholarly research were divine inspiration. What he produced would be a way of getting to know God better, and he would also receive just reward for his endeavor. His great book created the systematic foundation of Muslim geography, as he introduced geographic terminology, the various methods of division of the Earth, and the value of empirical observation.

One of the earliest Turkish geographers was Mahmud al-Kashghari, who was also a well-known lexicographer. He drew a world map, which looked unusual and circular, on a linguistic basis. It appears in his masterpiece work, a treatise on grammar called *Compendium of Turkish Dialects*, completed in 1073. A considerable portion of Central Asia as well as China and North Africa were also included, but little beyond the Volga in the west. This is perhaps because it was drawn before the Turks began to move west.

In the 11th and 12th centuries, two Muslim writers, Al-Bakri and Ibn Jubayr, collected and collated the information assembled by their predecessors into an easily digestible format. The first of them was the son of the governor of the province of Huelva and Saltes in Spain. Al-Bakri was an important minister at the court of Seville who undertook several diplomatic missions. Despite his busy official duties, he was an accomplished scholar and

SURVEYING

Surveying is a vital part of construction today.

Roman surveyors used a simple triangular level with a plumb line to "balance or equalize the land," a technique continued in Muslim and Christian Spain.

What the Romans did not have was triangulation, which is a method still used today in surveying and mapmaking. It was introduced from the East in the astrolabe treatises of two Muslim Spanish scholars, Maslama and Ibn al-Saffar, and Maslama's work was translated into Latin by John of Seville in the 12th century.

A tenth-century book called *Geometria* gave details of a variety of triangulation procedures that could be used with an astrolabe, especially for producing straight boundaries to large areas of land.

As today, teams of surveyors carry out the challenging projects like surveying irrigation canals. In al-Andalus, these teams were called *muhandis*, and in eastern Spain they were known as *soguejador*.

Today, triangulation is still used to determine the location of an unknown point by using the laws of plane trigonometry, but with the help of advanced technology, such as the Global Positioning System.

An artist's rendering shows Al-Idrisi in the court of Roger II of Sicily, with his circular map that showed he knew that the world was spherical.

essayist. He wrote an important geographical work devoted to the Arabian Peninsula, including the names of various places. *Route and Kingdoms* was alphabetically arranged, including the names of villages, towns, valleys, and monuments. His other major work was an encyclopedic treatment of the entire known world.

Ibn Jubayr of Valencia, who was secretary to the governor of Granada, Spain, was one of those who habitually recorded his hajj journeys to Mecca. These 700-year-old travel books were journals, giving a detailed account of the eastern Mediterranean world. His itineraries and road books all went well beyond the science of geography to include botany, culinary information, and travel advice.

In Muslim Spain, the passion for keeping travelogues thrived, and this inspired the compilation of the most comprehensive world atlas of the time, by the highly celebrated scholar Al-Idrisi. He was commissioned by the Norman king of Sicily, Roger II, in 1139 to come from Córdoba to Sicily and make a map for the king. He spent 15 years on this, enjoying exalted status at the king's Palermo court, interviewing thousands of travelers and producing 70 accurate maps, including some territories previously uncharted.

His work was based on that done by previous authors as well as on information he gathered in Sicily. As well as reiterating the fact that the Earth was a globe, he also said that the Earth remained "stable in space like the yolk of an egg," while giving accounts of the hemispheres, climates, seas, and gulfs. His work specifically contained a mine of excellent information about the more remote parts of Asia and Africa.

In the 13th century, Yaqut al-Hamawi toured from Mosul in Iraq to Aleppo in Syria, and then Palestine, Egypt, and Persia. Only four of his works have survived until today. The best known is his *Dictionary of Countries*. It is a vast geographical encyclopedia, which summed up nearly all medieval knowledge of the known world, including archaeology, ethnography, history, anthropology, natural sciences, and geography, and gave coordinates for every place. He described and named most towns and cities, giving details of their every monument, and their economy, history, population, and leading figures.

Like many areas of science, technology, and art, the list of personalities dedicated to the study of geography is immense. Many of them struck out into the world to gather information firsthand, to quench a thirst for knowledge and understanding, to sate their curiosity, and to leave information that would help others. Today, we have glossy magazines and TV satellite channels to experience our world. Although some people undoubtedly travel for pleasure, we often learn and understand through "professionals" from our armchairs, unlike those from the last millennium who were guided by curiosity and faith to make sense of their surroundings.

A SPHERICAL GLOBE

"Arab scientists had long known this [that the earth was round] but Europeans still clung to the belief that it was flat . . . He [Al-Idrisi] also included a travel guide and map, surprisingly accurate for 350 years before Columbus. It described England as 'gripped in perpetual winter.' . . . It is an essential ingredient in this Islamic scholarship that helped shape European civilization."

RAGEH OMAAR IN THE BBC'S *AN ISLAMIC HISTORY OF EUROPE* ON 12TH-CENTURY GEOGRAPHER AL-IDRISI

Muslim geographers agreed the world was round and made detailed measurements of the globe.

> *"Widely recognised as being the greatest single work of geography in the medieval era, [Al-Idrisi's book] also included a travel guide and map, surprisingly accurate for 350 years before Columbus. It described England as 'gripped in perpetual winter'... It is an essential ingredient in this Islamic scholarship that helped shape European civilization."*
>
> RAGEH OMAAR IN THE BBC'S *AN ISLAMIC HISTORY OF EUROPE*

VIEW FROM A CULTURAL CROSSROADS

The First Map to Show Europe, Asia, and North Africa

LEGACY: With his map the most accurate of his day, Al-Idrisi joined a long line of skilled mapmakers in Muslim civilization

LOCATION: Sicily

DATE: 12th century

KEY FIGURES: Al-Idrisi, Muslim scholar, and King Roger II, Norman king of Sicily

A thousand years ago, accurate plans of countries, continents, and waterways were unknown. But as more people began to travel the world for trade, exploration, and religious reasons, the demand for good maps increased.

Some of the world's most precious maps were drawn by great scholars of Muslim civilization, who assembled all the geographical knowledge available to them. They also drew on eyewitness accounts of the medieval world, which often came from Muslim geographers and travelers who kept detailed diaries as they journeyed.

In the 12th century, scholar Al-Idrisi produced an atlas showing most of Europe, Asia, and North Africa for the first time. Almost 850 years old, the map was created centuries before Marco Polo or Columbus explored the world. Al-Idrisi ranged widely, drawing on older knowledge and interviewing thousands of travelers to make his map the most accurate of its day. The Arabic text shows that Al-Idrisi drew the map with the south to the top and north to the bottom, as was customary then.

The map was commissioned by Roger II of Sicily, a Norman king who had recently overthrown the Muslim rulers. Nonetheless, he invited Al-Idrisi, who was at that time living in Spain, to make the map for him—a task that would take 15 years. Twelfth-century Sicily was a global crossroads for culture and ideas, and Al-Idrisi became a respected member of the king's court, before finally returning to his home in Morocco.

Al-Idrisi drew India, Arabia, Asia, the Mediterranean, Europe, and northern Africa on a circular map, and made a large, silver planisphere for King Roger. People had accepted that the Earth was spherical since ancient Greek times, and Muslim scholars made detailed measurements of the globe. In Al-Idrisi's writings, he calculated that the Earth's circumference was 22,900 miles at the Equator, about 10 percent adrift from the modern figure. He also wrote about the hemispheres of the Earth, its climatic zones, the seas, and gulfs.

In early Muslim civilization, measuring and charting the Earth's features was a key aim of mathematical geography and scholars used sophisticated astrolabes to help assess height and distance. In the ninth century, Caliph Al-Ma'mun had commissioned his astronomers to determine the Earth's circumference, which they did to within 125 miles of the figure we accept today.

Al-Idrisi's 12th-century map was the first to show most of Europe, Asia, and North Africa. The Norman King Roger of Sicily commissioned it from Al-Idrisi.

05 Maps

Maps have been helping people find their way for about 3,500 years, with the earliest inscribed on clay tablets. The introduction of paper enabled a leap forward in mapmaking, but the most recent cartographical revolution was with the development of Geographic Information Systems, or GIS. This meant that in 1973 the first computerized, large-scale, digitized maps appeared in the United Kingdom, and by 1995, most of the industrialized world was completely digitized.

Before this modern technology, which uses a system of satellites and receivers to compute positions on the Earth, maps were being made from travelers' and pilgrims' accounts.

The travel bug bit seventh-century Muslims, and they began to leave their homes for trade and religious reasons, to explore the world they lived in. They walked routes, sometimes simply gathering knowledge about new places, and when they returned gave accounts of the ways they had trodden and the people and sights they had encountered. First this was by word of mouth, but with the introduction of paper to eighth-century Baghdad, the first maps and travel guides could be produced.

Reports were commissioned by the Abbasid caliphs to help their postmasters deliver messages to addresses within their empire. These accounts made up the *Book of Routes,* and this encouraged more intensive information gathering about faraway places and foreign lands, including their physical landscapes, production capabilities, and commercial activities.

While Muslims were exploring the world, in Europe, few, excepting the Vikings (eighth-11th centuries), were traveling such distances and the average European's knowledge of the world around them was limited to their local area, with maps usually produced by religious authorities. The great European explorers of the 15th and 16th centuries would probably not have set off, were it not for the geographers and mapmakers of the Islamic world.

The maps we have today are in the style of European maps, but they are only a few centuries old. The "north" that is conventionally at the top of a map is artificial, because European navigators started using the North Star and the magnetic compass for navigation. Before that, the top of the map on European maps was to the east, which is where the word "orientation" comes from. In medieval Europe, Jerusalem was usually placed at the top or in the center, because that was the Holy Land.

A significant difference between Islamic maps and European ones was that Muslims usually drew them with the south facing upward and north downward. With Muslim development of more accurate astronomy and mathematics, map plotting became a respected branch of science, and as far as Muslims were concerned, the maps Western cartographers drew later on were upside down, with the north facing upward and south downward.

In 1929, scholars working in Turkey's Topkapi Palace Museum discovered a section of an early 16th-century map of the world signed by a Turkish captain named Piri ibn Hajji Mohammed Reis (meaning admiral), dated Muharram 919 A.H., or

> *"Columbus studied Arabic maps . . . without Jewish or Muslim expertise, Spain would not have become the greatest colonial power in 16th-century Europe."*
>
> RAGEH OMAAR ON THE BBC'S *AN ISLAMIC HISTORY OF EUROPE*

1513 C.E. This map is known as the famous "Map of America," and was made only 21 years after Columbus reached the New World.

When the Piri Reis map was discovered, there was great excitement worldwide, because of its connection with a now lost map made by Columbus during his third voyage to the New World and sent to Spain in 1498. In an inscription in the area of Brazil, Piri Reis said: "This section explains how the present map was composed. No one has ever possessed such a map. This poor man [himself] constructed it with his own hands, using 20 regional maps and some world maps, the latter including . . . one Arab map of India, four maps recently made by the Portuguese that show Pakistan, India, and China drawn by means of mathematical projection, as well as a map of the Western Parts drawn by Columbus . . . The coasts and islands [of the New World] on this map are taken from Columbus's map." No other traces of the maps made by Columbus have been found.

Very recently, a world map by the Muslim Chinese admiral Zheng He was discovered. It dates back to 1418. We do not know whether Piri Reis had come across it.

Charles Hapgood, in 1966, suggested that the Piri Reis map shows Antarctica (307 years before it was "discovered"). Now though, this theory has been thoroughly discredited, and it seems more likely that it is the South American coastline, which has been bent to conform to the animal skin parchment on which the map was drawn. Also shown on the map are the Andes Mountains of South America, which were again "first seen" by Spaniards in 1527, 14 years after the map's production. This fragment of the 1513 world map shows the adjacent coasts of Spain, and West Africa with the New World, and was drawn on a gazelle skin.

Piri Reis did not stop there, but made a second world map in 1528, of which about one-sixth has survived. This covers the northwestern part of the Atlantic, showing the New World from Venezuela to Newfoundland, as well as the southern tip of Greenland. Historians have been amazed by the richness of the map, and regret that only a fragment of the first world map was found. The search for the other parts has remained fruitless.

From the seventh century on, Muslims have undertaken the hajj, or pilgrimage to Mecca, journeying thousands of miles on horseback, by camel, or on foot. With the introduction of paper, eighth-century hajj pilgrims could produce maps to guide others.

LEFT: *Christopher Columbus.* RIGHT: *A replica of Christopher Columbus's flagship, the* Santa María.

So who was this Piri Reis, and why is his contribution to mapmaking absent from so many history books?

Piri Reis was born around 1465 in Gallipoli, and began his maritime life under the command of his illustrious uncle, Kemal Reis, toward the end of the 15th century. He fought many naval battles alongside his uncle, and became a naval commander, leading the Ottoman fleet that fought the Portuguese in the Red Sea and Indian Ocean.

In between his wars, he retired to Gallipoli to devise his first world map, his *The Book of Sea Lore* (a manual of sailing directions), and a second world map in 1528. Mystery surrounds his long silence between 1528, when he made the second of the two maps, and his reappearance in the mid-16th century as a captain of the Ottoman fleet in the Red Sea and the Indian Ocean. A sad end came to Piri Reis, as he was executed by the Ottoman sultan for losing a critical naval battle.

Like a lot of the information in this book about 1001 Inventions, not much of it has reached the present public, perhaps because Europe has concentrated on its own history, unspooling its own dramatic stories of ocean voyages, discoveries, and commercial and colonial empires. Turkish maps were given little attention, or wrongly called Italian.

But in actual fact, Turkish nautical science was ahead of its time. With Piri Reis presenting his New World map to the Ottoman sultan in 1517, the Turks had an accurate description of the Americas and the circumnavigation of Africa well before many seafaring Europeans.

Perhaps the most amazing map of the world is that of Ali Macar Reis, made in 1567, which depicts the world in such fine detail that it resembles modern-day maps and we can almost wonder if Ali Macar Reis was looking at the Earth from the moon.

Other important maps include 70 regional maps that Al-Idrisi made for the Norman king, Roger II, in Sicily, which together made up an atlas of the world as it was then known. He interviewed thousands of travelers, producing accurate maps charting previously undocumented territories. For three centuries, geographers copied his maps without alteration. More can be read about this fascinating man in the Navigation section of this chapter.

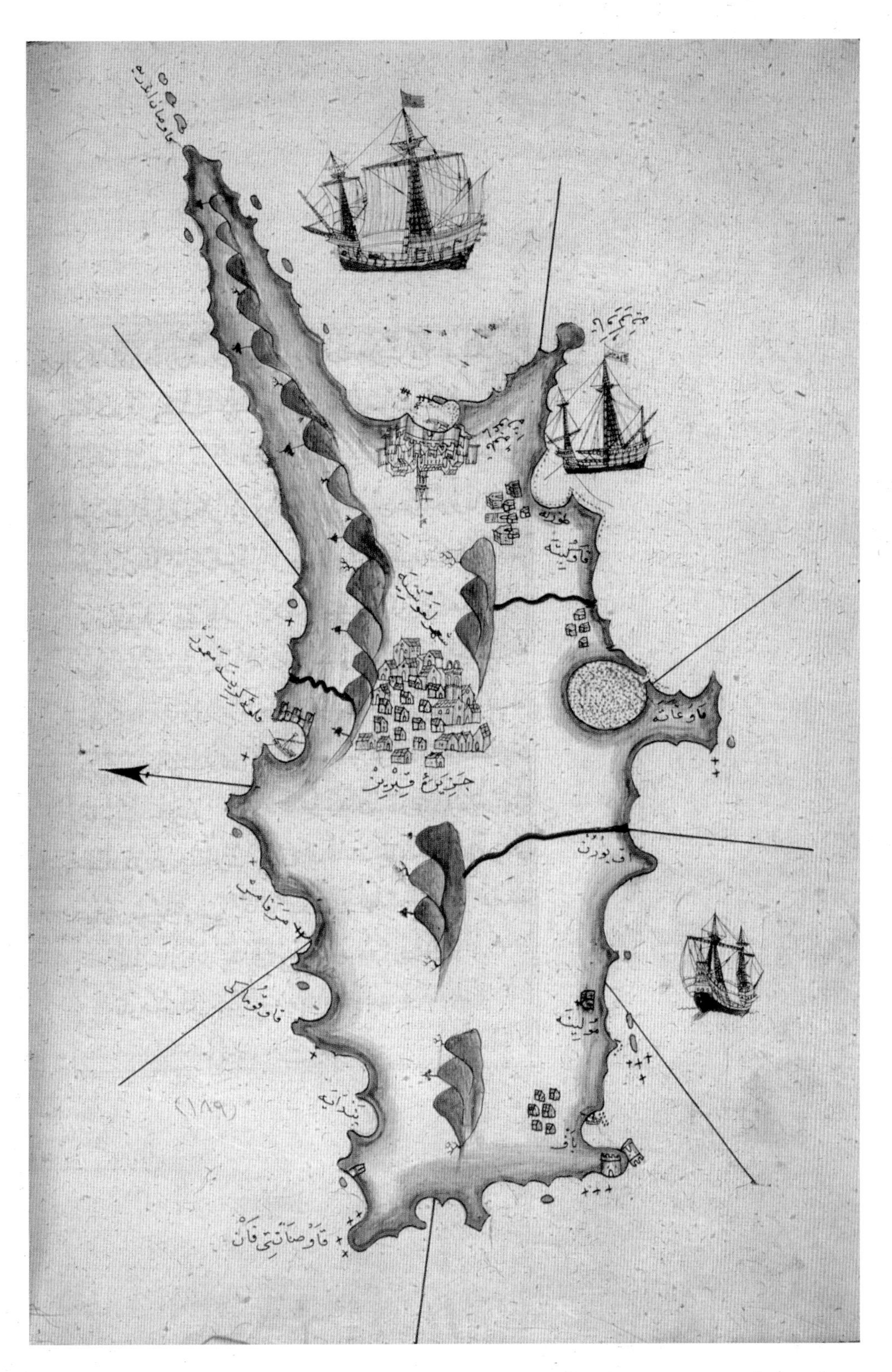

A map by Turkish admiral Piri Reis, from his 16th-century book Kitab-i-bahriyye, *shows Cyprus.*

A MAP TO INTRIGUE EAST AND WEST ALIKE

The Oldest Surviving Detailed Map Showing the Americas

LEGACY: Exciting in itself, Piri Reis's outstanding map showing America is also the only record we have today of a lost map of Christopher Columbus

LOCATION: Turkey

DATE: 16th century

KEY FIGURE: Piri Reiss, a Turkish admiral

In 1929, scholars working in Turkey's Topkapi Palace Museum discovered a section of an early 16th-century Turkish world map. It was signed by a captain named Piri ibn Hajji Mohammed Reis (meaning "admiral"), and it was dated 1513. Now known as the famous "Map of America," it was made only 21 years after Christopher Columbus had reached the New World.

Piri Reis was born in the second half of the 15th century in Gallipoli, and began his maritime life under the command of his illustrious uncle, Kemal Reis. He fought many naval battles alongside his uncle, and was later a naval commander, leading the Ottoman fleet that fought the Portuguese in the Red Sea and Indian Ocean. To draw the map rediscovered in the Palace Museum, Piri Reis did not rely on personal experience of traveling or sailing, as you might expect. He assembled it by referring to 20 regional maps; an Arab map of India; four Portuguese maps showing India and China; and a map "of the Western Parts," the coasts, and islands, drawn by Columbus.

It was this last fact that made the rediscovery of Piri Reis's map so exciting to historians worldwide. The last map he mentions was made by Columbus during his third voyage to the New World. Columbus sent it to Spain in 1498—but it has been lost for years. Now we can look at Piri Reis's map and see what Columbus must have recorded.

We know little more of Piri Reis himself. In between fighting sea battles, he retired to Gallipoli where he devised his map of 1513, and wrote *The Book of Sea Lore,* a manual of sailing directions. In 1528, he made a second world map of which about one-sixth has survived. This covers the northwestern part of the Atlantic, and the New World from Venezuela to Newfoundland, as well as the southern tip of Greenland. Historians have been amazed by the richness of the map, and regret that only a fragment of the first world map was found. The search for the other parts has remained fruitless. Mystery surrounds his long silence between 1528, when he made the second of the two maps, and his reappearance in the mid-16th century as a captain of the Ottoman fleet in the Red Sea and the Indian Ocean. A sad end came to Piri Reis, as he was executed by the Ottoman sultan for losing a critical naval battle.

Today, Piri Reis's map is the oldest-surviving detailed map showing the Americas. It shows that early 16th-century Turkish nautical science was way ahead of its time. When Piri Reis presented his New World map to the Ottoman sultan in 1517, the Turks had an accurate description of the Americas and the circumnavigation of Africa well before many European rulers.

The oldest-surviving detailed map showing America was drawn by Turkish admiral Piri Reis in 1513. He used one of Christopher Columbus's maps, now lost, for reference. Piri Reis's map shows Brazil's coastline to the left, and the coast of Spain and North Africa to the right.

06 Travelers and Explorers

In the early 1300s, Dar al-Islam, the Muslim world, was one of the greatest lands, stretching over much of the known globe, bound together by the principles of Islam. Al-Biruni, an 11th-century polymath, wrote in his *The Book of the Demarcation of the Limits of the Areas* that "Islam has already penetrated from the Eastern countries of the earth to the Western. It spreads westward to Spain [Al-Andalus], eastward to the borderland of China and to the middle of India, southward to Abyssinia and the countries of Zanj Zanj [meaning black Africa from Mali to Kilwa (Tanzania) and Mauritania to Ghana], eastward to the Malay Archipelago and Java, and northward to the countries of the Turks and Slavs. Thus the different people are brought together in mutual understanding, which only God's own art can bring to pass."

The arteries that coursed through this great body, giving it life, were trading and pilgrim routes. Within this intermeshing system, individual Muslim sultans ruled, and although there were military campaigns between them after the 13th century, an ordinary Muslim could pass through, albeit sometimes only with a passport.

Ibn Battuta said, when going into Syria, "No one may pass this place ... without a passport from Egypt, as a measure of protection for a person's property and of precaution against spies from Iraq [a Mongol-conquered country] ... This road is under the Bedouins. At nightfall they smooth down the sand so that no mark is left on it, then the governor comes in the morning and examines the sand. If he finds any track on it he requires the Arabs to fetch the person who made it, and they set out in pursuit of him and never fail to catch him."

The Muslims were natural explorers, since the Quran said every able-bodied person should make a pilgrimage, or hajj, to Mecca at least once in their lifetime. Thousands traveled from the farthest reaches of the Islamic empire to Mecca from the seventh century, even though transport was on foot, with only the fortunate ones riding in tents on camels, on ox-driven carriages, or astride horses and donkeys. As they traveled, they made descriptions of the lands that they passed through. Some of these were the first accounts in Arabic of many places, including China.

The first descriptions of China were from the early ninth century, when trade with the Chinese was recorded in the Persian Gulf. Abu Zayd Hasan was a Muslim from Siraf, and told that boats were sailing for China from Basra in Iraq and from Siraf on the Persian Gulf. Chinese boats, much larger than Muslim boats, also visited Siraf, where they loaded merchandise bought from Basra.

For thousands of years, these boats sailed along the Arabian coast, to Muscat, then Oman, and from there to India. All along the way, trade and exchanges were made, until the boats reached China and the city of Guangzhou (Khanfu in Arabic), where an important Muslim colony grew. Here, Muslim traders had their own establishments, and exchanges took place involving the emperor's officials, who had first choice for what

"And He [Allah] has set up on the earth mountains standing firm, lest it should shake with you; and rivers and roads; that ye may guide yourselves; And marks and sign-posts; and by the stars [men] guide themselves."

QURAN (16:15-16)

suited him. From Khanfu, some Muslim traders traveled as far as the empire's capital, Khomda, which was a two-month journey.

Ibn Wahhab was a ninth-century trader from Basra who sailed to China and described the Chinese capital as divided into two halves, separated by a long, wide road. On one side the emperor, his entourage, and administration resided, and on the other lived the merchants and ordinary people. Early in the day, officials and servants from the emperor's side entered the other, bought goods, left, and did not mingle again.

China, according to Muslim merchants, was a safe country and well administered, with laws concerning travelers achieving both good surveillance and security. Ibn Battuta said: "China is the safest and best country for the traveler. A man may travel for nine months alone with great wealth and have nothing to fear."

Al-Muqaddasi (circa 945-1000) was a geographer who set off from his home in Jerusalem many centuries before Ibn Battuta. He also visited nearly every part of the Muslim world and wrote a book called *Best Divisions for Knowledge of the Regions*, completed around 985.

There were many other travelers who toured the world of Islam and beyond. Al-Ya'qubi wrote the *Book of Countries,* which he completed in 891 after a long time spent traveling, and he gave the names of towns and countries, their people, rulers, distances between cities and towns, taxes,

A 13th-century manuscript shows a caravan en route to Mecca.

topography, and water resources; Ibn Khurradadhbih, who died in 912, wrote the *Book of Roads and Provinces,* which gave a description of the main trade routes of the Muslim world, referring to China, Korea, and Japan, and describing the southern Asian coast as far as the Brahmaputra River, the Andaman Islands, Malaya, and Java. The 13th-century geographer Yaqut al-Hamawi wrote the encyclopedic *Dictionary of Countries* about every country, region, town, and city that he visited, all in alphabetical order, giving their exact location, and even describing a town's monuments and wealth, history, population, and leading figures. Abu al-Fida' wrote *The Survey of Countries* in the 13th century, and this had a huge reputation in western Europe, so that by 1650 extracts about Khwarazm and Transoxonia (two regions of Central Asia) were published in London.

A 13th-century miniature depicts an eastern Muslim boat from the classical Arabic work of literature Maqamat al-Hariri. *The Arabic writing refers to a sea voyage, and mentions a verse from the Quran referring to Noah's ark. This is usually used as a blessing: "In the name of Allah, the one who protects the ship's sailing, seafaring and berthing."*

Muslim travelers and the works they left have not been completely ignored by the West, as Gabriel Ferrand compiled in the early 20th century a great study of accounts by Muslim travelers of the Far East between the seventh and eighteenth centuries. This contained 39 texts, 33 in Arabic, 5 in Persian, and 1 in Turkish. One of the early travelers covered is the ninth-century al-Ya'qubi, who wrote that "China is an immense country that can be reached by crossing seven seas; each of these with its own color, wind, fish, and breeze, which could not be found in another, the seventh of such, the Sea of Cankhay [which surrounds the Malay Archipelago] only sailable by a southern wind."

> *"If anyone travels on a road in search of knowledge, Allah will cause him to travel on one of the roads of Paradise."*
>
> PROPHET MUHAMMAD, NARRATED BY ABU AL-DARDAH

Travelers from the ninth and tenth centuries include Ibn al-Faqih, who compares the customs, food diets, codes of dress, rituals, and also some of the flora and fauna of China and India. Ibn Rustah focuses on a Khmer king, surrounded by 80 judges, and his ferocious treatment of his subjects while indulging himself in drinking alcohol and wine, but also his kind and generous treatment of the Muslims. Abu Zayd also deals with the Khmer land and its vast population, a land in which indecency, he notes, is absent. Abu al-Faraj dwells on India and its people, customs, and religious observations. He also talks of China, saying it has 300 cities, and that whoever travels in China has to register his name, the date of his journey, his genealogy, his description, age, what he carries with him, and his attendants. Such a register is kept until the journey is safely completed. The reasoning behind this was a fear that something might harm the traveler and thereby bring shame to the ruler.

Ferrand also referred to 13th-century travelers such as Zakariya' ibn Muhammad al-Qazwini, who has left accounts of the marvelous creatures that thrive in the China Sea, notably very large fish (possibly whales), giant tortoises, and monstrous snakes, which land on the shores to swallow whole buffaloes and elephants. And Ibn Sa'id al-Maghribi, who gave the latitude and longitude of each place he visited, wrote much on the Indian Ocean islands and Indian coastal towns and cities.

A 14th-century traveler, Al-Dimashqi, gives very detailed accounts of the island of Al-Qumr, also called Malay Island or Malay Archipelago. He says there are many towns and cities, rich, dense forests with huge, tall trees, and white elephants. Also there lives the giant bird called the *Rukh*, a bird whose eggs are like cupolas. The Rukh features in a story about some sailors breaking and eating the contents of its egg, so the giant bird chased after them on the sea, carrying huge

It is reported that Prophet Muhammad said, "Seek knowledge even from as far as China."

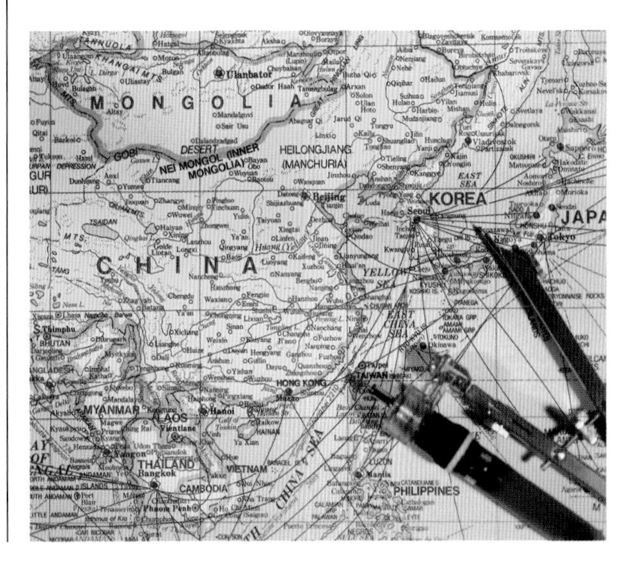

FROM LEFT: *An artist's rendering shows Ibn Battuta dictating his* Rihla, *passing through a dangerous gorge, and walking with his camel.*

rocks, which it hurled at them relentlessly. The sailors only escaped with their lives under the cover of night. This story, like other accounts by travelers, formed the basis of many of the tales that enrich Islamic literature, such as *The Adventures of Sinbad the Sailor* and *The Thousand and One Nights.* The richness of these thousand-year-old accounts has inspired many writers and filmmakers. Ibn Fadlan was an Arab chronicler, and in 921 the caliph of Baghdad sent him with an embassy to the king of the Bulgars of the Middle Volga. He wrote an account of his journey, and this was called *Risalah.* Like Ibn Battuta's *Rihla,* the *Risalah* is of great value because it describes the places and people of northern Europe, in particular a people called the *Rus* from Sweden.

He wrote, "I have seen the Rus as they came on their merchant journeys and encamped by the Volga. I have never seen more perfect physical specimens, tall as date palms, blonde and ruddy; they wear neither tunics nor caftans, but the men wear a garment which covers one side of the body and leaves a hand free."

This book inspired novelist Michael Crichton to write *Eaters of the Dead,* on which the film *The 13th Warrior* was based. Many other Muslim travelers have inspired people in modern times. Ibn Battuta's legacy now includes the world's largest shopping mall named after him in Dubai, as well as a music CD by the German band Embryo with tracks including "Beat from Baghdad."

Ibn Battuta

Ibn Battuta was only 21 on June 13, 1325, when he set out alone on his donkey at the beginning of a 3,000-mile overland journey to Mecca from Tangier in Morocco. He left his family, friends, and hometown, and would not see them again for 29 years. Some he never saw, because the plague reached them before he returned. He went to the four corners of the Muslim world by walking, riding, and sailing over 75,000 miles, through more than 40 modern countries; many know him as the Muslim Marco Polo.

His accounts have placed the medieval world before us, so we know that gold traveled from south of the African Sahara into Egypt and Syria; pilgrims continuously flowed to and from Mecca; shells from the Maldives went to West Africa; pottery and paper money came west from China. Ibn Battuta also flowed along with the wool and the wax, gold and melons, ivory and silk, sheikhs and sultans, wise men and fellow pilgrims. He worked as a *qadi*, a judge, for sultans and emperors, and as a pious Muslim his driving force was faith and learning on the road of life, in great Islamic cities like Cairo and Damascus, and from the great minds of his time.

It is reported, albeit some contest it, that Prophet Muhammad had said, "Seek knowledge, even as far as China," and Ibn Battuta took this literally. His journey became a kind of grand tour, mixing prayer, business, and adventure, and as a Muslim he understood codes of conduct throughout 14th-century Eurasia, which included equality, charity, trade, good citizenship, the pursuit of knowledge, and faith.

When he returned to his native city three decades later, he was a famous wayfarer, recounting stories of distant, exotic lands. Some simply would not believe him when he talked of these places. It was then that the sultan of Fez, Abu 'Inan, asked him to write down his experiences in a *Rihla*, a travel book, and with a royal scribe, Ibn Juzayy, he completed the task in two years.

He has left us with one of the greatest historical books ever; in particular, his account of medieval Mali in West Africa is the only record we have of it today. Now we can see his world of the 14th century with our own eyes.

LEFT: *A manuscript shows tenth-century Ibn Fadlan's* Risalah, *which was an account of his journey into northern Europe.* RIGHT: *An artistic re-creation shows Ibn Battuta making supplications after reading the Quran. Muslims usually make these supplications, or requests, to God, after reading the Quran or finishing prayers.*

07 NAVIGATION

It is widely believed that the Chinese developed the compass for use in feng shui, and mariners later developed it further for use in navigation. The earliest evidence of the magnetic compass is found in the Persian work called *Collection of Stories* by Muhammad al-Awfi.

The year was 1233, and the voyage was over the Red Sea and the Persian or Arabian Gulf. The compass was described as follows: "a fish made of iron is rubbed with a magnetic stone and then put in a bowl filled with water; it rotates until it stops, pointing to the south."

The first full description of the use of the magnetic compass for navigation in the Islamic world was by Baylak al-Qibjaqi in his *The Book of Treasure for Merchants Who Seek Knowledge of Stones,* written in Egypt in 1282. He described the use of a floating compass during a sea voyage from Tripoli in Syria to Alexandria in 1242. He wrote that "an iron needle is joined crosswise with a rush and put in a bowl filled with water. Then a magnetic stone is brought close to this device, and the hand holding the magnetic stone describes a circle clockwise above it. The cross of the needle and the rush follows this move. When the magnetic stone is suddenly removed, the needle is supposed to be aligned with the meridian."

Willow wood or pumpkin "fish" designs that had magnetic needles were also mentioned. These were sealed with tar or wax to make them waterproof, as they floated on the water. They were known as wet compasses, but there was also a dry compass. Here, two magnetized needles are on opposite sides of a disc of paper, and in the middle is something like a funnel. This funnel rotates on an axis, which is pivoted in the middle of a box sealed with a plate of glass to prevent the disc of paper from dropping.

These designs and uses of the compass were taken to Europe by Muslim traders, who developed them further.

Master Navigators

As well as having developed navigation instruments, Muslims were also master navigators. Ibn Majid was such a person from Najd, in Arabia, in the 15th century. It ran in the family, as both his father and grandfather were *Mu'allim,* or masters of navigation, knowing the Red Sea in detail. He knew almost all the sea routes from the Red Sea to East Africa, and from East Africa to China. He wrote at least 38 treatises about those, some in prose, others in poetry, of which 25 are still available. These talk about astronomical and nautical subjects, including lunar mansions, sea routes, and the latitudes of harbors.

But the most important navigator was the 16th-century admiral Piri Reis, whose 450-year-old book of sailing instructions, *Kitab-i-bahriyye*, is known in translation by three names, *The Book of the Mariner*, *The Naval Handbook,* and *The Book of Sea Lore*. It was recently published in 1991 by the Turkish Ministry of Culture and Tourism, and this new printing includes a color copy of the original manuscript, with the Ottoman text translated into Latin, modern Turkish, and English.

The Book of Sea Lore by Piri Reis was a mariner's guide to the coasts and islands of the Mediterranean, and paved the way for modern sea travel. It was also known as a portolan, and was a comprehensive guide to nautical instructions for sailors, containing maps covering coastlines, waterways, ports, and distances of the Mediterranean coast. It gave sailors instructions and good knowledge of the Mediterranean coast, islands, passes, straits, bays, where to shelter in the face of perilous seas, and how to approach ports and anchor. It also provided directions and precise distances between places.

It is the only full and comprehensive manual covering the Mediterranean and Aegean Seas ever made, with 219 detailed charts. It was the pinnacle of more than 200 years of development by Mediterranean mariners and scholars.

There were two editions of the handbook; the first came out in 1521, the second five years later. The first was primarily aimed at sailors; the second, on the other hand, was a gift from Piri Reis to the sultan. It was full of craft designs, its maps drawn by master calligraphers and painters, and already in the 16th century it had become a collector's item. For more than a century, copies were produced, becoming even more luxurious as they gave good descriptions of storms, the compass, portolan charts, astronomical navigation, the world's oceans, and the lands surrounding them. Interestingly, it also referred to European voyages of discovery, including the Portuguese expedition to the Indian Ocean and Columbus's discovery of the New World.

"City views, costumes, ships, flora and fauna, and even portraits found on certain maps have brought them close to works of art."

PROFESSOR GUNSEL RENDA
HACETTEPE UNIVERSITY, ANKARA, TURKEY

There are around 30 manuscripts of this *Book of Sea Lore* scattered all over libraries in Europe, but most are of the first version.

More can be read about Piri Reis in the Maps section of this chapter, and also about Zheng He, the Chinese Muslim sea explorer.

Muslim sailors developed navigation instruments and became master navigators at sea.

08 Naval Exploration

More than 630 years ago, a man was born who would revolutionize exploration by sea. His name was Zheng He, and he became the "Admiral of the Chinese Fleet." According to Gavin Menzies, author of *1421*, the recent book on Zheng He, he sailed throughout the Indian Ocean. He navigated to Mecca, the Persian Gulf, East Africa, Ceylon (Sri Lanka), and Arabia. He did so decades before the journeys of Christopher Columbus or Vasco da Gama, whose ships were less than a quarter of the size of those of Zheng He.

Zheng He was a Muslim who helped present China to the world. During 28 years of travel, his junk ships visited 37 countries, making seven monumental sea voyages in the name of trade and diplomacy. The expeditions covered a distance of more than 50,000 kilometers (31,000 miles), and his first fleet included 27,870 men on 317 ships. It was like a small town or an entire football stadium on the move. Sailing with such a large fleet into largely unknown waters required great skill in management as well as sailing. There was no margin for error, and what he achieved is comparable to us going to the moon today.

Zheng He was born and named Ma He, and his Muslim father and grandfather made pilgrimages to Mecca, which enabled him to grow up speaking both Arabic and Chinese. As a boy, he was taken from his town of Kunming, in Mongolia, by the invading Chinese Ming dynasty. He was castrated and became a eunuch, employed as a functionary

A stone statue of Zheng He stands in Nanjing's Zheng He Memorial Hall.

in the imperial household, assigned to the retinue of Duke Yan or Zhu Di, a prince. Zhu Di later seized the throne and became the Emperor Yong Le.

Gavin Menzies said that "Zheng He was a devout Muslim besides being a formidable soldier, and he became Zhu Di's closest advisor. He was a powerful figure, towering above Zhu Di; some accounts say he was over two metres tall and weighed over a hundred kilograms, with a 'stride like a tiger's.' "

Through dedicated service and for accompanying the duke on successful military campaigns, Ma He was awarded the supreme command of the Imperial Household Agency and was given the surname Zheng. He was also known as the Three-Jewel Eunuch *(San-Pao Thai-Chien)*, which has Buddhist connotations (even though he was a Muslim) and was a mark of honor for this high palace official.

There were quite a number of reasons he went on the seven great "Treasure Ship" voyages. There was scientific discovery and the search for gems, minerals, plants, animals, drugs, and medicine, which became increasingly important as the voyages multiplied. The emperor wanted to improve navigational and cartographical knowledge of the world and had a desire to show all foreign countries that China was the leading cultural and economic power. So overseas trade was encouraged; this meant other countries seeing the massive Chinese ships, thereby boosting their prestige. Other nations swore allegiance to China through diplomacy, with local and regional leaders acknowledging "overlordship" of the imperial power. A country would then send envoys to pay tribute to the emperor.

Zheng He made these voyages between 1405 and 1433, and he was joined by two other able eunuch leaders, Hou Hsien and Wang Ching-Hung.

What was amazing about the voyages was that they were long and highly organized. Zheng He wrote, "Sixty-two of the largest ships were 440 feet

Fifteenth-century Zheng He and his crew used this navigation chart to chart the routes taken during his voyages.

long, and at broadest beam 180 feet." These are Ming units (0.31 meter, or 1.02 feet), so it would be 137 meters (449 feet) long and 56 meters (184 feet) wide in our measurements. Zheng He also wrote that they were manned by 450 to 500 men each, including sailors, clerks, interpreters, soldiers, artisans, medical men, and meteorologists. On the fourth voyage, he set out with 30,000 men to Arabia and the mouth of the Red Sea.

The Chinese shipbuilders realized that the gigantic size of their ships would make maneuvering difficult, so they installed a "balanced rudder" that could be raised and lowered for greater stability. Shipbuilders today do not know how the Chinese built a framework without iron that could carry a 400-foot-long vessel. Some doubted the ships ever existed, but in 1962, the rudder post of a treasure

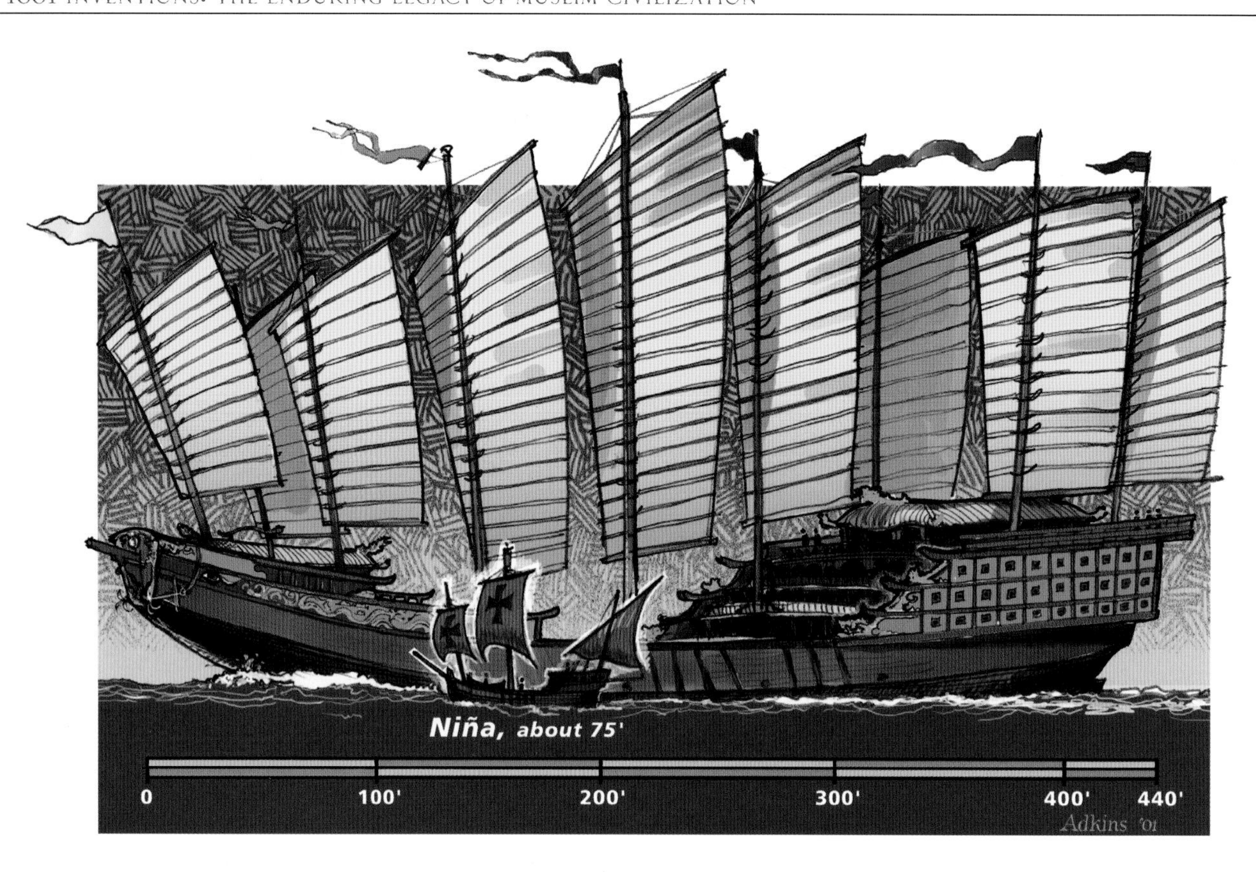

A rendering compares Christopher Columbus's boat, which was 23 meters (75 feet) long, with Zheng He's boat, which was 134 meters (440 feet) long.

was found in the ruins of one of the Ming boatyards in Nanjing. It was 36 feet long. Doing reverse calculations using the proportions of a typical traditional junk, the estimated hull for this rudder would be 152 meters (500 feet).

On board these mighty vessels were large quantities of cargo including silk goods, porcelain, gold and silverware, copper utensils, iron implements, and cotton goods. They also carried live animals—including giraffes; zebras, or "celestial horses"; oryx, or "celestial stags"; and ostriches, or "camel-birds"—watertight bulkheads to hold live fish and also provide bathhouses, and otters that were sent out to round up fish into large nets. The ships were able to transfer water from floating water tankers to their holds, and they could communicate by means of flags, lanterns, bells, carrier pigeons, gongs, and banners.

A Ming account of the voyages said: "The ships which sail the Southern Sea are like houses. When their sails are spread they are like great clouds in the sky," and they were described collectively as "swimming dragons," because all were dotted with dragons' eyes to help them "see."

By the end of Zheng He's fleet's seven voyages, China was unrivaled in naval technology and power, and China also benefited from many exotic species being introduced, such as the first giraffe from Africa. It was initially misidentified as the qilin, the unicorn central to Chinese mythology. According to Confucian tradition, a qilin was a sage of the utmost wisdom, and benevolence was felt in its presence.

It is believed that Zheng He died in India on his way home in 1433. With his death and the revival of Confucian philosophy, the Chinese Empire became inward-looking and eventually seagoing trade was banned. In less than a hundred years, it was a capital offense to set sail from China in a multi-masted ship. In 1525, the Chinese emperor ordered the destruction of all oceangoing ships. The greatest navy in history, which once had 3,500 ships (the U.S. Navy today has around 300), was gone.

In 1985, at the 580th anniversary of Zheng He's voyages, his tomb was restored. The new tomb was built on the site of the original one in Nanjing, and reconstructed according to the customs of Islamic teachings. At the entrance to the tomb is a Ming-style structure, which houses the memorial hall. Inside are paintings of the man himself and his navigation maps.

> *"We have . . . beheld in the ocean huge waves like mountains rising sky high, and we have set eyes on barbarian regions far away, hidden in a blue transparency of light vapors, while our sails, loftily unfurled like clouds, day and night continued their course rapid like that of a star, transversing the savage waves as if we were treading a public thoroughfare."*
>
> ZHENG HE IN HIS BIOGRAPHY, *MING SHIH*

To get to the tomb, there are newly laid stone terraces and steps. The stairway to the tomb is of 28 stone steps divided into four equal sections. This represents Zheng He's seven journeys westward. Inscribed on top of the tomb are the Arabic words *Allahu Akbar,* or God Is Great.

There were no other ships in the world as big as or with as many masts as Zheng He's. These were floating cities on the move. Most of the ships were built at the Dragon Bay shipyard near Nanjing, the remains of which can still be seen today.

Zheng He's Seven Epic Voyages

1. 1405-1407: Visited Champa (Indochina), Java and Sumatra, Ceylon and Calcutta, India.

2. 1407-1409: Sailed to Siam and India, stopping at Cochin.

3. 1409-1411: Went to numerous places in the East Indies using Malacca as a base, and visited Quilon in India for the first time.

4. 1413-1415: The fleet split up. Some went to the East Indies again, others (based in Ceylon) went to Bengal, the Maldives, and the Persian sultanate of Ormuz. This voyage provoked so much interest that a huge number of envoys visited Nanjing in 1416. A large fleet the following year had to take them home again.

5. 1416-1419: The Pacific squadrons went to Java, Ryukyu, and Brunei. The Indian-based ones went to Ormuz, Aden, Mogadishu, Mombasa, and other East African ports. It was from this trip that the giraffe was brought back.

6. 1421-1422: Sailed the same seas as before, including more ports in southern Arabia and East Africa. The fleet visited 36 states in two years from Borneo in the east to Zanguebar (Zanzibar) in the west. This suggests the squadrons had split up again, using Malacca as the main rendezvous port, which, before the advent of radio, was an incredible achievement.

7. 1431-1433: This final voyage, when Zheng He was 60, established relations with more than 20 realms and sultanates from Java to Mecca to East Africa. No one knows how far down the East African coast the Chinese went, but there are accounts that they rounded the Cape.

09 GLOBAL COMMUNICATION

Communicating information—whether electronic or paper-based—can be a risky process. To avoid vital secrets falling into the wrong hands, messages are scrambled (encrypted) so that only someone with the right code can unscramble it.

A famous case of encryption was during World War II when the Germans used a typewriter-like machine, called Enigma, to encrypt military messages before playing them on the radio. These were decrypted by a group of Polish code breakers from the Cipher Bureau and British code breakers from Bletchley Park.

> *"The birth of cryptanalysis required a society which has reached a high standard of development in three disciplines, namely linguistics, statistics and mathematics. These conditions became available at the time of al-Kindi who had command of these three disciplines and more."*
>
> DR. SIMON SINGH, *THE CODE BOOK,* 1999

These 20th-century problem solvers were carrying on the code-breaking tradition first written about by ninth-century polymath Al-Kindi from Baghdad. At this time the mail was delivered by birds, so messages had to be light in weight, and the confidential ones were encrypted. He revolutionized cryptography when he wrote *A Manuscript on Deciphering Cryptographic Messages*. Part of this included a description of the method of frequency analysis, which means he noticed that if a normal letter is replaced with a different letter or symbol, the new letter will take on all the characteristics of the original one. If we look at the English language the letter *e* is the most common, accounting for 13 percent of all letters. So, if *e* is replaced by the symbol #, # would become the most common symbol, accounting for 13 percent of the new symbols. A cryptanalyst can then work out that # actually represents *e*.

From studying the Arabic text of the Quran closely, Al-Kindi noticed the characteristic letter frequency, and laid cryptography's foundations, which led many cryptographers from European Renaissance states to devise other schemes to combat it. Even though Al-Kindi developed methods that enabled greater encryption and code breaking 1,100 years ago, the actual word "cryptanalysis" is relatively recent and was first coined by a man named William Friedman in 1920.

Frequency analysis is now the basic tool for breaking classical ciphers or codes that use the basic, plain text alphabet. It relies on linguistic and statistical knowledge of plain text language, and good problem-solving skills.

Modern ciphers are a lot more complex, but back in the days of World War II, Britain and America recruited code breakers by placing crossword puzzles in major newspapers and running contests for those who could solve them the fastest.

Air Mail

Today, we take postal communication for granted, but once, the pigeon post was the only "air mail" available.

In a book about carrier pigeons, the Muslim scholar Ibn 'Abd al-Dhahir wrote that normally there would be about 1,900 pigeons in the lofts of the citadel of Cairo, the communication nerve center of the time.

Al-Nuwayri, a Muslim chronicler, tells the story of a tenth-century Fatimid caliph called 'Aziz who one day, in Cairo, felt a desire to eat fresh cherries of a kind grown in Antioch. The order was sent by carrier pigeon to Baalbek, near Antioch, and from there, 600 pigeons were released, each with one cherry in a silk bag tied to each leg. Just three days after expressing his desire, the caliph was served a large bowl containing 1,200 fresh cherries from Lebanon, which had arrived by special "air mail" delivery.

An Enigma machine was used to encrypt military messages in World War II. It was Al-Kindi in the ninth century who laid the foundation of cryptography.

BREAKING CODE

"One way to solve an encrypted message, if we know its language, is to find a different plain text of the same language long enough to fill one sheet or so, and then we count the occurrences of each letter. We call the most frequently occurring letter the 'first,' the next most occurring letter the 'second,' the following most occurring the 'third,' and so on, until we account for all the different letters in the plaintext sample . . . Then we look at the cipher text we want to solve and we also classify its symbols. We find the most occurring symbol and change it to the form of the 'first' letter of the plain-text sample, the next most common symbol is changed to the form of the 'second' letter, and so on, until we account for all symbols of the cryptogram we want to solve."

AL-KINDI IN HIS NINTH-CENTURY *A MANUSCRIPT ON DECIPHERING CRYPTOGRAPHIC MESSAGES*

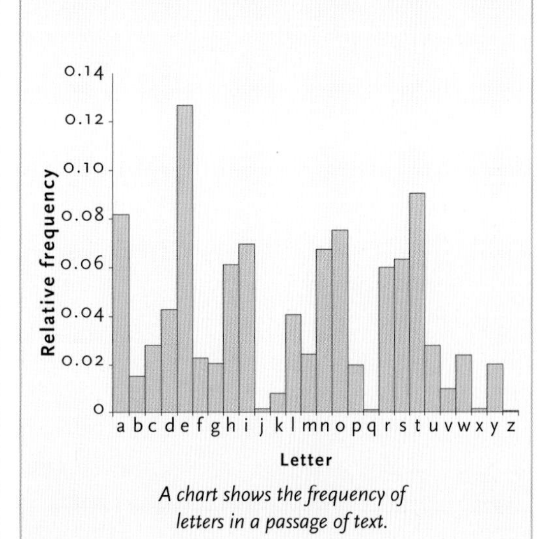

A chart shows the frequency of letters in a passage of text.

10 War and Weaponry

Military talk in the 13th century was sophisticated, and discussions included grenades, sulfur bombs, cannons, rockets, and torpedoes. One of the most important books on military technology was *The Book of Horsemanship and Ingenious War Devices* by the Syrian scholar Hasan al-Rammah, written around 1270. It was packed full of weapon diagrams, including the first documented rocket, a model of which is exhibited at the National Air and Space Museum in Washington, D.C.

The Chinese invented the first gunpowder. They developed saltpeter, one of gunpowder's ingredients, but probably only used it in fireworks. As Amani Zain in the BBC's *What the Ancients Did for Us* said, "research has shown that Muslim chemists did develop a powerful formula for gunpowder and may well have used it in the first firearms."

The Chinese did not use it in explosions because they could not get the right proportions, nor could they purify potassium nitrate. It was not until 1412 that Huo Lung Ching wrote the first Chinese book detailing proportions for explosives. About a hundred years earlier, Hasan al-Rammah's book was the first to explain the purification procedure for potassium nitrate, and it describes many recipes for making exploding gunpowder.

LEFT: *A 13th-century manuscript of Al-Rammah depicts a trebuchet for flinging missiles.* CENTER: *Another trebuchet is shown in a 14th-century* Manual on Armory *by Ibn Aranbugha al-Zardkash.* RIGHT: *A 14th-century* Manual on Armory *by Ibn Aranbugha al-Zardkash shows a pedestal crossbow.*

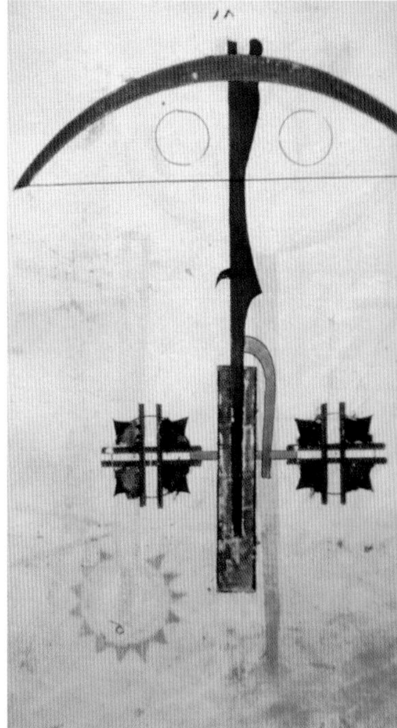

"For the Islamic armies led by Baybars in 1249 the use of gunpowder in war proved decisive against the invading Crusaders. At the battle of al-Mansura in Egypt, Muslim incendiary devices were so terrifying and destructive that the French Crusader Army was routed and King Louis IX was taken prisoner," said Amani Zain on the BBC's *What the Ancients Did for Us*.

By the 15th century, the cannons used by the Ottomans were awesome and today the Fort Nelson Museum in Portsmouth, U.K., has a huge bronze cannon weighing 18 tons. It was originally cast in two pieces, and screwed together, to make it easier to transport because its overall length is more 5 meters (16.4 feet), with a diameter of 0.635 meter (about 2 feet). The length of the barrel alone is more than 3 meters (9.8 feet) and the gunpowder reservoir is 0.248 meter (0.814 feet) in diameter. No such large split guns existed in Europe before this one.

This novel cannon was cast in 1464 by the order of Sultan Mehmed II. He was very interested in firearms, especially in cannons. During his siege of Constantinople, he ordered his cannon master to cast a larger cannon than had ever been seen before and this one could fire cannon balls up to a mile.

On the muzzle is inscribed in Arabic: Help, O Allah, The Sultan Mohammed Khan son of Murad. The work of Kamina Ali in the month of Rajab. In the year 868 (of the *Hejira* calendar, or 1464.)

Sultan Mehmed's cannon ended up in a Portsmouth museum because, after unsuccessful attempts by the English for 60 years to persuade the Ottomans to give it, Queen Victoria personally asked Sultan Abdul Aziz for it during his visit to Europe. One year later, the sultan sent it as a gift. It was transported from the Dardanelles to London and placed in the museum in 1868. Queen Victoria perhaps wanted it because it was known as the "most important cannon in Europe."

Muslims also built rockets and the first torpedo. The rocket was the so-called self-moving and combusting egg and the torpedo was a cleverly modified rocket designed to skim along the surface of the water. It was called "the egg, which moves itself and burns," and Hasan al-Rammah's illustrations and text show two sheet-iron pans were fastened together and made tight with felt. This made a flattened pear-shaped vessel that was filled with "naphtha, metal filings, and good mixtures [probably containing saltpeter], and the apparatus was provided with two rods and propelled by a large rocket." The two rods would probably have acted as tail rudders, while a spear at the front would lodge into the wooden hull of an enemy ship to secure it before exploding.

GUNPOWDER

To make gunpowder: "Take from white, clean and bright [or fiery] *barud* [saltpeter] as much as you like and two new [earthen] jars. Put the saltpeter into one of them and add water to submerge it. Put the jar on a gentle fire until it gets warm. Skim off the scum that rises [and] throw it away. Make the fire stronger until the liquid becomes quite clear. Then pour the clear liquid into the other jar in such a way that no sediment or scum remains attached to it. Place this jar on a low fire until the contents begin to coagulate. Then take it off the fire and grind it finely."

HASAN AL-RAMMAH DESCRIBES A COMPLETE PROCESS FOR THE PURIFICATION OF POTASSIUM NITRATE IN HIS BOOK *THE BOOK OF HORSEMANSHIP AND INGENIOUS WAR DEVICES*

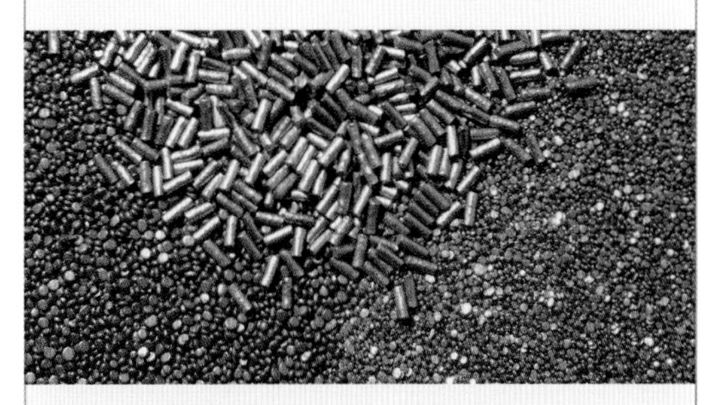

Many gunpowder recipes are included in Hasan al-Rammah's 14th-century writings.

11 Social Science and Economics

Ibn Khaldun was one of the last scholars of the classic medieval Muslim civilization. In many ways his writing, family story, and life reflect the changes that caused the decline, and eventual fall, of medieval Islamic civilization. Born in Tunis in 1332 and died in Cairo in 1406, he explained how Islamic civilization was undone.

He began by looking at the various invaders who undermined it, and how his ancestors were themselves affected by such invasions. Up until 1248, they had lived in Seville—until the Christian Spanish advanced to reoccupy their ancestral home and his family's home was lost as they fled. His ancestors escaped to North Africa, where his parents died of the plague.

Ibn Khaldun then left his native Tunisia for Egypt, in 1382, and his family followed him, but they fell victim to yet another of the scourges of the day, piracy. His family were killed or taken captive, and he never saw them again, nor did he ever say another word about them.

Ibn Khaldun wrote Al-Muqaddimah, *or An Introduction to History, during a period of enforced exile in Algeria.*

"Human beings require cooperation for the preservation of the species, and they are by nature equipped for it. Their labor is the only means at their disposal for creating the material basis for their individual and group existence. Where human beings exist in large numbers, a division of activities becomes possible and permits greater specialization and refinement in all spheres of life. The result is umran *[civilization or culture], with its great material and intellectual achievements, but also with a tendency toward luxury and leisure, which carries within itself the seeds of destruction."*

FROM 14TH-CENTURY IBN KHALDUN'S *AL-MUQADDIMAH*

The last years of Ibn Khaldun's life correspond to possibly the last years of classical Muslim scholarship and bright civilization. By the early 15th century, having lost Spain and Sicily, and having endured the Crusades and the Mongol invasions, the Muslim world now suffered the most devastating onslaught of Timur the Lame, also known as Tamurlane, the effects of which were in part witnessed by Ibn Khaldun.

Despite the demands of his work as a judge and diplomat, he managed to continue his academic research, and produced his world history called *Book of the Lessons and Archive of Early and Subsequent History*. This became known as *Al-Muqaddimah* or *Introduction*.

The *Al-Muqaddimah* is a gigantic endeavor, a discourse on universal history. Ibn Khaldun explored and implemented the idea that the documentation of history is not just a list of correct facts, but is dependent on who is interpreting them, what region they come from and when, as well as their impartiality. This was a revolutionary approach and his methodology is still used by historians today. He completely rejected partiality and unchecked facts. In this way, he brought in a rigorous new dimension to scholarship and the social sciences, which provided the basis for arguments before they could become accepted as fact.

One of his best-known studies relates to the rise and decline of civilizations, and it is this that laid down the foundations of social science, the science of civilization and sociology. He explains how civilization and culture breed their own decline following a cyclical nature. They rise up because of a common need for protection and domination, reach a peak when the social bonds are at their strongest, before declining, and perish when group support and social bonds become diluted because of unhealthy competition and corruption at times of prosperity.

In Ibn Khaldun's mind, the only thing that could counteract the disintegrative forces, inherent in every nation, was religion. He said that Islam gave a community a lasting spiritual content, a complete answer to all problems of life; that it furnished the complete answer to his empirical inquiry into the organization of the human race.

IBN KHALDUN

Ibn Khaldun was a judge, university scholar, and diplomat, known for his works of sociology, economics, commerce, history, philosophy, political science, and anthropology. He wrote his famous *Al-Muqaddimah* or *Introduction to a History of the World* during a period of enforced exile, taking refuge in Algeria, while running from Fez because of political unrest. The first volume gave a profound and detailed analysis of Islamic society, referring and comparing it to other cultures, and he traced the rise and fall of human societies in a science of civilization.

A contemporary artist's rendering of Ibn Khaldun

He was ahead of his time in economic theory. Four centuries before Adam Smith, Ibn Khaldun had already concluded that labor was the source of prosperity. He had also distinguished between the direct source of income in agriculture, industry, and commerce, and the indirect source of income of civil servants and private employees. These concepts may seem like second nature today, but they were groundbreaking 700 years ago, and have paved the way for classical economics and models relating to consumption, production, demand, cost, and utility.

Chapter Eight

"It is He who created the Night and the Day,
and the sun and the moon:
all [the celestial bodies] swim along,
each in its orbit."

QURAN (21:33)

UNIVERSE

ASTRONOMY • OBSERVATORIES • ASTRONOMICAL INSTRUMENTS
ASTROLABE • ARMILLARY SPHERE • SIGNS FOR WISE PEOPLE
THE MOON • LUNAR FORMATIONS • CONSTELLATIONS • FLIGHT

The night sky and the concept of the universe have inspired poetry, music, philosophy, and science for thousands of years, and it was no different in the Muslim world of a millennium ago.

The wonders of the heavens inspired the first successful manned flight 1,200 years ago as others were keeping a watchful eye on the night sky. Muslims needed to know the times of the daily prayers that depend on the sun's position, the direction of Mecca from every geographical location, and the moon's cycle for the Muslim lunar calendar.

From these impetuses Muslims also made epoch-making discoveries such as the first record of a star system outside our own galaxy and the third inequality of the moon's motion, and they developed instruments that laid the foundation for modern-day astronomy. These included celestial globes, armillary spheres, universal astrolabes, and sextants. This all began in the eighth century with the first observatory and accurate astronomy tables.

Today, these stargazing scholars, along with other eminent Muslims, are remembered as we look up because areas of the moon bear their names and more than 165 stars have Arabic titles.

OPPOSITE: *A manuscript depicts the sextant in Taqi al-Din's famous 16th-century Istanbul observatory.*

01 Astronomy

Why did Muslims spend so much time looking at the sky? First, there was a practical need to determine the times of the daily prayers throughout the year, and these times depended on the sun's position in the sky, as prayers were at dawn, midday, afternoon, sunset, and evening. Muslims also needed to know the direction of Mecca from every geographical location, and this could be done by observing the position of the sun and moon. And the Quran had some major revelations about the heavens that needed to be explored. A final motivation was the calendar.

The Muslim calendar is a lunar calendar, so the months change according to the phases and position of the moon. Each month begins with the first sighting of the crescent moon. This is especially important in the Muslim holy month of Ramadan when Muslims fast during the day for one month.

From all these religious motivations, astronomy became a main concern for Muslim scholars a thousand years ago, and what they produced lasted for centuries. During the Renaissance, Regiomontanus, a celebrated 15th-century mathematician and astronomer, had to rely on Muslim books for his sources, while Nicolaus Copernicus referred repeatedly in his book *De Revolutionibus* to Al-Zarqali and Al-Battani, Muslim astronomers of the eleventh and tenth centuries.

Most of the great astronomical discoveries happened in observatories in the East, but for the 300 years that Muslims ruled Toledo in Spain, it was the center of world astronomy. The new astronomical tables made here were used in Europe for two centuries.

Observing the sky was an intense activity, and it happened on a daily basis when the sun and moon would be studied as they tracked across the heavens. This helped to determine solar parameters and produced information on the longitudes and latitudes of the planets whose measurements were made at intervals of two weeks.

In ninth-century Baghdad Caliph Al-Ma'mun set up an intellectual academy, the House of Wisdom, to translate manuscripts, which you can read about in the School chapter. Among the first works translated into Arabic was the Alexandrian astronomer Ptolemy's *Great Work*, which

TOP: *Muslim astronomers were skillful instrument makers, paving the way for the development of modern astronomy.* BOTTOM: *Scholars made daily observations of the sun and moon, along with measurements of the positions of the planets every two weeks.*

described a universe in which the sun, moon, planets, and stars revolved around Earth. *Almagest*, as the work was known to Arabic scholars, became the basis for cosmology for the next 500 years. Yet the Muslims developed and went far beyond the Greek mathematical methods found in this treatise. In particular in the field of trigonometry, the advances made in Muslim lands provided the essential tools for the creation of Western Renaissance astronomy.

There were many Muslim astronomers who contributed hugely to the field of studying the heavens, laying the foundation for astronomers in the future, but these eminent individuals stand out:

■ **Al-Battani,** known in the West as Albategnius, who died in 929, wrote *The Sabian Tables,* which was a very influential work for centuries. His work also included timing of the new moons, calculation of the length of the solar and sidereal year, the prediction of eclipses, and the phenomenon of parallax. He also popularized, if not discovered, the first notions of trigonometric ratios used today, and made serious alterations to Ptolemy's theories, which had been used as the main astronomical works until then. He made the important discovery that the motion of the solar apogee, or the position of the sun among the stars at the time of its greatest distance from the Earth, was not what it had been in the time of Ptolemy. The Greek astronomer placed the sun at longitude 65 degrees, but Al-Battani found it at longitude 82 degrees. This discrepancy was a distance too great to be accounted for by any inaccuracy of measurement, and today we know it is because the solar system is moving through space. Then, though, it was still believed that the Earth was the center of the universe, so this conclusion could not be made.

■ **Al-Biruni** lived between 973 and 1050. He stated that the Earth rotated around its own axis, calculated the Earth's circumference, and fixed scientifically the direction of Mecca from any point of the globe. He also wrote about 150 works, including 35 treatises on pure astronomy, but only six astronomical writings have survived.

"After having lengthily applied myself in the study of this science, I have noticed that the works on the movements of the planets differed consistently with each other, and that many authors made errors in the manner of undertaking their observation, and establishing their rules. I also noticed that with time, the position of the planets changed according to recent and older observations; changes caused by the obliquity of the ecliptic, affecting the calculation of the years and that of eclipses. Continuous focus on these things drove me to perfect and confirm such a science."

AL-BATTANI, ASTRONOMER AND MATHEMATICIAN (858-929)

■ **Ibn Yunus** made observations for nearly 30 years from 977 using a large astrolabe nearly 1.4 meters (4.6 feet) in diameter. He recorded more than 10,000 entries of the sun's position throughout all these decades.

■ **'Abd al-Rahman al-Sufi** was a Persian astronomer who lived during the tenth century. You can read more about him in the Constellations section of this chapter.

Al-Farghani was one of Caliph Al-Ma'mun's astronomers who wrote about the astrolabe, explaining the mathematical theory behind the instrument and correcting the faulty geometrical constructions of the central disc that were current then. His most famous work, the *Book on Sun Movement and Encyclopedia of Star Science,* on cosmography, contains 30 chapters, including a description of the inhabited part of the Earth, its size, and the distances of the heavenly bodies from the Earth and their sizes.

Al-Zarqali, known as Arzachel or Azarquiel in Europe, died in 1087. He prepared the Toledan Tables and also made a sophisticated astrolabe that could be used at any geographical location.

Jabir ibn Aflah, who died in 1145, was the first to design a portable celestial sphere to measure celestial coordinates. Jabir is specially noted for his work on spherical trigonometry.

Ibn Rushd, from 12th-century Córdoba, was known in the West as Averroes. He was one of the most famous doctors in Córdoba, but he was also an astronomer, and he is believed to have discovered sunspots.

Ibn al-Shatir was a 14th-century astronomer. In the case of lunar motion, he corrected Ptolemy, whose imagined moon approached far closer to the Earth than did the actual moon. After noting, as did other Muslim astronomers before him, the shortcomings of the Greeks' planetary theory, Ibn al-Shatir said, "I therefore asked Almighty God to give me inspiration and help me invent models that would achieve what was required, and God, may He be praised and exalted, all praise and gratitude to Him—did enable me to devise universal models for the planetary motions in longitude and latitude and all other observable features of their motions, models that were free—thank God—from the doubts surrounding previous models."

Traces of medieval Islamic astronomy are still seen today. The words zenith, azimuth, and the names of stars in the Summer Triangle, Vega, Altair, Deneb, are all of Arabic origin.

THE BIRTH OF MODERN ASTRONOMY

Nicolaus Copernicus

Many believe that astronomy died with the Greeks, and was brought to life again by Nicolaus Copernicus, the 15th-century Polish astronomer who is famous for introducing the sun-centered theory of the solar system, which marked the beginning of modern astronomy.

However, many historians now think it is not a coincidence that his models of planetary theory are mathematically identical to those prepared by Ibn al-Shatir more than a century before him. It is known that Copernicus relied heavily on the comprehensive astronomical treatise by Al-Battani, which included star catalogs and planetary tables.

The mathematical devices discovered by Muslims before Copernicus, referred to in modern terms as linkages of constant length vectors rotating at constant angular velocities, are exactly the same as those used by Copernicus. The only, but important, difference between the two was that the Muslims' Earth was fixed in space, whereas Copernicus had it orbiting around the sun. Copernicus also used instruments that were particular to astronomy in the East, like the parallactic ruler, which had previously only been used in Samarkand and Maragha observatories.

A 15th-century Persian manuscript of Nasir al-Din al-Tusi's observatory at Maragha depicts astronomers at work teaching astronomy, including how to use an astrolabe. The instrument hangs on the observatory's wall.

02 OBSERVATORIES

From the beginnings of human awakening, people have marveled at the amazing canopy of stars and at the movement of everything in the sky. Clearly, there was order in the heavens, and many attempts were made to identify the patterns in this order.

This had great significance for life since through these observations came the beginnings of predictive science, and now we can predict the position of the sun in the sky, the moon, the timing of eclipses, the changing position of the planets and the stars.

Muslims were not the first to study astronomy, but they were the first to do it on a large scale, with massive instruments in observatories. Astronomical research was expensive, needing costly equipment and the cooperation of many astronomers. Good work was done previously with small, portable instruments, and Ptolemy carried out his observational work with these.

There was one man, Caliph Al-Ma'mun, who ruled from Baghdad from 813 to 833, who really gave astronomy the patronage and impetus it needed to become a major science. He was the first person to set up observatories. The concept of a fixed location with large and fixed instruments, programs of work, scientific staff made up of several astronomers, and royal patronage or affiliation from the state were all novel ideas introduced by Al-Ma'mun. Nothing comparable can be found before the Muslim astronomers. Not only did Al-Ma'mun build the first observatory in Islam, but he also arguably built the first observatory in the world or in history. Al-Ma'mun was an enlightened leader who played a major part in setting up the House of Wisdom, one of the greatest intellectual academies in history, which you can read about in the School chapter.

The earliest observatories were in the Al-Shammasiyah quarter of Baghdad and on Mount Qasiyun at Damascus, and these led to the emergence of fixed places for specialized and collective work. A major task of such observatories was to construct astronomical tables. These helped in the calculation of planetary positions, lunar phases, eclipses, and information for calendars. They often included explanations of astronomical instruments. Al-Ma'mun's observatories prepared solar and lunar tables and had a star catalog, plus some planetary observations.

At Al-Shammasiyah, astronomers observed the sun, the moon, the planets, and some fixed stars. The results of the work done here were presented in a book called the *Mumtahan Zij*, or *Verified Tables*, whose author is said to have been Ibn Abi Mansour.

Other observatories were built all over the Muslim world, such as the Maragha Observatory founded by Hulagu Khan, the Samarkand Observatory of Ulugh Beg, the Malik Shah Observatory at Isfahan, and the Tabriz Observatory of Ghazan Khan.

The Maragha Observatory was completed in 1263 in Iran to the south of Tabriz, and its foundations are still there. The main work done in Maragha was the preparation of new astronomical tables and the observatory library contained more than 40,000 books. Among the eminent

Many observatories were built in the Muslim world during the golden age of discovery.

astronomers associated with the observatory were Nasir al-Din al-Tusi and Qutb al-Din al-Shirazi, who is credited with the discovery of the true cause of the rainbow. Nasir al-Din al-Tusi prepared the Ilkhanid Tables and the catalog of fixed stars, which remained in use for several centuries throughout the world. An astronomer from Maragha was sent to China, and the dynastic chronicles of the Yuan bear record of how he designed an instrument for observing the heavens, which was erected on the Great Wall.

Ulugh Beg was the 15th-century ruler of the Timurid Empire, which stretched over central and Southwest Asia. As well as being the ruling sultan, he was an astronomer and mathematician, which led him to build a three-story observatory for solar, lunar, and planetary observations in Samarkand.

The Samarkand Observatory was a monumental building equipped with a huge meridian, made of masonry, which became the symbol of the observatory as a long-lasting institution. A trench about 2 meters wide (6.6 feet) was dug in a hill, along the line of the meridian, and in it was placed the segment of the arc of the instrument. The radius of that meridian arc was equal to the height of the dome of the Hagia Sofia Mosque in Istanbul, which was about 50 meters (164 feet). Built for solar and planetary observations, it was equipped with the finest instruments available, including a Fakhri sextant with a radius of 40.4 meters (132.5 feet). This was the largest astronomical instrument of its type. The main use of the sextant was to determine the basics of astronomy, such as the length of the tropical year. Other instruments included an armillary sphere and an astrolabe, which you can read about in this chapter.

Ulugh Beg's work was very advanced for his time and surprisingly accurate. His calculation that the stellar year was 365 days, 6 hours, 10 minutes, and 8 seconds long was only 62 seconds more than the present estimation: an accuracy of 0.0002 percent, which is remarkable.

Observatories were massive, with continuous observational programs, which needed organization and efficiency of administration, so astronomers directed and supervised other members of the staff in their work. Later observatories are

OBSERVATORIES IN THE WEST

Spanish King Alfonso X

Alfonso X, a Spanish king of the second half of the 13th century, tried to carry on the Islamic tradition of building observatories in western Europe, but he didn't succeed—maybe because astrology was frowned upon by the Church and the usefulness of astronomy was questioned. Four centuries later, however, the situation gradually changed and knowledge of astronomy gained depth and breadth, with Europe absorbing all that had gone on in the Islamic world. So much so that the instruments used by the famous 16th-century observational astronomer Tycho Brahe were very similar to those used earlier by Muslim astronomers. His famous mural quadrant was like those developed in eastern Islam.

Astronomers study the moon and the stars in this Ottoman-style painting.

known to have had directors, treasurers, clerks, librarians, and other administrative officers, as well as their staff of scientists.

Even though the main work carried out at Al-Ma'mun's observatories in Al-Shammasiyah and on Mount Qasiyun was the construction of astronomical tables, other original and epoch-making discoveries also occurred, resulting, for example, in the discovery of the movement of the solar apogee. Other remarkable discoveries can be read about in other sections of this chapter.

A magnificent but short-lived observatory was built in the 16th century for Taqi al-Din, one of the most notable scientists in the Muslim world. He had convinced the new sultan, Murad III, to fund the building of the Istanbul Observatory, and it was completed in 1577.

With two outstanding buildings, built high on a hill overlooking the Anatolian section of Istanbul, it had an unobstructed view of the night sky. Like a modern observatory today, the main building held the library and housed the technical staff, while the smaller building contained an impressive collection of instruments built by Taqi al-Din. They included a giant armillary sphere and a mechanical clock for measuring the position and movement of the planets.

Taqi al-Din wanted to update the old astronomical tables describing the motion of the planets, sun, and moon. However, his observatory was destroyed by the sultan for numerous sociopolitical reasons linked to the plague and internal palace rivalry. Despite this, Taqi al-Din left an enormous legacy of books on astronomy, mathematics, and engineering.

As well as building the first observatories, Muslims had among them a pioneering ninth-century Cordoban who built a planetarium. Unlike an observatory where the heavens are studied, a planetarium is a room where images of the stars, planets, and other celestial bodies are projected. Ibn Firnas, better known for his experiments in flight, made a planetarium out of glass in one room of his house, showing the night sky as it was then. This very much resembled today's planetariums, and he even added artificial thunder noise and lightning.

LEFT: *A giant marble sextant sits inside Ulugh Beg Observatory in Uzbekistan.* RIGHT: *The radius of the meridian arc at Ulugh Beg Observatory was equal to the height of the dome of the Hagia Sophia in Istanbul.*

THE HEAVENLY REALMS

A Rich Shared Heritage of Astronomy from East and West

LEGACY: **A wealth of astronomical data, a six-cylinder pump, and mechanical clocks**

LOCATION: **Istanbul, Turkey**

DATE: **16th century**

KEY FIGURE: **Taqi al-Din**

The wonder and glory of the starry skies impressed the scholars of Muslim civilization—but they also looked for order and logic in what they saw. With sophisticated instruments and new mathematical techniques, astronomers made great leaps forward in understanding the universe. They built on ancient Greek ideas, leaving a joint astronomical heritage that we see today in the Greek and Arabic names of many stars.

Caliph Al-Ma'mun began the Muslim tradition of observatory building in the ninth century, when he founded facilities in Baghdad and on Mount Qasiyun in Damascus. Sultan Malikshah built the first large-scale observatory in the Muslim world in Isfahan in the late 11th century.

The most important Islamic observatory was built by the 13th-century astronomer Nasir al-Din al-Tusi in Maragha, Iran. It was once considered the most prestigious observatory in the world, and its foundations are still visible today. Chinese records inform us that an astronomer called Jamal al-Din, who was linked to the 13th-century Maragha Observatory, visited the imperial court in Beijing in 1267 and brought with him several astronomical instruments. He became famous in China and was known as Cha-ma-lu-ting.

Key astronomical tools of Muslim civilization were quadrants and sextants, which scholars used for measuring the altitude of heavenly bodies. Once confirmed, detailed stellar coordinates could be mapped on celestial globes such as those made by Al-Battani. In a tenth-century work he described an instrument called *al-baydha,* or "the egg." It combined elements of a solid celestial sphere with others derived from the tradition of the armillary sphere, allowing him to plot the positions of a large number of stars. Accuracy was essential for maintaining an observatory's reputation. Astronomers thus built larger tools to reduce the percentage of error in measurements. A famous observatory built in the 1420s by the astronomer Ulugh Beg in Uzbekistan had a sextant set into a trench more than three stories deep to protect it from earthquakes.

The 16th-century astronomer Taqi al-Din built a magnificent observatory in Istanbul. It contained an impressive array of instruments for making observations and updating astronomical tables. Huge versions of some tools, including a sextant, were installed.

The astronomers of Muslim civilization challenged some of the ideas they had inherited from ancient Greek scholars. Their observations and mathematical research allowed them to develop new models to which, through translation, Western astronomers gained access. Scientific texts, tables of data, and descriptions of instruments from Arabic into Latin provided the foundations for Western pre-modern and early modern astronomy, paving the way for Nicolaus Copernicus's ideas of the sun-centered solar system published in 1543.

■ *A 16th-century Turkish manuscript from the* Book of the Kings *shows Taqi al-Din and other astronomers with their sophisticated devices at the Istanbul Observatory.*

03 Astronomical Instruments

What Muslims really pioneered were huge observational instruments designed and built to study the heavens, and by using large instruments they reduced the percentage of error in their measurements. The observatory at Damascus had a 6-meter (20-foot) quadrant and a 17-meter (56-foot) sextant, which more than four average-length cars end to end. The Maragha Observatory also had many large instruments including quadrants, armillary spheres, and astrolabes.

Other instruments included celestial globes, quadrants, and sextants. (You can read about astrolabes and armillary spheres in more detail in separate sections in this chapter.) All these instruments used in observatories had to be accurate because the observatories' reputations depended on the results they produced.

Jabir ibn Aflah from Spain, who died in 1145, was the first to design a portable celestial sphere to measure celestial coordinates (called a torquetum), but it was tenth-century astronomer Al-Battani, working in Iraq, who was the main astronomer writing on celestial globes. He did not use his globes as observational instruments; instead he wanted to precisely record celestial data. He described one that was suspended from five rings, which he called *al-baydah* or "the egg," giving detailed directions on how to plot the coordinates of each of 1,022 stars. His treatise was very influential, as it gave details of how stars should be marked onto the globe. This meant that instrument makers around that time could produce a globe to this particular standard.

Al-Battani's treatise was very different from the pre-Ptolemaic design of a celestial globe, which used five parallel equatorial rings and constellation outlines. Instead, al-Battani had a more precise method of charting the stars using the ecliptic and the Equator, and dividing them into small divisions. This method allowed the stars to be given exact coordinates, and of course this increased precision.

The Muslims were skillful tool and instrument makers. An important maker of celestial globes was 'Abd al-Rahman al-Sufi, who was born in 903. He wrote a treatise on the design of constellation images for celestial globe makers that had great influence in the Muslim world as well as in Europe. His other treatises included one on the astrolabe and one on how to use celestial globes.

Many globes were constructed up to the 16th century, and many still exist today, but none prior to the 11th century has survived.

There are many scholars who wrote about astronomical instruments, and one of these was Abu Bakr ibn al-Sarraj al-Hamawi, who died in Syria in circa 1329. He wrote several books on scientific instruments and

Muhammad ibn Hilal made this 13th-century brass celestial globe in Maragha, Iran.

FROM LEFT: *The reverse and obverse of a 14th-century astrolabic quadrant were created by Ibn Ahmad al-Mizzi, the official timekeeper of the Great Umayyad Mosque in Damascus, Syria.*

geometrical problems, while also inventing a quadrant called *al-muqantarat al-yusra*. He dedicated much time to writing about the quadrant, and his books include *Treatise on Operations with the Hidden Quadrant* and an opulent-sounding work called *Rare Pearls on Operations with the Circle for Finding Sines*. Despite his accomplishments, especially in the field of scientific instrument making, there has been no single study of him and his works.

Another was Ahmad al-Halabi, who died in 1455, an astronomer from Aleppo in Syria. He wrote on instruments in *Aims of Pupils on Operations with the Astrolabe Quadrant*.

His contemporary 'Izz al-Din al-Wafa'i was primarily a mathematician, muezzin, and *muwaqqit*, or timekeeper, at the Umayyad Mosque in Cairo, and he wrote a staggering number of 40 treatises on mathematics including arithmetic, operations with the sexagesimal ratio, and a large number of works dealing with instruments. Among these was *Brilliant Stars on Operations with the Almucantar Quadrant*.

Sextants and quadrants were used to measure the altitude of celestial objects above the horizon. The quadrant, in particular, was used extensively by Islamic astronomers, who had greatly improved its design.

Muslim astronomers invented quite a few quadrants. There was the sine quadrant, used for solving trigonometric problems and developed in ninth-century Baghdad; the universal quadrant, used for solving astronomical problems for any latitude and developed in 14th-century Syria; the horary quadrant, used for finding the time with the sun; and the astrolabe/almucantar quadrant, a quadrant developed from the astrolabe. Most of these were used in conjunction with the astrolabe.

To measure the obliquity of the ecliptic, the angle between the plane of the Earth's Equator and the plane of the sun's ecliptic, Al-Khujandi in 994 used a device that he claimed was his own invention. It was called the Fakhri sextant because his patron was Fakhr al-Dawla, the Buwayhid ruler of Isfahan. Al-Khujandi claimed to have vastly improved on similar past instruments, because these could only be read in degrees and minutes, while with his instrument, seconds could also be read.

This instrument incorporated a 60-degree arc on a wall aligned along a meridian, the north–south line. Al-Khujandi's instrument was larger than previous such instruments, and had a radius of about 20 meters (65.6 feet).

Taqi al-Din preferred to use a fifth type of quadrant called a mural quadrant rather than Al-Khujandi's Fakhri sextant. This mural quadrant had two graduated brass arcs with a total radius of 6 meters (19.7 feet) only, a staggering 20 meters (65.6 feet) smaller than Al-Khujandi's. These arcs were placed on a wall along a meridian. In order to take a reading, the astronomers aligned the rod or cord on the quadrant with the celestial body, a moon, or the sun, and read off the angle from the mural quadrant.

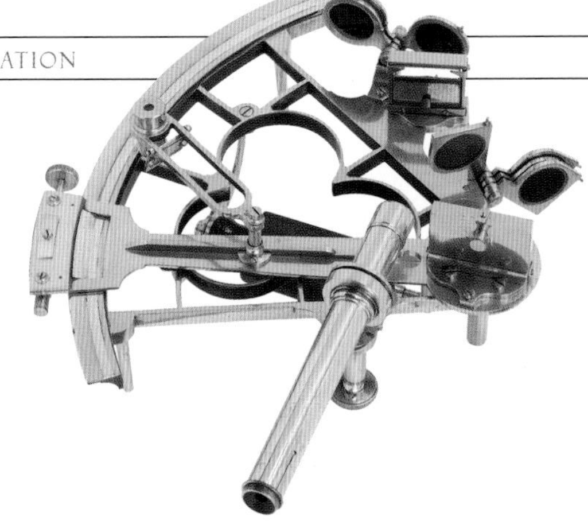

Used for navigation, a sextant is an instrument primarily used to find the angle between a star and the horizon.

These enormous astronomical observation instruments have been significantly downsized in modern times, but their technology laid the foundations for the modern-day sextant, a portable instrument, and before Global Positioning Systems existed, they were the main navigational instruments.

INNOVATIVE INSTRUMENTS

A mural quadrant by Tycho Brahe, 1598

Two of the most influential astronomers in the 16th century were Taqi al-Din from Istanbul, and Tycho Brahe, who built an observatory under the sponsorship of King Frederic II of Denmark in 1576. This observatory was equipped with the best possible and refined instruments of his time, helping him make accurate observations and aiding the discoveries of Johannes Kepler, who was Tycho Brahe's assistant.

Recent research has shown that there is an exact correlation between most of the instruments of Tycho Brahe and Taqi al-Din's observatories (you can read more about this in the Observatories section), but both men were not satisfied with the instruments of the previous astronomers. They had newly discovered instruments to use, such as the sextant, the wooden quadrant, and the astronomical clock.

Taqi al-Din's sextant was called *mushabbahah bil-manatiq* or "replication by areas" and was made from three ruled scales. Two of the scales formed the edges of the three-edged sextant. At the end was an arc, which was attached to one of the rules and was used to determine the distances between the stars. The sextants of these two men should be considered among the finest achievements of the 16th century.

An illustration depicts Taqi al-Din's mushabbahah bil-manatiq, *or sextant, in his observatory in 1580 in Istanbul. The image is taken from the manuscript* Alat-i Rasadiya li Zij-i Shahinshahiya, *which means "Astronomical Observational Instruments."*

04 ASTROLABE

Since Islam began, the muezzin has called the faithful to prayer five times a day. These prayer times are astronomically determined, changing from day to day, so it is vital to know exactly when they were. Before modern technology, Muslims developed an extraordinarily accurate device called the astrolabe to help them do this.

The astrolabe is described by Dr. Harold Williams, an American astrophysicist, as "the most important astronomical calculating device before the invention of digital computers and the most important astronomical observational device before the invention of the telescope."

The earliest origins of the astrolabe are unknown. We know Theon of Alexandria wrote on the astrolabe in the fourth century C.E., and the earliest preserved Greek treatise on the subject is from the sixth century. The origin of the word "astrolabe" is in the Arabic word *asturlāb*, which is said to be a transliteration of a Greek word. Whatever its origins, the instrument was fully developed, and its uses expanded, by Muslim astronomers because they needed to determine prayer times and the direction of Mecca. In the Islamic world, astrolabes remained popular until 1800.

New treatises on the astrolabe were produced, the earliest by Masha'Allah 'Ali ibn 'Isa' and Al-Khwarizmi in the early ninth century, while the earliest surviving Islamic instrument dates from the middle of the tenth century, built by an apprentice to 'Ali ibn 'Isa' in Baghdad. With the Muslim presence in Spain from the eighth century, Islamic learning, including that on the astrolabe, seeped into western Europe, so that the earliest surviving Christian or Western instruments are from the 13th century onward.

Quite a few types were made, and the most popular was the planispheric astrolabe, where the celestial sphere was projected onto the plane of the Equator.

Astrolabes were two-dimensional models of the heavens, showing how the sky looked at a specific place at a given time. This was done by drawing the sky on the face of the astrolabe and marking it so that positions in the sky were easy to find. Some astrolabes were small, palm-size, and portable; others were huge, with diameters of a few meters.

> *"The astrolabe is the most important astronomical calculating device before the invention of digital computers and the most important astronomical observational device before the invention of the telescope."*
>
> ASTROPHYSICIST HAROLD WILLIAMS

They were the astronomical and analog computers of their time, solving problems relating to the position of celestial bodies, like the sun and stars, and time. In effect, they were the pocket watches of medieval astronomers. They could take altitude measurements of the sun; could tell the time during the day or night; or find the time of a celestial event such as sunrise, sunset, or culmination of a star. This was made possible by the use of ingenious tables printed on the back of the astrolabe. These tables could contain information about curves for time conversions, a calendar for converting the day of the month to the sun's position on the ecliptic, trigonometric scales, and a graduation of 360 degrees.

They were based on the model of the Earth being at the center of a spherical universe, with an imaginary observer positioned at a particular latitude and time outside this sphere and looking down upon it. On the astrolabe that the astronomer was holding, the major stars in the sky were represented on a pierced metal plate, which was set into a larger flat circular holder called a mater. Because the plate with the stars was pierced, the astronomer could see through it onto another plate beneath, which would have lines representing his particular geographical location. Several plates would be included in an astrolabe, so that the astronomer could move about from one latitude to another. After using the sighting device on the back of the plate to determine the altitude of a star or the sun, the astronomer would rotate the pierced star map over the plate for his location so as to coincide with the sky at that time. Then, all sorts of calculations could be made. For the more accurate coordinates of celestial bodies necessary for detailed astronomical tables, astrolabes had to be used in conjunction with other instruments, such as large quadrants and observational armillary spheres.

Astrolabes worked with fixed and rotating parts. The mater was a hollow disc holding the

A working astrolabe, created by Mohamed Zakariya, requires a wealth of knowledge to build. Using ancient techniques, such an astrolabe can take three to six months to complete; it requires extensive geometrical calculations and precision engraving in order to work accurately.

rete (the pierced star map) and the rotating plates were placed on top of each other. On the back of the mater were the alidade (the sighting device) and various trigonometric tables. In this respect, the astrolabe was a graphical computer.

"TREATISE ON THE ASTROLABE"

Chaucer, author of the *Canterbury Tales,* also wrote a "Treatise on the Astrolabe" for his ten-year-old son, Lewis, in 1387. Here is what he said about it:

"Little Lewis my son, I have . . . considered your anxious and special request to learn the Treatise of the Astrolabe . . . therefore have I given you an astrolabe for our horizon, constructed for the latitude of Oxford. And with this little treatise, I propose to teach you some conclusions pertaining to the same instrument. I say some conclusions, for three reasons. The first is this: you can be sure that all the conclusions that have been found, or possibly might be found in so noble an instrument as an astrolabe, are not known perfectly to any mortal man in this region, as I suppose."

Geoffrey Chaucer

Islamic makers attempted to develop different types of astrolabes, such as the spherical astrolabe and the linear astrolabe, neither of which was widely adopted. Mariner's astrolabes were developed in the late 15th and 16th centuries by the Portuguese.

A highly sophisticated form of the astrolabe, the universal astrolabe, was developed in Toledo in the 11th century, and it revolutionized star mapping. Two individuals, Ali ibn Khalaf al-Shakkaz, an apothecary or herbalist, and Al-Zarqali, an astronomer, were important in this new development. The universal astrolabe was a major breakthrough because it could be used at any location. Ordinary astrolabes needed different latitude plates if they were moved, because they were designed for a certain place and were latitude dependent.

An important aspect of the universal astrolabe was that its stereographic projection used the vernal or autumnal equinox as the center of projection onto the plane of the solstitial colure.

Dr. Julio Samso, of Barcelona University, talking with Rageh Omaar on the BBC's *An Islamic History of Europe*, said that Muslims used new computing devices and "the universal astrolabe was designed that had applications that were impossible with the standard astrolabe."

Any discussion of astrolabes would be incomplete without mentioning a young engineer astronomer, Maryam al-Ijliya al-Astrulabi. Born in Aleppo, Syria, in 944, she was raised in a family of astronomers and instrument makers. She became a skilled astrolabe maker and worked at the famous Aleppo castle under the auspices of the ruler Saif al-Dawla. She died in 967.

Astrolabes, and in particular universal astrolabes, were the cutting edge of technology, used and developed prolifically by Muslim astronomers who were intrigued and fascinated by the heavens. It was through these hardworking scholars that the astrolabe made it into Europe, where modern astronomy was born.

Two close-ups of an astrolabe made by Mohamed Zakariya show its intricate construction.

MAGICAL GADGET

A Device That Puts Time and Space in the Palm of Your Hand

LEGACY: Until the invention of the sextant in the 17th century, astrolabes were essential tools

LOCATION: Baghdad, Iraq

DATE: Late 13th century

MAKER: Ibn Shawka al-Baghdadi

Crafted in gleaming brass and engraved with the names of stars, astrolabes look as if they might belong to a wizard. And in a way, they are almost magical. Long before clocks and watches were widespread, Muslim engineers and astronomers built astrolabes that brought time and space together into a single gadget.

The scholar Theon of Alexandria had described astrolabes in the fourth century, and they were mentioned in other early Greek writings. But the need to calculate accurate prayer times, and determine the direction of Mecca, drove the development of more sophisticated instruments in Muslim civilization. The oldest surviving astrolabe made in the Muslim world is from tenth-century Baghdad.

Astrolabes told the time during the day or night, helped people navigate on land, and were designed specifically for calculating times of sunrise and sunset.

With the most important stars marked on a pierced metal plate, a sighting device for calculating the angle of a star or the sun, and data tables for accurate astronomical calculations, astrolabes were very sophisticated calculating devices, centuries before computers.

According to tenth-century astronomer Al-Sufi, an astrolabe could perform a thousand tasks useful in astronomy, astrology, navigation, and surveying.

Earlier versions required calibration for particular latitudes on Earth, but 11th-century scholar Al-Zarqali invented a universal astrolabe. Known as *Saphea Arzachelis* from the Arabic *safiha* (plate), this astrolabe worked anywhere.

Universal astrolabes became essential tools for navigation, in use until the invention of the sextant in the 17th century. They were eventually superseded by mechanical clocks and more advanced methods of calculation, but simplified astrolabes for stargazers are still made today.

Ibn Shawka al-Baghdadi made this 13th century astrolabe. Each astrolabe includes several interchangeable plates, which are stored inside the mater, the device's hollow disc. Each plate corresponds to different geographical locations.

■ *Data tables engraved on the astrolabe allow you to perform a variety of different calculations. For example, to tell the time at night, you line up a rule on the back of the astrolabe with a star to find its altitude. You rotate the rete until the star's pointer sits on the correct altitude line on the plate, and read the time off the rim.*

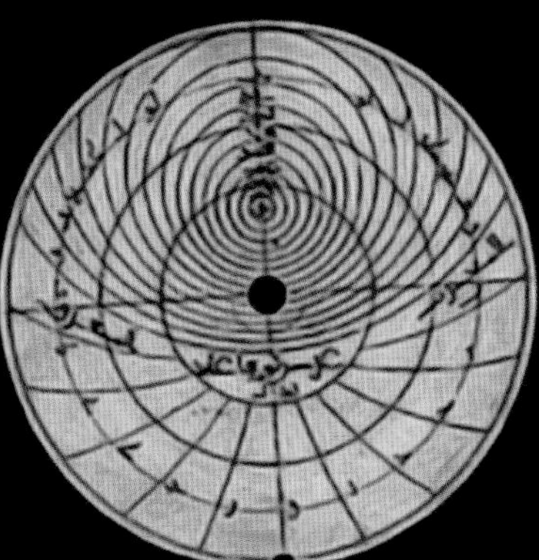

■ LEFT, TOP, AND RIGHT: *The lines engraved on each plate are projections of the sphere of the sky overhead. Each plate covers a narrow range of latitudes (the north-south position on Earth).*

■ CENTER: *The mater of the astrolabe is a hollow disc deep enough to hold several flat plates.*

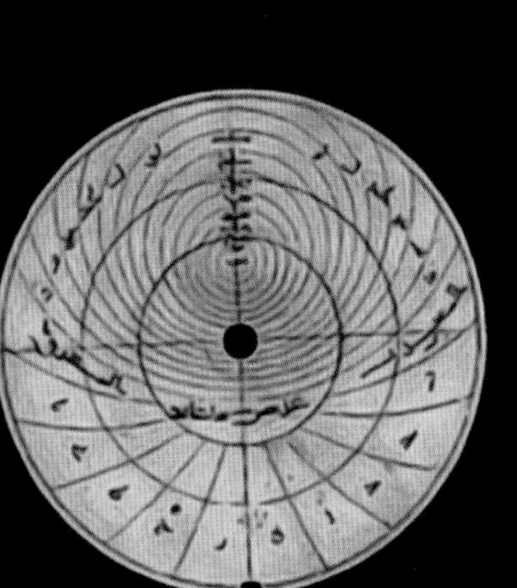

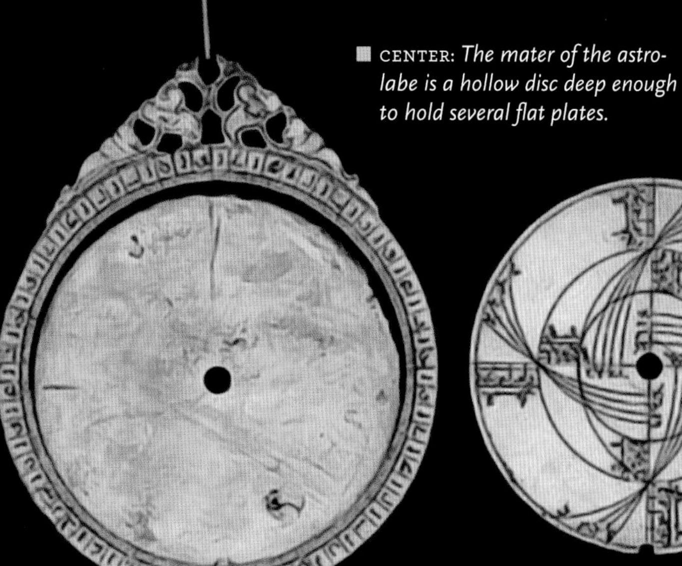

■ BOTTOM: *The rete has a circle to track the sun's path across the sky, and pointers correspond to bright stars. Dagger-shaped pointers were characteristic of early Islamic astrolabes.*

05 Armillary Sphere

In an attempt to make predictions of the movement of heavenly bodies easier, people from many great civilizations built different kinds of models to represent in physical form what they saw in the sky. These models were built based on the idea of the Earth having a sphere of stars surrounding it. One of these models was the armillary sphere.

Armillary spheres modeled the heavens and planetary motions, showing medieval Muslim astronomers how the universe worked in three dimensions, and they came very close to the model we know today. They were not solid globes, but were made up of concentric rings, with the Earth at the center and other bodies surrounding it.

The construction and use of the armillary sphere started in the eighth century when they were first written about in Baghdad in the treatise of *The Instrument with the Rings* by Al-Fazari.

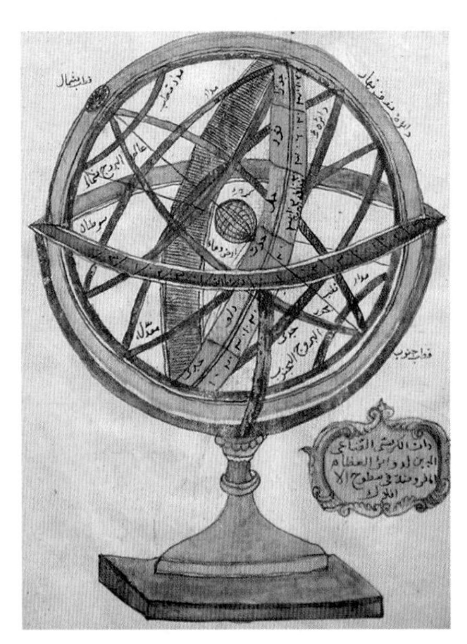

An armillary sphere is depicted in an engraving from the Jihannuma *or* Universal Geography, *Istanbul, 1732—a reprint of the original* Jihannuma *written in the 17th century by scholar Katib Celebi (Hajji Khalifa).*

But it was in the tenth century that they reached an advanced level of sophistication, and were produced in two main varieties.

Demonstrational armillary spheres concentrated on the Earth, and a tiny model of the globe was surrounded by the rings of ecliptic (the apparent path of the sun around the Earth), the circle of the Equator, tropics, and polar circles. These were all held in place by a graduated meridian ring, and pivoted about the equatorial axis. The moon, planets, and stars did not appear in these models, but they did give the relative motions of bodies around the Earth.

The second type was the observational armillary sphere, which was different because it did not have the Earth globe in the center, but had mounted sighting devices on the rings. These spheres were larger and were tools used to determine coordinates and other values.

There were many Muslim astronomers who wrote about observational armillary spheres, like Jabir ibn Aflah from Seville, known in the West as Geber (not to be confused with Geber the chemist), from the mid-12th century. They referred to the descriptive work of Ptolemy's *Syntaxis*, written in the second century, known as *Almagest* in the Islamic world.

Armillary spheres to study the Earth and skies were found in observatories such as the 13th-century Maragha Observatory, the 15th-century Samarkand Observatory, and the 16th-century observatory at Istanbul.

A 16th-century manuscript shows astronomers lining up various parts of the armillary sphere with specific stars to produce flat charts of the heavens, which were plotted and made into astrolabes. These would guide people, using the stars. The central pendulum is used to trace the trajectories of the stars and planets on the flat ground so as to create these charts.

06 SIGNS FOR WISE PEOPLE

The Quran often refers to various natural phenomena in a very inspiring manner, and challenges mankind to ponder these phenomena using reason.

For example: **2:164:** "Verily, in the creation of the heavens and of the earth, and the succession of night and day: and in the ships that speed through the sea with what is useful to Man: and in the waters which God sends down from the sky, giving life thereby to the earth after it had been lifeless, and causing all manner of living creatures to multiply thereon: and in the change of the winds, and the clouds that run their appointed courses between sky and earth: [in all this] there are messages indeed for people who use their reason."

Astronomical phenomena are frequently cited in the Quran and often put in the context of their use to mankind as in timekeeping and navigation. The Quran talks about precise orbits and courses, thus passing on the message that behind these phenomena lies a coherent system that people are invited to explore. Here are some examples:

6:97: "[God] is the One Who has set out for you the stars, that you may guide yourselves by them through the darkness of the land and of the sea. We have detailed the signs for people who know."

16:12: "For you [God] subjected the night and the day, the sun and the moon; the stars are in subjection to His Command. Verily in this are signs for people who are wise."

21:33: "[God is] the One Who created the night, the day, the sun and the moon. Each one is traveling in an orbit with its own motion."

55:5: "The sun and the moon follow courses [exactly] computed."

Verses like these cited above formed an intellectual challenge to people to build the required knowledge to explore a universe abundant with God's wonders.

Not only that, but in one verse, humans are even encouraged to make their way out of the Earth in order to explore space, but with a warning that this should only be done when they have enough power and control.

55:33: "O you assembly of Jinns and Humans! If it be you can pass beyond the zones of the heavens and the earth, pass you! not without authority [power] shall you be able to pass!"

The coherent systems behind astronomical phenomena are explored in verses in the Quran.

"The Coiling of Day and Night," by Dr. Ahmed Moustafa

07 THE MOON

On July 20, 1969, Apollo 11 landed on the lunar surface, and Neil Armstrong became the first person on the moon. However, long before Armstrong made his first lunar step and uttered his now famous line, a number of great Muslims had become associated with Earth's closest astronomical neighbor.

For Muslims, the moon is extremely important because the calendar that is used, the *Hegira* calendar, is determined by the cycle of the moon. A problem they faced was that the approximately 29.5 days of a lunar month were not in sync with the 365 days of a solar year; 12 lunar months add up to only 354 days.

Christians and Jews had confronted the same problem, and they had adopted a scheme based on a discovery made in about 430 B.C.E. by Athenian astronomer Meton. He developed the Metonic cycle of 19 years. This was made of 12 years of 12 lunar months and 7 years of 13 lunar months. Periodically a 13th month was added to keep the calendar dates in step with the seasons.

Muslims would use this cycle, but unscrupulous rulers sometimes added a 13th month when it suited their own interests, so the second caliph, Umar ibn al-Khattab, who reigned for ten years from 634, introduced the *Hegira* calendar that is still used in Islamic countries today.

This strictly follows a lunar cycle. The *Hegira* year is about 11 days shorter than the solar year, and holidays such as Ramadan, the month of fasting, slowly cycle through the seasons. So each year Ramadan is about 11 days earlier than the last, and the month of fasting falls on the same date only about every 33 solar years.

Ramadan and the other Islamic months also begin when the crescent moon is sighted, so no one knows exactly when Ramadan will start until the crescent moon appears in the night sky.

Predicting just when the crescent moon would become visible was a special challenge to Muslim mathematical astronomers. Although Ptolemy's theory about the moon's movements was accurate near the time of the new moon, the invisible moon, it looked only at the lunar path as part of the eclipse or the sun's path on the moon.

Muslims realized that to predict the sighting of the crescent moon, its movement in relation to the horizon had to be studied, and this problem

The phases of the moon are used to determine the Muslim calendar, the Hegira *calendar.*

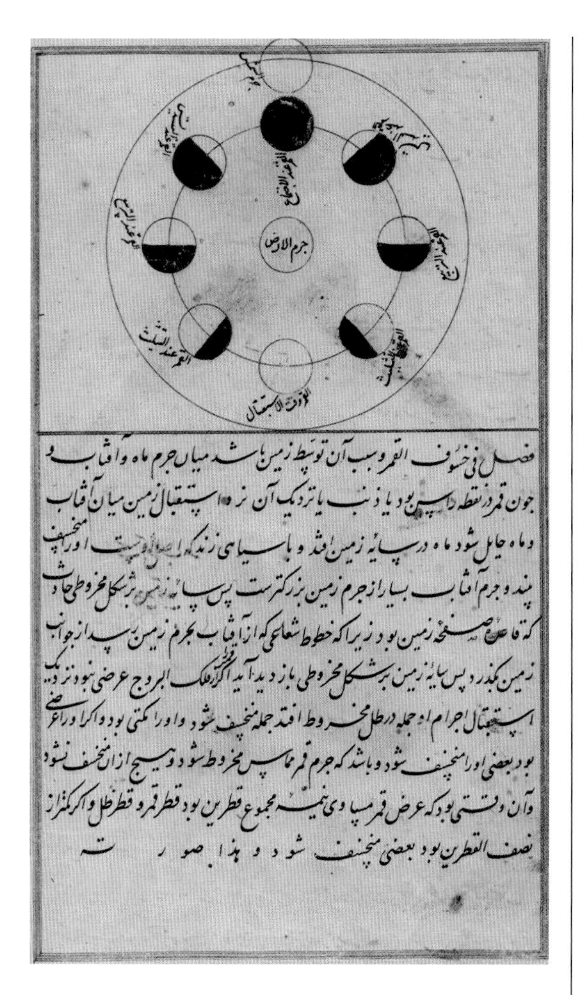

A diagram from an early 16th-century Persian translation of Ajaib al Makhluqat', *or* Wonders of Creation, *by al-Qazwini*

demanded fairly sophisticated spherical geometry, or geometry that deals with shapes on the surface of the sphere. It was Al-Kindi, working in Baghdad in the ninth century, who was the first to develop spherical geometry, which he used extensively in his astronomical works.

Spherical geometry was also needed when Muslims were finding the *Qibla*, the direction of Mecca, which they prayed toward and which their mosques faced, and it was Al-Biruni who worked this out from any location on the globe. Al-Biruni was interested in literally everything, and sometimes he is referred to as the Leonardo da Vinci of his day. Concerning the moon, he described the eclipse of May 24, 997, while he was at Kath, in today's Uzbekistan. This event was also visible in Baghdad, and he arranged with Abu al-Wafa' al-Buzjani, a fellow astronomer, that the latter would observe it there. When they compared their timings, they were able to calculate the difference in longitude between the cities.

So moon watching and recording was a serious business. Back then, as now, the moon was a constant source of fascination, since logging the order of its movements supported the idea that there was order in the heavens, too. And these observations produced the structure for the Muslim calendar that has been used for more than 1,400 *Hegira* years.

MOON'S VARIATION

A Muslim astronomer who lived in Cairo and observed at Baghdad in 975 discovered the third inequality of the moon's motion called the moon's variation. Ptolemy knew about the first and second. He bore the formidable name of Abu al-Wafa al-Buzjani.

In Europe, this third inequality of motion, that the moon moves quickest when it is new or full, and slowest in the first and third quarters, was "rediscovered" by Tycho Brahe six centuries later in about 1580.

08 LUNAR FORMATIONS

AD
MAGNANIMVM PRINCIPEM
HONORATVM II.
MONOECI PRINCIPEM
DVCEM VALENTINVM
PAREM FRANCIÆ
MARCHIONEM BAVCII, COMITEM CARLADESII
BARONEM BVISSII, ET CALVINETI
DOMINVM S. REMIGII &c.
ALMAGESTI NOVI
PARS POSTERIOR
TOMI PRIMI.

When viewed with the naked eye, the surface of the moon appears unevenly bright, with dark and light patches. These features are called "lunar formations."

In 1651 Joannes Baptista Riccioli, a Jesuit professor of astronomy and philosophy in Bologna, Italy, compiled a comprehensive work on astronomy, called *Almagestum Novum*, with a complete map of the moon. He named the lunar formations after distinguished astronomers of the Middle Ages. Ten were given the names of Muslim astronomers and mathematicians.

These names were finally agreed upon at a conference of the International Astronomical Union in 1935. Of the 672 lunar formations, 13 were given the names of major Muslim astronomers, and since then more have been added. These names include:

■ **Messala** is a plain in the 13th section of the moon named after Masha'Allah, who was active in 809. He was a Jew of Egypt who embraced Islam during the time of the Abbasid caliph Al-Mansur. Two of his books on astronomy were translated into Latin in the 16th century: *De Scientia Motus Orbis* and *De compositione et utilitate astrolabii.*

■ **Almanon** is a crater in the ninth section named after Caliph Al-Ma'mun, the son of Harun al-Rashid, famous from *The Thousand and One Nights*. In 829, Al-Ma'mun built an observatory in Baghdad. In his academy, Bayt al-Hikmah, the House of Wisdom, the greatest scientists and philosophers of his age carried out their research.

■ **Alfraganus** is a crater in the second section named after Al-Farghani, who died around 861. He was one of Al-Ma'mun's team of researchers in astronomy. His most famous book was the *Book of the Summary of Astronomy*, and this was the main influence for the Italian poet Dante.

■ **Albategnius** is a plain in the first section named after Al-Battani, who was born in 858. He determined many astronomical measurements with great accuracy.

■ **Thabit** is a prominent circular plain in the eighth section named after Thabit ibn Qurra, who died in Baghdad in 901. He translated into Arabic a large number of Greek and Syrian works on science. He also made major contributions of his own to pure mathematics.

■ **Azophi** is a mountainous ring in the ninth section named after the tenth-century 'Abd al-Rahman al-Sufi. He was one of the most outstanding practical astronomers of the Middle Ages. Al-Sufi's illustrated book *The Book of Fixed Stars* is a masterpiece on stellar astronomy.

■ **Alhazen** is a ring-shaped plain in the 12th section named after Abu Ali al-Hasan ibn al-Haytham, usually known as Ibn al-Haytham, He was born in Basra around 965 and spent most of his working life in Egypt, where he died in 1039. He composed almost a hundred works, of which about 55 are preserved today, all concerned with mathematics, astronomy, and optics. He was one of the foremost investigators of optics in the world, and his *Book of Optics* had an enormous influence on European science.

Arzachel is a plain in the eighth section named after Al-Zarqali, who died in 1100. He worked in Muslim Spain in collaboration with other Muslim and Jewish astronomers and prepared the famous Toledan Tables. His work may have influenced that of Copernicus.

Geber is a circular, flat plain in the ninth section named after Jabir ibn Aflah, who died in 1145. He was a Spanish Arab who was the first to design a portable celestial sphere to measure celestial coordinates, today called a torquetum.

Nasireddin is a crater 30 miles in diameter named after Nasir al-Din al-Tusi, who was born in 1201. He was a minister to Hulagu Khan, Ilkhanid ruler of Persia from 1256 to 1265. He was put in charge of the observatory installed at Maraghah by Hulagu, and of preparing the Ilkhanid Tables and the catalog of fixed stars, which remained in use for several centuries throughout the world, from China to western Europe.

Alpetragius is a crater in the eighth section named after Nur al-Din ibn Ishaq al-Bitruji, who was born in Morocco, lived in Ishbiliah (Seville), and died around 1204. He worked hard, unsuccessfully, at modifying Ptolemy's system of planetary motions. Al-Bitruji's book *On Astronomy* was popular in 13th-century Europe in its Latin translation.

Abulfeda is a circular plain in the ninth section named after Abu al-Fida', who was born in 1273 in Syria. He was the last Muslim geographer and astronomer trained and nurtured on the traditions established by Caliph Al-Ma'mun. He was also a great historian, the most famous of his works being *Survey of Countries*.

Ulugh Beigh is a prominent elliptical ring in the 18th section named after Ulugh Beg, who was born in 1394 and founded in 1420 a magnificent observatory in Samarkand, which was equipped with excellent and accurate astronomical instruments. His most commendable and enduring work was a new catalog of stars.

So when you glance at the moon tonight, remember all those individuals who have been immortalized in craters, plains, and elliptical rings, people who have brought greater understanding and knowledge into our lives.

BOTTOM: *A lunar map shows the formations named after eminent Muslim scholars.* OPPOSITE TOP: The Almagestum Novum, *compiled in 1651, by Johannes Baptista Ricioli, contained a detailed map of the moon.*

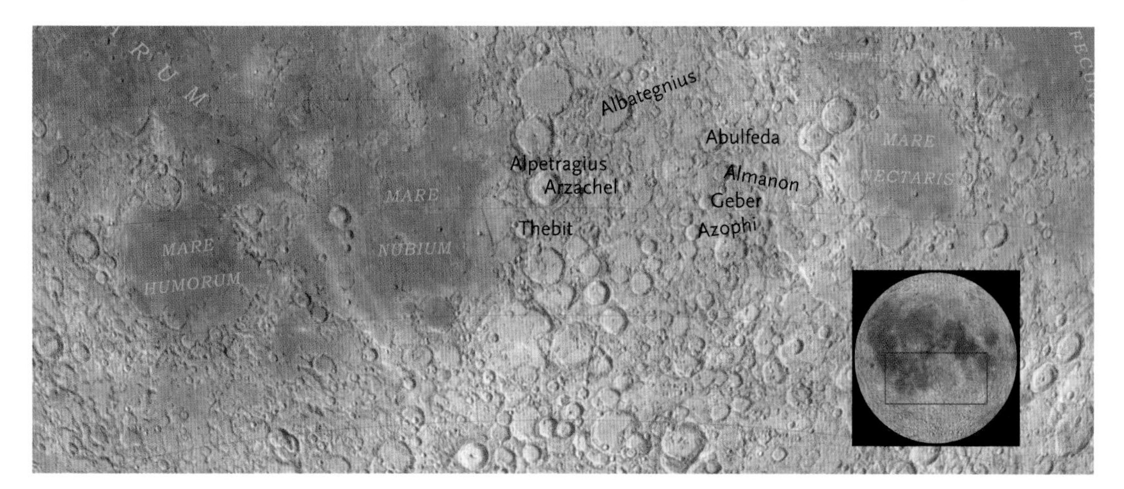

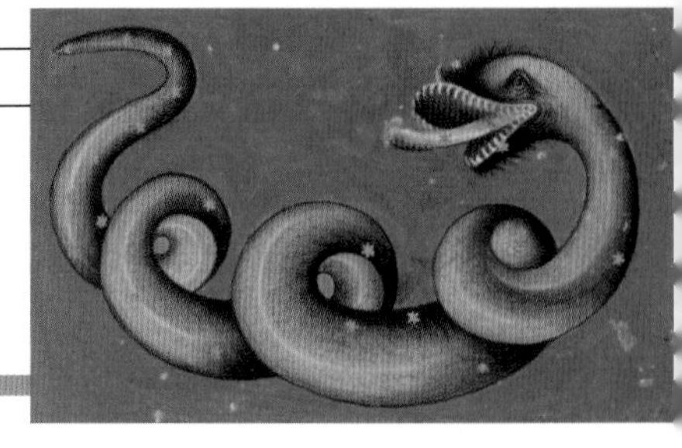

09 CONSTELLATIONS

With the rise of observatories and a greater interest in the night sky, Muslim astronomers from the ninth century onward were fascinated by the night sky and carried out substantial work on stars and constellations. These included 'Abd al-Rahman al-Sufi, a Persian astronomer who lived during the tenth century; he was a real stargazer and in 964 described the Andromeda galaxy, our closest neighbor, calling it "little cloud."

This was the first written record of a star system outside our own galaxy. He set out his results constellation by constellation, discussing the stars' positions, sizes and colors, and for each constellation he produced two drawings, one from the outside of a celestial globe and the other from the inside. He also wrote about the astrolabe and its thousand or so uses.

The result of this hard work was the recording of many stars and constellations, which are still known by their original Arabic names. In fact, astronomers gave names and assigned magnitudes to 1,022 in all. Today, more than 165 stars still have names that reflect their original Arabic names, such as Aldebaran, meaning "Follower" of the Pleiades, and Altair, meaning "The Flying Eagle."

Muslims also devised star maps and astronomical tables, and both of these would be used in Europe and the Far East for centuries. Maps of the heavens also appeared in art, such as on the dome of a bathhouse at Qusayr 'Amra, a Jordanian palace built in the eighth century, which has a unique hemispherical celestial map. The surviving fragments of the fresco show parts of 37 constellations and 400 stars.

TOP: *The constellation the Dragon, or al-tinnin in Arabic.* BOTTOM LEFT: *The constellation Cepheus, or qifa'us in Arabic.* BOTTOM RIGHT: *The manual of cosmography in Turkish by Mustafa ibn Abdallah.*

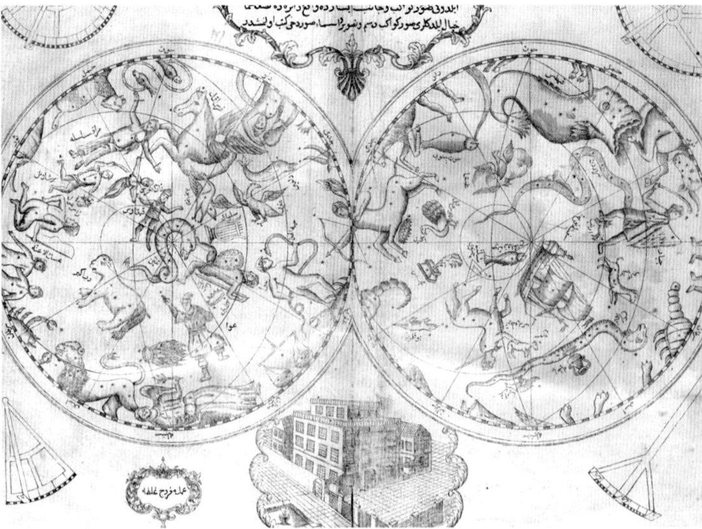

The Pleiades star group, known in Arabic as Al-Thurayya

10 FLIGHT

The concept of flight has fascinated and challenged humans for thousands of years. The ancient Egyptians left behind many paintings demonstrating their desire to fly, depicting pharaohs soaring with wings. The Chinese and the Greeks had mythical stories and legends about flying, as did the Sassanians.

The most popular story is one recounted by Al-Firdawsi in his *Book of Kings*, written around 1000. It says a certain King Kai Kawus was persuaded by evil spirits to invade heaven with the help of a flying craft that was a throne, and attached to its corners were four long poles pointing upward. Pieces of meat were placed at the top of each pole and ravenous eagles were chained to the feet of the throne. As the eagles attempted to fly up to the meat, they carried the throne up, but, inevitably, they grew tired and the throne came crashing down.

Al-Firdawsi's Book of Kings *includes a popular story of a proud king who tried to fly. The book's cover is illustrated here.*

Pre-Islamic Arabic legends also have stories about flying magicians and sorcerers, supernatural powers, birds, or just feathers. For Muslims, flight has a spiritual dimension. The pious soul reaches for goodness until it attains a certain level, then it rises above.

The first Muslim, and perhaps person, to make a real attempt to construct a flying machine and fly was Cordoban 'Abbas ibn Firnas in the ninth century. He was the usual polymath of the time, becoming a renowned poet, astrologer, musician, astronomer, and engineer. But his greatest fame was for constructing a flying machine, the first of its kind capable of carrying a human into the air. He flew successfully a number of times over desert regions, improving his designs before attempting his two famous flights in Córdoba in Spain.

The first flight took place in 852, when he wrapped himself in a loose cloak stiffened with wooden struts and jumped from the minaret of the Great Mosque of Córdoba. The attempt was unsuccessful, but his fall was slowed enough that he got off with only minor injuries, making it at least one of the earliest examples of a parachute jump. Western sources wrongly gave him a Latin name calling him Armen Firman, instead of 'Abbas ibn Firnas.

A swan lands on the surface of water. After observing how birds land, 'Abbas ibn Firnas realized that a tail was needed to land accurately and safely. He had not noticed this before his earlier attempt at flying, during which he crash-landed.

Ibn Firnas was one to learn from experience, and he worked hard to improve his next design. Accounts from various eyewitnesses and medieval manuscripts described it as machine consisting of large wings. So about 1,200 years ago, the nearly 70-year-old 'Abbas ibn Firnas made a flight machine from silk and eagle feathers.

In the Rusafa area on the outskirts of Córdoba, Ibn Firnas mounted a hill and appeared before the crowd in his bird costume, made from silk covered with eagle feathers, which he tightened with fine strips of silk. Ibn Firnas explained with a piece of paper how he planned to fly using the wings fitted on his arms: "Presently, I shall take leave of you. By guiding these wings up and down, I should ascend like the birds. If all goes well, after soaring for a time I should be able to return safely to your side."

He flew to a significant height and hung in the air for more than ten minutes before plummeting to the ground, breaking the wings and one of his vertebrae. After the event, Ibn Firnas understood the role played by the tail, telling his close friends that when birds land, they normally land on the root of the tail, which did not happen for him because he did not have one.

All modern airplanes land on their rear wheels first, which makes Ibn Firnas's comment ahead of its time. Recording the event, one witness wrote: "He flew a considerable distance as if he had been a bird, but in alighting again on the place where he started from, his back was very much hurt. For not knowing that birds when they alight come down upon their tails, he forgot to provide himself with one."

It would be centuries until Leonardo da Vinci's drawings of a flying machine and the Wright brothers' first flight.

Unfortunately, the injury Ibn Firnas sustained in the flight prevented him from carrying out further experiments. However, he was an enterprising man, and he probably guided somebody, perhaps one of his apprentices, to create a newer version.

An artistic impression shows the first successful manned flight by 'Abbas ibn Firnas.

The existence of such a machine was mentioned in a manuscript by Roger Bacon, who described it as an ornithopter. In 1260, Bacon wrote *On the Marvelous Powers of Art and Nature*, including two possible ways a person might fly. One is a rough description of what was later to become known as an ornithopter. The other is a more detailed description of a globe filled with "ethereal air." Bacon claimed, "There is an instrument to fly with, which I never saw, nor know any man that hath seen it, but I full well know by name the learned man who invented the same." It is known that Bacon studied in Córdoba, the homeland of Ibn Firnas. It is likely that the description of the ornithopter could have been taken from Muslim contemporary manuscripts in Spain that have since disappeared.

Ibn Firnas died in 887, and none of his original works has survived to the present day. His life has been reconstructed from a few verses and from the information given by the chroniclers of the time.

After Ibn Firnas, Muslims and non-Muslims pursued the endeavor of flying, and many more flight attempts were made: Al-Juhari, a Turkistani teacher, launched himself from the minaret of Ulu Mosque in 1002, using wings made from wood and rope. He died instantly on impact. Eilmer of Malmesbury was an 11th-century English Benedictine monk who also forgot the use of a tail, and broke both his legs as he jumped from a tower in 1010 before gliding 183 meters (600 feet).

After these two flying attempts, aviation history is silent until the works of the famous Florentine artist and scientist Leonardo da Vinci. Leonardo remains the leading engineer

> *"By guiding these wings up and down, I should ascend like the birds. If all goes well, after soaring for a time I should be able to return safely to your side."*
>
> ABBAS IBN FIRNAS, PIONEER OF FLIGHT

to establish proper scientific thinking on the quest for flight. Although he did not attempt to fly himself, da Vinci discussed and drew on paper many sketches relating to flight and flying, including the bird-winged machine known as an ornithopter, which was designed to be strapped to a person's back. Other sketches included a glider and, according to some interpretations, even a helicopter.

In 1633, a Turk named Lagari Hasan Celebi invented the first manned rocket, which he launched using about 300 pounds of gunpowder as the firing fuel. The event is recorded by an artist's sketch drawing. William E. Burrows in his book *This New Ocean: The Story of the First Space Age* says: "[T]here was a Turk named Lagari Hasan Celebi, who . . . was shot into the sky by fifty-four pounds of gunpowder to celebrate the birth of Sultan Murad IV's daughter, Kaya Sultan . . .The rocket then carried Celebi high into the air, where he opened several 'wings,' and then glided to a safe landing in front of the royal palace. Celebi was rewarded with a pouch of gold, made a cavalry officer, and is said to have been killed in combat in the Crimea."

Hazarfen Ahmed Celebi, another 17th-century Turk, used eagle feathers stitched to his wings to fly. After nine experimental attempts, he finally decided on the shape of his wings. His most famous flight took place in 1638 from the Galata tower near the Bosporus in Istanbul, and he successfully landed on the other side of the strait. According to Turkish historian Evliya Celebi, who witnessed the feat and recorded it in his book *A Book of Travel*, the famous Turkish flier used Al-Juhari's calculations with some corrections and balancing adjustments, derived from studying the eagle in flight. Hazarfen earned a reward of a thousand gold pieces for his achievement, and a Turkish postal stamp bears tribute to his historic flight.

After the successful flight over the Bosporus, the Montgolfier brothers were the next to

LEFT: *An illustration depicts the flight of Hazarfen Ahmed Celebi in 1638 from Galata Tower near the Bosporus in Istanbul.* RIGHT: *An artistic impression illustrates the first manned rocket flight flown by Lagari Hasan Celebi in 1633. Lagari Hasan was launched into the sky by a seven-winged rocket, which used a mixture of gunpowder paste.*

publicly air their hopes for flight with a model hot air balloon whose passengers were a sheep, a duck, and a cockerel. A few weeks later, Jean François Pilâtre de Rozier, a science teacher, and the Marquis d'Arlandes, an infantry officer, became the first human air travelers when they flew for 9 kilometers (5.6 miles) over Paris in a hot air balloon.

Nineteenth-century aeronautics was dominated by German Otto Lilienthal, who studied the lifting power of surfaces, the best form of wing curvature, and the movement of the center of pressure with different wing angles, an important factor in the stability of aircraft. He was a great hang glider, but died in flight in the Berlin hills in 1896 when a gust of wind stalled his machine and he was unable to regain control.

The Wright brothers with their famous flight of December 1, 1903, are probably the best known names today in aviation history. Wilbur Wright's key insight was to study birds, a lesson Ibn Firnas learned, too. Wilbur realized that birds keep their lateral balance, or control when banking, by twisting their wings. He devised a kite that reproduced the same effect mechanically, allowing it to roll one way or the other as desired.

Before developing a powered aircraft, the Wright brothers used gliders, aiming "to escape accident long enough to acquire skill sufficient to prevent accident." They also hit on the essential principle of combining rudder control and roll for smoother balanced turns. In 1908, Wilbur Wright demonstrated his airplane in France, and within the year Henri Farman and Louis Blériot were making extended flights.

All this history of aviation, and even space travel, started with the humble beginnings of one man, 'Abbas ibn Firnas, who was one of the first to try out his ideas when he glided with his eagle feathers and silk.

A photograph depicts the first flight by the Wright brothers in 1903.

An illustration shows the balloon "Le Flesselles" ascending over Lyon, France, on January 19, 1784. It carried seven passengers, including Joseph Montgolfier and Jean-François Pilâtre de Rozier.

بفرخ ترین ساعت آمد بتخت
خدیو جهانگیر فیروزبخت
آمد شه کامران بر سریر
چو بر آسمان آفتاب منیر

"Sit down before fact as a little child, be prepared to give up every conceived notion, follow humbly wherever and to whatever abysses nature leads, or you will learn nothing."

THOMAS HUXLEY, ENGLISH BIOLOGIST

WEALTH OF KNOWLEDGE

PERSONALITIES FROM THE PAST • EUROPE'S LEADING MINDS
A THOUSAND YEARS OF SCHOLARSHIP • AUTHORS AND TREATISES
FURTHER READING • GLOSSARY • ILLUSTRATIONS CREDITS
INDEX • ACKNOWLEDGMENTS

THE FOLLOWING SECTION HAS BEEN COMPILED FOR YOUR EASE OF USE IN ACCESSING 1,000 years of missing history and getting to know the individual scholars. Many individuals are discussed throughout this book. We have selected 11 outstanding scholars from the Muslim world for you to get to know in greater detail; all others are presented in the A Thousand Years of Scholarship section. Scroll through to discover who was who, when, and where.

To discover more about the effect of the works, innovations, and inventions of Muslim scholars on European thought and the Renaissance, see Europe's Leading Minds, where you will read how Roger Bacon spoke Arabic and never tired of telling people that knowledge of Arabic and Arab science was the only way to true knowledge. For those who have been inspired to find out more, use Further Reading to guide you, and if you want to know more about the original manuscripts authored by the scholars mentioned in the book, see the Authors and Treatises section.

There is also a glossary to browse through and an index for you to locate whatever subject interests you.

OPPOSITE: *A Persian manuscript shows Timur the Lame on his throne in the ancient city of Balkh.*

01 PERSONALITIES FROM THE PAST

Throughout this book, you have read how men and women from Muslim civilization have contributed to our daily lives. Here now is a "Who's Who" of some of the biggest names from a thousand years ago.

'Abbas ibn Firnas

FULL NAME: 'Abbas Abu al-Qassim ibn Firnas ibn Wirdas al-Takurini

BORN: Ninth-century; Andalusian descendant from a Berber family residing in Takuronna (now Ronda)

DIED: 887

MOST INFLUENTIAL WORK: Producing a flying machine, crystal, and a planetarium

Go to: Fine Dining in Home; Glass Industry in Market; and Observatories and Flight in Universe

It is difficult to pin one profession on Cordoban 'Abbas ibn Firnas because he had numerous talents, including poetry, astrology, music, and astronomy. He was also fluent in Greek, and made translations of philosophical and musical manuscripts.

After perfecting the technique of cutting rock crystal (quartz) and producing glass, he made a kind of glass planetarium, complete with artificial thunder and lightning.

His most famous achievement is the construction of a flying wing, the first known to be capable of allowing a human to glide through the air.

Unfortunately he left no trace of his original works, and his biography was reconstructed only from a few verses and information from eyewitnesses left to us in numerous documents.

Al-Jazari

FULL NAME: Badi'al-Zaman Abu al-'Izz Isma'il b al-Razzaz al-Jazari

BORN: Birth date not known; he served the Artuq kings of Diyarbakir (now in southeast Turkey) from 1174 to 1200

DIED: Date not known

MOST INFLUENTIAL WORK: *Al-Jami Bain al-Ilm Wal-Amal al-Nafi fi sina'at al-Hiyal*, or *The Book of Knowledge of Ingenious Mechanical Devices*

GO TO: Clocks and Cleanliness in Home, and Water Supply in Market

Today, we might call Al-Jazari a mechanical engineer, and he was an outstanding one at that. There is little known about his life, but we do know that he was in the service of Nasir al-Din, the Artuqid, king of Diyarbakir, who asked him to document his inventions in a manual, *The Book of Knowledge of Ingenious Mechanical Devices,* which he completed in 1206.

Before this, he had built many machines, including clocks and water-raising machines, and a large number of mechanical devices that revolutionized engineering, like the crankshaft. He is possibly the first to use robotics, as many of his machines incorporated moving figures.

Al-Kindi

Al-Kindi

FULL NAME: Abu Yusuf Yaqub ibn Ishaq al-Sabbah al-Kindi
BORN: About 801 in Kufa, Iraq
DIED: 873
MOST INFLUENTIAL WORK: Wrote more than 361 works on a variety of subjects including *The Book of the Chemistry of Perfume and Distillations*
GO TO: Music, Cleanliness, Vision and Cameras in Home; Chemistry, Commercial Chemistry, House of Wisdom, Translating Knowledge in School; Pharmacy in Hospital; Earth Science and Natural Phenomena in World

Al-Kindi was an encyclopedic man, working as a physician, philosopher, mathematician, geometer, chemist, logician, musician, and astronomer. A son of the governor of Kufa, he studied there and at Baghdad's House of Wisdom, where he gained a high reputation at the caliph's court for translation, science, and philosophy. Caliph al-Mutassim also chose him as tutor to his son Ahmad.

His contributions include an introduction to arithmetic, eight manuscripts on the theory of numbers, and two on measuring proportions and time. He was the first to develop spherical geometry, and used this in his astronomical works. He wrote on spherics, the construction of an azimuth on a sphere, and how to level a sphere. As a musician he used musical notation and played a part in the development of the *'ud*, or lute.

Al-Zahrawi

FULL NAME: Abul Qasim Khalaf ibn al-Abbas al-Zahrawi, known in the West as Abulcasis
BORN: 936 in Medinat al-Zahra, near Córdoba, Spain
DIED: 1013
MOST INFLUENTIAL WORK: *Al-Tasrif liman 'Ajiza 'an al-Ta'lif*, shortened to *Al-Tasrif*, and translated as *The Method of Medicine*, which became a central part of the medical curriculum in European countries for many centuries
GO TO: Cleanliness in Home; Translating Knowledge in School; Medical Knowledge, Instruments

Al-Zahrawi

of Perfection, Pharmacy, and Surgery in Hospital

Al-Zahrawi was a revolutionary physician and surgeon of Umayyad Spain. His 30-volume work, *Al-Tasrif*, gave detailed accounts of dental, pharmaceutical, and surgical practices, and it was one of the most influential medical encyclopedias of the time.

His surgical breakthroughs included his discovery of the use of catgut for internal stitching, and administering drugs by storing them in catgut parcels that were ready for swallowing, known today as capsules.

He also designed and illustrated more than 200 surgical instruments such as syringes, droppers, scalpels, and forceps, and his detailed diagrams of these figured prominently in medieval medical texts and journals in Europe and the Muslim world for centuries. Many modern surgical instruments have changed little from his original designs.

Ibn al-Haytham

Fatima al-Fihri

Fatima al-Fihri

FULL NAME: Fatima al-Fihri
BORN: Ninth century
DIED: 880
MOST INFLUENTIAL WORK: Building the college mosque complex of Al-Qarawiyin in Fez, Morocco, in 859
GO TO: Universities in School

Fatima al-Fihri was a young, well-educated pious woman who received a large amount from her father, a successful businessman. She vowed to spend her entire inheritance on building a mosque and learning center for her Qairawaniyyin community. It was completed in 859 and developed into Morocco's number one university.

Studies included astronomy, the Quran and theology, law, rhetoric, prose and verse writing, logic, arithmetic, geography, medicine, grammar, Muslim history, and elements of chemistry and mathematics. This variety of topics and the high quality of the teaching drew scholars and students from all over.

Fatima's sister, Maryam, had simultaneously constructed Al-Andalus Mosque in the vicinity of Qairawaniyyin. These two neighborhoods became the nuclei of the city of Fez.

Ibn al-Haytham

FULL NAME: Abu 'Ali al-Hasan Ibn al-Haytham, known in the West as Alhazen
BORN: 965 in Basra, Iraq
DIED: 1039 in Cairo, Egypt
MOST INFLUENTIAL WORK: *Kitab al-Manazir*, or *Book of Optics*, which formed the foundations for the science of optics. The Latin translation had an enormous impact on Roger Bacon, Witelo, Leonardo da Vinci, Descartes, and Johannes Kepler, centuries later.
GO TO: Vision and Cameras in Home; Translating Knowledge in School; Natural Phenomena in World; The Moon in Universe

Ibn al-Haytham revolutionized optics, taking

the subject from one discussed philosophically to a science based on experiments. He rejected the Greek idea that an invisible light emitting from the eye caused sight, and instead rightly stated that vision was caused by light reflecting off an object and entering the eye.

By using a dark room with a pinhole on one side and a white sheet on the other, he provided the evidence for his theory. Light came through the hole and projected an inverted image of the objects outside the room on the sheet opposite. He called this the *qamara*, and it was the world's first camera obscura.

Ibn Battuta

Ibn Battuta

FULL NAME: Abu Abdullah Muhammad ibn Battuta
BORN: 1304 in Tangier, Morocco
DIED: 1368 or 1370
MOST INFLUENTIAL WORK: The *Rihla*, or his travel book, narrated by him and written by Ibn Juzayy, a royal scribe, under the patronage of Abu 'Inan, the sultan of Fez and Morocco
GO TO: Trade, Jewels, and Currency in Market; Public Baths in Town; Travelers and Explorers in World

Ibn Battuta left his hometown of Tangier in Morocco as a 21-year-old, about 680 years ago. He set off as a lone pilgrim and did not return for 29 years. In this time, he covered more than 75,000 miles, through 44 modern-day countries traveling on horse, cart, camel, boat, and foot. This journey took him through North, West, and East Africa; Egypt; Syria; Persia; the Arabian Gulf; Anatolia; the steppe; Turkistan; Afghanistan; India; Maldives; Ceylon (Sri Lanka); Bengal; Sumatra; China; Sardinia; and Spain. By the end he had visited Mecca four times, and had met and could name more than 1,500 people, including 60 heads of state.

He was asked by the sultan of Fez and Morocco to record all this in his *Rihla*, and this is our window into the 14th-century world because he has left some of the best eyewitness accounts of culture, customs, people, animals, and plants of the medieval world stretching from Córdoba to Canton.

Ijliya al-Astrulabi

FULL NAME: Maryam al-Ijliya al-Astrulabi
BORN: 944 in Aleppo, Syria, in the era of Saif al-Dawla
DIED: 967
MOST INFLUENTIAL WORK: The daughter of Al-Ijili al-Astrulabi, she continued her father's work of making astrolabes.
GO TO: Astrolabe in Universe

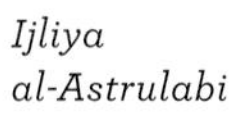

Ijliya al-Astrulabi

Jabir ibn Hayyan

Jabir ibn Hayyan

Full name: Abu Musa Jabir ibn Hayyan, known in the West as Geber
Born: 722 in Tus, Iran
Died: 815 in Kufa, Iraq
Most influential work: Devising and perfecting the processes of sublimation, liquefaction, crystallization, distillation, purification, amalgamation, oxidation, evaporation, and filtration; and producing sulfuric acid by distilling alum
Go to: Chemistry and Commercial Chemistry in School

Jabir ibn Hayyan is generally known as the father of chemistry. The son of a druggist and perfume maker, he worked under the patronage of the Barmaki vizier during the Abbasid Caliphate of Harun al-Rashid. This meant that he shared some of the effects of the downfall of the Barmakis and was placed under house arrest in Kufa, where he died.

His work was not all in the lab and it had practical applications, as he described processes for the preparation of steel, hair dyes, metal refinement, dyeing cloth and leather, making varnishes to waterproof cloth, and illuminating manuscript ink. Some of his most groundbreaking work was in acids and in discovering sulfuric and hydrochloric acid.

Sinan

Full name: Koca Mimar Sinan
Born: 1489
Died: 1588
Most influential work: Designing and building more than 477 buildings, including the Selimiye Mosque in Edirne, which has the tallest earthquake-defying minarets in Turkey
Go to: Architecture in Town

Sinan was the son of Greek Orthodox Christian parents who embraced Islam. His father was a stonemason and a carpenter, and from an early

Sinan

age Sinan followed in his footsteps, learning the skills of his trade. At 21, he was recruited into the Janissary Corps and as a conscript he mentioned that he wanted to learn carpentry. This led to him eventually building ships, wooden bridges, and many kinds of temporary wooden constructions.

Through military service he participated in a number of Ottoman campaigns and gained experience building and repairing bridges, defenses, and castles. The Ottoman sultans noticed his talents and he became their chief architect, constructing mosques, schools, and other civic buildings all around the Muslim (Ottoman) world, from Turkey to Damascus, Mecca, and Bosnia.

He is also honored with a crater on Mercury named after him.

Zheng He

FULL NAME: Born as Ma He; then his name changed to Zheng He as he was awarded the supreme command of the Chinese Imperial Household Agency
BORN: 1371 in Kunming, China
DIED: 1433 in India
MOST INFLUENTIAL WORK: Transformed China into the 15th century's regional, and perhaps world, superpower by making seven monumental sea voyages
GO TO: Naval Exploration in World

Zheng He was the admiral of the Chinese fleet, and within 28 years of travel he had visited 37 countries in the name of trade and diplomacy. The expeditions covered more than 50,000 kilometers, and his first fleet included 27,870 men on 317 ships. Today, it is not known how his ships, which were more than 400 feet long, were built without metal. These massive vessels were five times as big as the vessels of other European explorers like Vasco da Gama and were described as "swimming dragons" because they were dotted with dragon's eyes to help them "see."

Some of the lands the great fleet visited included Java, Sumatra, Ceylon, Siam, the East Indies, Bengal, the Maldives, the Persian Sultanate of Ormuz, Ryukyu and Brunei, Borneo, Mogadishu, Mombasa, and other East African ports. They even possibly rounded the Cape of Good Hope.

These voyages fostered scientific discovery and the search for gems, minerals, plants, animals, drugs, and medicine. They also improved navigational and cartographical knowledge of the world; developed international relations; and traded large quantities of cargo, including silk and cotton goods, porcelain, gold and silverware, copper utensils, and iron implements. They also carried live animals including giraffes and ostriches, had watertight bulkheads to hold live fish and also make bathhouses, and used otters to round up fish into large nets.

Zheng He

02 Europe's Leading Minds

There are numerous European scholars who had groundbreaking achievements and are remembered for their outstanding contributions to modern-day science and discovery. We selected the following, listed chronologically, whose genius rose above the knowledge of their day, who had long-lasting impact on science and technology, and who were in harmony with or may have been influenced by Muslims.

Roger Bacon (1214-1292)
This Oxford scholar is known as the originator of the experimental method in western Europe, and he received his training from the pupils of Spanish Moors. He spoke Arabic and never tired of telling people that knowledge of Arabic and Arab science was the only way to true knowledge.

Bacon quotes Ibn al-Haytham or refers to him at almost every step in the optics section of his *Opus Maius*. Part VI of this work also rests almost entirely on the findings of Ibn al-Haytham, especially in areas relating to the intromission theory of vision. It was Ibn al-Haytham who introduced the scientific, experimental method, and it was this that Bacon picked up on.

Al-Kindi was another source of inspiration for Bacon, and his two treatises on geometrical and physiological optics were used by the Englishman.

Ibn Firnas's flying machine inspired Bacon's flying machine or ornithopter, which he described in his manuscript *De Mirabili Potestate Artis et Naturae* or *On the Marvelous Powers of Art and Nature* from 1260. It is known that Bacon studied in Córdoba, the homeland of Ibn Firnas.

Bacon's writing on gunpowder was based on Muslim sources, and the Latin book *Liber Ignium of Marcus Graecus*, which gives many recipes for making gunpowder, was originally in Arabic and translated in Spain.

Roger Bacon became acquainted with Muslim chemistry from the Latin translations of Arabic works, and believed in the great importance of alchemy, and in transmutation.

His chief guide in medicine was the *Canon* of Ibn Sina, which he cites as frequently as all these other writers combined.

The book that had the greatest impact upon Bacon's method of thinking, and made him different from his Western contemporaries, was *The Book of the Secret of the Secrets* by ninth-century Zakariya' al-Razi, known in the West as Rhazes. In Latin, this was *Secretum Secretorum*.

Leonardo da Vinci (1452-1519)
Leonardo da Vinci was an Italian painter, sculptor, architect, musician, engineer, mathematician, and scientist, and a key figure in late Renaissance Europe. Recent research suggested that his mother was of North African origin.

Da Vinci drew "The Vitruvian Man," a man of perfect proportions in two superimposed positions with his arms apart, appearing in a circle and square, which illustrated the text of the *Roman Canon of Vitruvius*. Da Vinci's drawing was seen as innovative because he said the man's center, when drawn in a square, was not his navel, as the *Canon* stated, but lower. But five centuries earlier in the tenth-century Muslim scholars *Ikhwan al-Safa'*, or the Brothers of Purity, had come to the same conclusion, saying that the center of the figure was the navel only for a child under seven, and after this the center moved to the groin area. Historians have acknowledged Ibn Sina's *Book of Cure, Healing or Remedy from Ignorance* as an inspiring source of thought for the founders of geometrical thought in Europe, including Leonardo da Vinci.

Ibn al-Haytham invented a camera obscura long before Leonardo da Vinci produced the full and developed camera design.

Da Vinci found arabesque designs fascinating and worked out his own complicated patterns. The Muslim knot, in particular, intrigued him so he produced two plates of six knots, which were later reproduced in circular copper engravings by one of his followers in Milan around 1483 and 1499.

Nicolaus Copernicus (1473-1543)
Polish scientist Nicolaus Copernicus is said to be the founder of modern astronomy.

Many of his theories were based on those of Nasir al-Din al-Tusi and Ibn al-Shatir. Ibn al-Shatir's planetary theory and models are mathematically identical to those prepared by Copernicus more than a century later. Copernicus would have come into contact with these in Italy, where he studied.

Another influence on Copernicus is believed to have been the famous Toledan Tables written by Al-Zarqali, who was born in 1028.

It is known that Copernicus relied heavily on the comprehensive astronomical treatise of Al-Battani that included star catalogs and planetary tables. He relies extensively on Al-Zarqali and Al-Battani in his book *De Revolutionibus*.

Tycho Brahe (1546-1601)
This leading Danish Renaissance astronomer was credited with many influential works, including the production of the quadrant and one of Europe's leading observatories.

He is renowned for rediscovering the moon's variation, which was first discovered by a Muslim astronomer Mohammed Abu al-Wafa' al-Buzjani about 600 years earlier.

Tycho's famous mural quadrant was like those developed in eastern regions of the Muslim world, especially by astronomer Taqi al-Din.

Johannes Kepler (1571-1630)
Johannes Kepler is renowned in the West for discovering the laws of planetary motion and for his work on optics; as the founder of the first correct mathematical theory of the camera obscura; and for providing the first correct explanation of the working of the human eye, with an upside-down picture formed on the retina.

Ibn al-Haytham's influence can easily be detected in Kepler's work, as the former had revolutionized optics 600 years earlier. His *Kitab al-Manazir* or *Book of Optics* was translated into Latin by Gerard of Cremona and called *Perspectiva* or *De aspectibus*.

Both Kepler and Descartes relied upon Ibn al-Haytham's studies on the refraction of light, and Kepler took up where Ibn al-Haytham left off.

Kepler developed the camera obscura after its first discovery by Ibn al-Haytham, improving it with a negative lens behind the positive lens, which enlarged the projected image—the principle used in the modern telephoto lens.

Robert Boyle (1627-1691)
Robert Boyle is widely regarded as one of the founders of modern chemistry. He is best known for Boyle's gas law, which relates the pressure and volume of a gas in a closed system. He studied Eastern languages like Arabic and Syriac because of the manuscripts of natural philosophy written in those languages. He had access to works such as that of Al-Iraqi, 13th-century Muslim chemist, and the tables of longitude and latitude compiled by the 14th-century Syrian geographer Abulfeda. He often turned to the ancient practices of Muslim chemists like Jabir ibn Hayyan (Geber). Boyle and Geber both championed the experimental approach to chemistry, despite the nine centuries between them.

03 A Thousand Years of Scholarship

Many of the individuals mentioned in this book are listed below for your reference, including their names, birth and death dates, places of birth and work, and profession. The names in bold refer to how these people were commonly known, as many of the scholars, coming from distinguished families, had long names.

'Abd al-Rahman III (891-961); the caliph of Córdoba, Spain (912-961); a man of wisdom and patron of arts; founder of Madinat al-Zahra (a city now in ruins) on the outskirts of Córdoba.

Yahya ibn Abi Mansour (ninth century); Baghdad, Iraq; astronomer at the court of Al-Ma'mun. He compiled the so-called *Al-Zij al-Mumtahan* or *The Validated Zij.*

Abu Abdullah al-Bakri (1014-1094); Huelva, Spain; geographer and historian.

Abu al-Fida' (1273-1331); Hama, Syria; geographer and astronomer.

Abu al-Wafa', Mohammed al-Bouzjani (940-998); Buzjani, Baghdad, Iraq; mathematician, astronomer, and geometrician.

Ad-Dakhwar (early 13th century); Aleppo, Syria; physician at Al-Nuri Hospital.

Adelard of Bath (ca 1080–ca 1160); Bath, England; mathematician and philosopher.

Albertus Magnus, also known as Albert the Great (1206-1280); Bavaria; scientist, philosopher, and theologian.

Alfonso X, also known as Alfonso the Wise (1221-1284); Spanish king of Castile and León (1252-1284); son and successor of Ferdinand III.

Archimedes (287-212 B.C.E.); Syracuse, Sicily; astronomer.

Aristotle (383-322 B.C.E.); Stagirus, Greece; philosopher.

Roger Bacon (1214-1292); Ilchester, England; physicist, chemist, and mathematician.

Al-Baghdadi, real name: Abu Mansur Abr al-Qahir ibn Tahir ibn Muhammad ibn Abdallah al-Tamini al-Shaffi, known as Ibn Tahir (980-1037); Baghdad, Iraq; mathematician.

Banu Musa brothers (ninth century); Baghdad, Iraq; Ibn Musa, Jafar Muhammad ibn Shakir (800-873); geometry and astronomy; Ibn Musa, Ahmed ibn Shakir (805-873); mechanics; Ibn Musa, Al-Hasan ibn Shakir (810-873); geometry.

Al-Battani, Abu 'Abdallah Muhammad ibn Jabir, known as Albategnius (858-929); born in Harran, Turkey, and worked in Baghdad, Iraq; astronomer and mathematician.

Baybars, al-Malik al-Zahir Rukn al-Din Baybars al-Bunduqdari (1223-1277); Solhat, Turkey; Mamluk sultan who rose to power from being a slave, ruled Egypt and Syria (1260-1277); defeated the Mongols at Battle of Ayn Jalut.

Al-Biruni, Mohammed ibn Ahmed Abul Rayhan (973-1050); born in Khwarizm, died in Gazna; mathematician, geographer, pharmacy, medicine, physics, and earth science scholar.

Al-Bitruji, Nur al-Din ibn Ishaq (d. 1204), also known as Alpetragius; Morocco and Seville; astronomer.

Tycho Brahe (1546-1601); Skane, Denmark; astronomer and engineer.

Robert Boyle (1627-1691); England; natural philosopher and chemist. One of the founders of modern chemistry, and one of the pioneers of the modern experimental scientific method.

Nicolaus Copernicus, Mikolaj Kopernik or Nicolaus Koppernigk (1473-1543); Thorn (Torun), Poland; astronomer and mathematician.

Al-Dimashqi (1256-1327); Damascus, Syria; traveler and explorer.

Al-Dinawari, Abu Hanifa (d. 895); Andalusia, Spain; botanist.

Edward I (1239-1307); king of England (1272-1309); went on Crusades to Acres (1271-1272); on his return he built castles on the Muslim plan, using the barbican design.

Euclid (325-265 B.C.E.); Alexandria, Egypt; Greek mathematician.

Al-Farabi, Abu Nasr (870-950), also known as Alpharabius; near Farab, Khazakhstan, but flourished and worked in Iraq; philosopher and music theorist.

Al-Farghani, Abu-al-Abbas Ahmad ibn Kathir, known as Alfraganus (d. 861); Farghana, Transoxiana; astronomer and surgeon.

Muhammad al-Fatih, known as Mehmed II or al-Fatih (1432-1481); Adrianople, Thrace, Turkey; Ottoman sultan who ruled from Constantinople (1451-1481); conqueror of Constantinople.

Al-Fazari, Abu Abdullah Muhammad ibn Ibrahim (d. ca 777); Kunduz, Afghanistan; mathematician, philosopher, poet, and astronomer. The first Muslim astronomer to construct astrolabes.

Leonardo Fibonacci (1170-1250); Pisa, Italy; mathematician.

Fatima al-Fihri (ninth century); nicknamed Um al-Banin or "the mother of children"; Fez, Morocco; art and building patron, founder of Al-Qarawiyin University, Fez.

Al-Firdawsi, Abu al-Qasim Mansur (940-1020); Korasan, Iran; historian and chronicler.

Frederick II (1194-1250); king of Sicily (1198-1250); holy roman emperor (1220-1250).

Galen, Claudius (ca 131–206); Pergamum/Bergama, Turkey; physician.

Gerard of Cremona (ca 1114-1187); Lombardy, Italy; translator.

Al-Ghafiqi, Muhammad ibn Qassum ibn Aslam (d. 1165); physician, eye surgeon, and herbalist.

Al-Ghazali, Abu Hamed, known in the West as Algazel (1058-1128); Khorasan, Iran; philosopher and theologian.

Al-Hakam I (796-823); ruled Córdoba.

Al-Hakam II (915-978); Córdoba, Spain; son of Abderrahamn III; ruled Al-Andalus from 961 to 978; famous for his library.

Ahmad al-Halabi (d. 1455); Aleppo, Syria; astronomer.

Abu Bakr ibn al-Sarraj al-Hamawi (d. 1328/9); Hama, Syria; geometer, astronomer, and engineer.

Al-Hanbali, Taqi al-Din (1236-1328); Harran, Turkey; theologian; Quranic exegesis *(tafsir)*; hadith and jurisprudence.

Abu Ishaq Ibrahim ibn Ishaq al-Harbi (d. 285); Baghdad, Iraq; prominent companion and theologian of the Hanbali School of Thought.

Harun al-Rashid (766-809); the fifth Abbasid caliph, who ruled from Baghdad (786-809); famously known for his good relations with Charlemagne, to whom he sent a delegation with gifts including a hydraulic organ.

Hazarfen Ahmed Celebi (17th century); Istanbul, Turkey; pilot flying in 1638 from the Galata tower near the Bosporus in Istanbul, landing on the other side of the Bosporus.

Hippocrates (ca 460–377 B.C.E.); Kos Island, Greece; physician.

Hunayn ibn Ishaq, al-'Ibadi (808-873); Baghdad, Iraq; member of the House of Wisdom; translator of Greek works into Arabic; physician.

Ibn Abi Usaybi'ah (d. 1270); Damascus, Syria (practiced in Egypt); chronicler of physicians and pharmacists; physician and oculist.

Ibn 'Aqil, Abu al-Wafa Ali (1040-1119); Baghdad, Iraq; theologian of the Hanbali School of Thought and humanist.

Ibn al-Awwam (12th century); Seville, Spain; agriculturist.

Ibn al-Baytar, Abu Muhammad Dia' al-Din Abdullah ibn Ahmad (1197-1248); Málaga, Spain; physician, herbalist, pharmacist, and botanist

Ibn al-Faqih, al-Hamadhani (tenth century); Baghdad, Iraq; geographer and traveler.

Ibn al-Haytham, Abu Ali al-Hasan (965-1039), also known as Alhazen; Syria, Egypt; physicist and mathematician.

Ibn al-Hajj, Muhammad ibn Muhammad, Abu Abdullah (1258-1336); Fez, Morocco; educationalist and theologian.

Ibn al-Jazzar, Abu Ja'far Ahmad ibn Abi Khalid (ca 855-955); Al-Qaryawan, Tunisia; physician.

Ibn al-Nadim, Abu al-Faraj Muhammad ibn Ishaq ibn Muhammad ibn Ishaq (tenth century); Baghdad, Iraq; bibliographer and the author of the *Kitab al-Fihrist*; bookseller and calligrapher.

Ibn al-Nafis, Abu-Alhassan Alauldin Ali ibn Abi Hazm al-Qurashi (1210-1288); Damascus, Syria, and flourished and worked in Cairo, Egypt; physician and discoverer of the circulation of the blood.

Ibn al-Quff, Abu'l-Faraj ibn Yacqub ibn Ishaq Amin al-Dawla al-Karaki (1233-1286); Damascus, Syria; physician.

Ibn al-Saffar, Abu al-Qasim Ahmed ibn Abdallah ibn Omar al-Ghafiqi, best known under the name of Ibn al-Saffar, meaning "son of coppersmith" (d. 1035); Córdoba, Spain; mathematician and astronomer.

Ibn al-Shatir al-Muwaqqit (1304-1375); Damascus, Syria; astronomer and timekeeper at the Umayyad Great Mosque of Damascus.

Ibn al-Thahabi, Abu Mohammed Abdellah ibn Mohammed al-Azdi (d. 1033); Suhar, Oman; physician and encyclopedist.

Ibn al-Wafid, Abu al-Mutarrif abd al-Rahman (1008-1074); also known as Abenguefit; Toledo, Spain; physician and pharmacologist.

Ibn Badis, al-Mu'izz (1007-1061); Tunisia; historian, scientist, chemist, and ruler of North Africa (1016-1062).

Ibn Bajjah, Abu Bakr Muhammad ibn Yahya ibn as-Say'igh, known as Avempace in the West (d. 1138); Saragossa, Spain; philosopher and physician.

Ibn Bassal, Abu 'Abd Allah Muhammad ibn Ibrahim al-Tulaytuli (1085); Toledo, Spain; botanist, agriculturist, and gardener.

Ibn Battuta, Abu Abdullah Muhammad (1304-1368/70); Tangier, Morocco; traveler, explorer, and chronicler.

Ibn Fadlan, Ahmed (tenth century); Baghdad, Iraq; explorer, traveler, and chronicler.

Ibn Firnas, 'Abbas (d. 887); Korah, Takrna, Spain; humanitarian, technologist, and chemist.

Ibn Hawqal, Abu Al-Qasim Muhammad (920-990); Nisibin, Iraq; explorer, traveler, and chronicler.

Ibn Hazm, Abu Muhammad 'Ali ibn Ahmad ibn Sa'id (994-1064); Córdoba, Spain; theologian and man of letters.

Ibn Isa, Ali (tenth century); also known as Jesu Haly; Baghdad, Iraq; physician and oculist.

Ibn Jubayr, Abu al-Husayn Muhammad ibn Ahmad ibn Jubayr (12th century); Granada, Spain; traveler, explorer, and chronicler.

Ibn Juljul al-Andalusi (ca 943); Córdoba, Spain; physician, herbalist, and pharmacist.

Ali ibn Khalaf al-Shakkaz (11th century); Toledo, Spain; an apothecary or herbalist, astronomer.

Ibn Khaldun, Abd al-Rahman ibn Mohammad (1332-1406); Tunis, Tunisia; sociologist, historian, philosopher, and economist.

Ibn Khurradadhbih (820-912); Baghdad, Iraq; geographer and director of the government postal service in Baghdad.

Ibn Majid, Shihab al-Din Ahmed al-Najdi (1432-1500); Najd, Saudi Arabia; navigator.

Ibn Muqla, Abu-Ali Mohammed (866-940); Baghdad, Iraq; Abbasid vizier, calligrapher, and one of the inventors of the *Naskhi* script.

Ibn Rushd, Abu'l Walid Muhammad al-Qurtubi, also known as Averroes (1126-1198); Córdoba,

Spain; philosopher, physician, humanist, and judge.

Ibn Rustah, Ahmed (tenth century); Isfahan, Iran; explorer and geographer.

Ibn Sa'id al-Maghribi (1214-1274); Granada, Spain; historian, poet, traveler, and geographer.

Ibn Samajun (d. 1002); Andalusia, Spain; herbalist, botanist, and pharmacologist.

Ibn Sarabiyun, Yuhanna, also known as Serapion (ninth century); Syria; physician and pharmacist.

Ibn Sina, also known as Avicenna (980-1037); Bukhara, Uzbekistan; physician, philosopher, mathematician, and astronomer.

Ibn Toloun, Ahmad (835-884); originally was in the service of the Abbasid caliph and moved to become governor of Egypt as part of the Abbasid Caliphate. He built the famous Ibn Tulun Mosque in Cairo.

Ibn Tufayl, Abu Bakr ibn Abd al-Malik ibn Muhammad ibn Muhammad ibn Tufayl al-Qaysi, also known as Abubacer (d. 1185); Granada, Spain; philosopher, physician, and politician.

Ibn Yunus, Abu'l-Hasan Ali ibn Abd al-Rahman ibn Ahmad al-Sadafi (950-1009); Fustat, Cairo, Egypt; mathematician and astronomer who compiled the Hakemite Tables.

Ibn Zuhr, Abu Marwan (1091-1161); also known as Avenzoar; Seville, Spain; physician, and surgeon.

Al-Idrisi (1099-1166); Ceuta (Morocco) and Palermo, Sicily; geographer and cartographer.

Ikhwan al-Safa', also known as Brothers of Purity (ca 983); Basra, Iraq; group of philosophers.

'Izz al-Din al-Wafa'i (d. 1469); Cairo, Egypt; astronomer and mathematician.

Jabir ibn Aflah (1100-1145); Seville, Spain; mathematician and astronomer.

Jabir ibn Hayyan, Abu Musa, also known as Geber (722-815); Tus, Iran, and lived and worked in Kufa, Iraq; chemist, druggist, and physician.

Al-Jahiz, Abu Uthman Amr ibn Bahr (ca 776-868); Basra, Iraq; philosopher and zoologist.

Al-Jazari, Badi'al-Zaman Abu al-'Izz Isma'il b al-Razzaz (early 13th century); Diyarbakir, Turkey; engineer.

Al-Jurjani, Abu Ruh Muhammad ibn Mansur ibn Abdullah (ca 1088); Astarabad, Iran; oculist and surgeon.

Kamal al-Din, Abu'l Hasan Muhammad al-Farisi (ca 1260-1319); Tabriz, Iran; mathematician and physicist.

Al-Karaji, Abu Bekr ibn Muhammad ibn al-Husayn, also known as Al-Karkhi (953-1029); Baghdad, Iraq; mathematician and engineer. He wrote the *Al-Fakhri*.

Al-Kashghari, Mahmud (1073); Turkey; geographer and lexicographer.

Al-Kashi, Ghiyat al-Din (1380-1429); Kashan, Iran; mathematician and astronomer.

Johannes Kepler (1571-1630); near Stuttgart, Germany; mathematician and astronomer.

Al-Khujandi, Abu Mahmud Hamid ibn al-Khidr (940-1000); Khudzhand, Tajikistan; astronomer, built an observatory in Ray, Iran, and constructed a sextant.

Al-Khwarizmi, Muhammad ibn Musa (780-850); Khwarizm, Iran; mathematician, astronomer, and geographer from whose name the word "algorithm" comes.

Al-Kindi, Abu Yusuf Yaqub ibn Ishaq al-Sabbah (801-873); Kufa, Iraq; cryptanalyst, mathematician, astronomer, physician, and geographer; also talented musician.

Al-Kuhi, Abu Sahl Wijan ibn Rustam (tenth century), born in Kuh in Tabaristan, in north Iran but worked and flourished in Baghdad around 988; mathematician and astronomer.

Leonardo da Vinci (1452-1519); Venice, Italy; painter, draftsman, sculptor, architect, and engineer.

Al-Majusi, 'Ali ibn al-'Abbas (tenth century); Ahwaz, Iran; geographer.

Al-Ma'mun, Abu Jafar al-Ma'mun ibn Harun (786-833); one of the most enlightened Abbasid caliphs, who ruled from 813 until 833; he expanded the House of Wisdom.

Al-Mansur, Abu Jafar Abdullah ibn Muhammad al-Mansur (712-775); Abbasid caliph who ruled from Baghdad (754–775); the founder of Baghdad in 762.

Al-Mansur, Yaqub (1160-1199); Marrakech, Morocco; Almohad sultan who ruled from Marrakech (1184-1199), succeeding his father, Abu Yaqub Yuf, who ruled from 1163 until 1184.

Al-Maqrizi, Taqi al-Din Ahmad ibn 'Ali ibn 'Abd al-Qadir ibn Muhammad (1364-1442); Cairo, Egypt; historian.

Yahya ibn Masawayh, Abu Zakariah (776-857); Baghdad, Iraq; physician, pharmacologist, earth scientist, and translator.

Masha'Allah 'Ali ibn 'Isa' (d. 815); Cairo, Egypt; astronomer and mathematician.

Maslama ibn Ahmad al-Majriti (d. 1007); Madrid, Spain; astronomer and mathematician.

Al-Masudi, Abul-Hassan Ali ibn al-Hussain (871–957); Baghdad, Iraq; explorer, geographer, and historian.

Michelangelo di Lodovico Buonarroti (1475-1564); Tuscany, Italy; Italian Renaissance sculptor, painter, architect, and poet.

Lady Mary Wortley Montagu (1689-1762); London, England; prominent member of society and wife of the British ambassador to the Ottoman Empire; brought smallpox inoculation from Turkey.

Al-Mawsili, Ammar ibn Ali (tenth century); Mosul, Iraq; eye surgeon and ophthalmologist.

Al-Mu'izz, Li-din Allah (930-975); a powerful Fatimid caliph who expanded the Fatimid rule from North Africa to Egypt; the founder of Islamic Cairo, Al-Qahirah, in 972/3 and the mosque of Al-Azhar.

Al-Muktafi (d. 908); Abbasid caliph who ruled from Baghdad (902-908).

Al-Muqaddasi, Muhammad ibn Ahmad Shams al-Din (ca 945-1000); Jerusalem; historian and geographer.

Al-Mutawakkil; Abbasid caliph who ruled from Samarra, Iraq (847-861), which was the short-lived Abbasid capital founded by his father, Al-Mu'tassim.

Muwaffaq, Abu al-Mansur (tenth century); Herat, Afghanistan; pharmacist.

Nur al-Din ibn Zangi (1118-1174); sultan who ruled Aleppo and Damascus, Syria; built one of the earliest hospitals, Al-Nuri Hospital.

Al-Nuwayri, Abu al-Abbas Ahmad (1278-332); Cairo, Egypt; historian.

Palladio, Andrea (1505-1580); Padua, Italy; architect and painter.

Piri Reis, Ibn Haji Muhammad (1465-1554); Gallipoli, Turkey; sea admiral, geographer, explorer, and cartographer.

Plato (427-347 B.C.E.); Athens, Greece; philosopher.

Claudius Ptolemaeus, also known as **Ptolemy** (85-165 C.E.); Alexandria, Egypt; geographer and astronomer.

Qalawun, Saif ad-Din al-Alfi al-Mansur (1222-1290); Mamluk sultan who ruled Egypt (1279-1290); in 1284 he built the famous and important Al-Mansuri Hospital.

Al-Qazwini, Zakariya' ibn Muhammad (1203-1283); Qazwin, Iran; traveler, explorer, and judge *(qadi)*.

Baylak al-Qibjaqi (c 1282); Istanbul, Turkey; explorer, seafarer, and geographer.

Qutb al-Din al-Shirazi (1236-1311); Shiraz, Iran; astronomer.

Al-Rammah, al-Hassan Najm al-Din (ca 1285); Syria; engineer and military historian.

Raphael, Raphaello (1483-1520); Urbino, Italy; painter and architect.

Al-Razi, Abu Bakr Muhammad ibn Zakariya (865-925); Ray, Iran; physician and chemist known in the West as Rhazes.

Roger II (1093-1154); Palermo, Sicily; Norman

king who ruled Sicily (1130-1154); son and successor of Roger I, famous for his interest in geography and support of the Muslim geographer Al-Idrisi.

Sabur ibn Sahl, also spelled Shapur (d. 869); Jundishapur, Iran; physician and pharmacist.

Saif al-Dawla, Abu al-Hasan ibn Hamdan (916-967); ruler of Aleppo and founder of the Hamadanid dynasty of Aleppo. He was famous for his patronage of scholars.

Al-Samawal, Ibn Yahia al-Maghribi (d. 1180); Baghdad, Iraq; mathematician and astronomer.

Michael Scott (ca 1175–ca 1236); Scotland; physician, astrologer, and translator.

Sibawaih (760-793); Bayza or Bayda, Iran; grammarian, considered the most important Arabic grammarian upon whose work all other Arabic grammars are based.

Sinan, Koca Mimar Sinan (1489-1588); Istanbul, Turkey; architect and designer.

Ibrahim ibn Sinan ibn Thabit ibn Qurra (908-946); Harran, Turkey; geometer, astronomer, and mathematician.

Al-Sufi, 'Abd al-Rahman (903-986); Isfahan, Iran; astronomer.

Al-Suli, Abu Bakr Muhammad (tenth century); great master of chess.

Pope Sylvester II, Gerbert of Aurillac (940/950-1003); Auvergne, France; pope (999-1003), philosopher, mathematician, and translator, brought Arabic numerals to Europe before being poisoned.

Umar ibn Farrukhan al-Tabari, also known as Omar Alfraganus (ninth century); Tabaristan, Iran; astrologer; compiled the *Liber universus*.

Taqi al-Din al-Rasid, Muhammad ibn Ma'rouf al-Shami al-Asadi (ca 1526-1585); Damascus, Syria; astronomer, engineer, and mechanic.

Thabit ibn Qurra (ca 836–901); Harran, Turkey; geometer, mathematician, astronomer, and translator of Greek work into Arabic.

Al-Tusi, Nasir al-Din (1201-1274); Maragha (Tus), Khorasan, Iran; astronomer, mathematician, and philosopher.

Ulugh Beg, Muhammed Taragai (1394-1449); Samarkand, Uzbekistan; astronomer.

Umar al-Khayyam, Ghiyath al-Din Abu'l-Fath Umar ibn Ibrahim Al-Nisaburi (1048-1122); Nishapur, Iran; astronomer and mathematician.

Umar ibn al-Khattab, ibn Nufayl ibn 'Abd al-'Uzza ibn Rayyah (ca 581-644); companion of Prophet Muhammad and second caliph, ruling from Medina, Saudi Arabia (634-644).

Uthman ibn Affan, ibn Abi Al-'As ibn Umayyah (577-656); companion of Prophet Muhammad and third caliph (644-656).

Vitruvius, Marcus Pollio (ca 70-ca 25 B.C.E.); Rome, Italy; architect and engineer.

Al-Walid ibn 'Abdulmalik ibn Marwan (668-715); Umayyad caliph who ruled from Damascus, Syria (705-715); he built the Umayyad Mosque in Damascus.

Sir Christopher Wren (1632-1723); London, United Kingdom; architect, astronomer, and mathematician.

Sanad ibn Ali al-Yahoudi (ninth century); Baghdad, Iraq; Jew converted into Islam, chief astronomer of Al-Ma'mun; distinguished member of the House of Wisdom.

Yaqut, Ibn-'Abdullah Rumi al-Hamawi (1179-1229); Arab biographer, historian, and geographer.

Al-Zahrawi, Abul Qasim Khalaf ibn al-Abbas, known in the West as Abulcasis (936-1013); Córdoba, Spain; physician and surgeon.

Al-Zarqali, Abu Ishaq Ibrahim ibn Yahya, also known as Arzachel (1029-1087); Toledo, Spain; astronomer who compiled the Toledan Tables.

Zheng He (1371-1433); Kunming, China; navigator and admiral.

Ziryab, Abul-Hasan Ali ibn Nafi' (789-857); Baghdad, Iraq; musician, astronomer, fashion designer, and gastronome.

04 AUTHORS AND TREATISES

Below are the titles of manuscripts, treatises, and books by and about some of the Muslim scholars mentioned in this book, and details of where the material can be found.

Locating original manuscripts is difficult. Since a thousand years have passed, they often do not exist anymore. Reasons for this vary, from libraries being burned in the Middle Ages, wars, and natural disasters to medieval scholarly rivalry that led to manuscripts being destroyed and a lack of preservation over the centuries. Thousands of original manuscripts also remain to be catalogued in many libraries, and some have yet to be located. Experts say as many as five million manuscripts exist and only about 60,000 of them have been edited.

Fortunately, copies and translations of many manuscripts have been preserved over the centuries in libraries such as the British Library (London), Topkapi Palace Museum Library (Turkey), Suleymaniye Library (Turkey), National Library of Medicine (U.S.), Princeton University Library (U.S.), Vatican Library, Leiden Library (Holland), and Cambridge and Oxford university libraries (U.K.).

HOME

On the Coffee Trail

'Abd al-Kadir ibn Muhammad al-Ansari al-Djaziri. *Umdat al-Safwa fi hill al-qahwa.* Partly edited in De Sacy, *Chrestomathie Arabe,* 2nd edition. Imprimerie royale, Paris, 1826.

'Abd al-Kadir ibn Shaykh ibn al-'Aydarus. *Safwat al-Safwa fi bayan hukm al-qahwa.* Ahlwardt, Verzeichnis, Bibliothek Berlin, MS 5479, 23 volumes, Berlin, Germany, 1853-1914.

Hattox, R. S. *Coffee and Coffeehouses: The Origins of a Social Beverage in the Medieval Near East.* University of Washington Press, Seattle and London, 1988.

Clocks

Al-Jazari. *Al-Jami' Bayn al-'Ilm al-Nafi' wa Sina'at al-Hiyal,* or *The Book of Knowledge of Ingenious Mechanical Devices.* Suleymaniye Library, Ayasofya Collection, MS 3606, Istanbul.

Al-Jazari. *Texts and Studies.* Collected and reprinted by Fuat Sezgin in collaboration with Farid Benfeghoul, Carl Ehrig-Eggert, and Eckhard Neubauer. Institute for the History of Arabic-Islamic Science at the Johann Wolfgang Goethe University, Frankfurt, 2001.

Hill, Donald R. *On the Construction of Water Clocks,* or *Kitab Arshimidas fi 'amal al-binkamat.* Turner and Devereaux, London, 1976.

Hill, Donald R. *Arabic Water-Clocks.* Institute for the History of Arabic Science, Aleppo, Syria, 1981.

Hill, Donald R. "Islamic Fine Technology and Its Influence on the Development of European Horology," in *Al-Abhath,* Vol. 35, pp. 8-28. American University of Beirut, Beirut, 1987.

Taqi al-Din. *Al-Kawakib al-durriyya fi al-binqamat al-dawriyya,* or *Pearl Stars on Cyclic Water Clocks.* Dar al-Kutub, MS Miqat 557/1, Cairo.

Taqi al-Din. *Alat al-rasadiya li-Zij al-shahinshahiyya.* Library of the Topkapi Palace Museum, MS Hazine 452, Istanbul.

Taqi al-Din. *Rayhanat al-ruh fi rasm al-sa'at 'ala mustawi'l-sutuh.* Vatican Library, MS 1424, Vatican City.

Tekeli, Sevim. *The Clocks in the Ottoman Empire in the 16th Century and Taqi al-Din's The Brightest Stars for the Construction of the Mechanical Clocks.* Ankara University Basimevi, Ankara, Turkey, 1966.

Chess

Al-Hanbali. *Kitab namudhaj al-qital fi naql al-'awal,* or *The Book of the Examples of Warfare in the Game of Chess.* Zuhayr Ahmad al-Qisi (editor). Dar-al-rashid, Baghdad, 1980.

Al-Suli. *Kitab al-shatranj,* or *Muntahab Kitab al-shatranj.* Suleymaniye Library, Lala Ismail Collection, MS 560, Istanbul.

Al-Suli. *Kitab al-shatranj.* Fuat Sezgin (publisher). Institut für Geschichte der Arabisch-Islamischen Wissenschaften, Frankfurt, 1986.

Murray, H. J. R. *A History of Chess.* Oxford University Press,

London, 1913; reprinted Benjamin Press, Northampton, Mass., 1985.

Music

Al-Farabi. *Kitab al-Musiqi al-Kabir,* or *The Great Book of Music.* Koprulu Library, MS 953, Istanbul.

Al-Farabi. *Kitab al-Musiqi al-Kabir,* or *The Great Book of Music.* Eckhard Neubauer (editor). Institut für Geschichte der Arabisch-Islamischen Wissenschaften, Frankfurt, 1998.

Safi al-Din al-Baghdadi al-Urmuwi. *Kitab al-Adwar.* Vatican Library, MS 319/3, Vatican City.

Chabrier, J. C. "Musical Science," in *Encyclopedia of the History of Arabic Science.* Roshdi Rashed (editor) with the collaboration of Régis Morelon. 3 volumes. Routledge, London/New York, 1996.

Farmer, Henry George. *Studies in Oriental Music.* Eckhard Neubauer (editor). 2 volumes. IGAIW, Frankfurt, 1997.

Maalouf, Shireen. *History of Arabic Music Theory: Change and Continuity in the Tone Systems, Genres, and Scales.* Université Saint-Esprit, Kaslik, Lebanon, 2002.

Neubauer, Eckhard. *Arabische Musiktheorie von den Anfangen bis zum 6./12. Jahrhundert.* The Science of Music in Islam, Vol. 3. IGAIW, Frankfurt, 1998.

Shiloah, Amnon. *The Theory of Music in Arabic Writings (ca 900-1900).* Henle, München, 1979.

Shiloah, Amnon. *Music in the World of Islam: A Socio-Cultural Study.* Wayne State University Press, Detroit, 1995.

Touma, Habib Hassan. *The Music of the Arabs.* Laurie Schwartz (translator). Amadeus Press, Portland, Ore., 1996.

Cleanliness

Al-Kindi. *Kitab Kimiya' al-'itr wa't-tas'idat,* or *Book of the Chemistry of Perfume and Distillations.* K. Garbers (German translation), *Buch uber die Chemie des Parfums und die Distillationen,* Leipzig, Germany, 1948.

Al-Zahrawi. *Kitab al-Tasrif Liman 'Ajaz An al-Ta'lif* or *Al-Tasrif,* or *The Method of Medicine.* Suleymaniye Library, Bashir Agha Collection, MS 502, Istanbul, Turkey; and Al-Khizana al-Hasaniyya, MS 134, Rabat, Morocco.

Al-Zahrawi. *Texts and Studies I.* Fuat Sezgin (editor); Mazen Amawi, Carl Ehrig-Eggert, and Eckhard Neubauer (publishers). IGAIW, Frankfurt, 1996.

Trick Devices

Banu Musa Brothers. *Kitab al-hiyal al-Handasiyah,* or *The Book of Ingenious Mechanical Devices.* Vatican Library, MS 317/1,Vatican City; and Dar al-Kutub, MS Taymur Sina'a 69, Cairo.

Bir, Atilla. *Kitab al-hiyal of Banu Musa bin Shakir Interpreted in the Sense of Modern System and Control Engineering.* Preface and edition by Ekmeleddin Ihsanoglu (Studies and Sources on the History of Science, 4). Research Centre for Islamic History, Art, and Culture IRCICA, Istanbul, 1990.

Al-Hassan, Ahmad Y., and **Hill, Donald R.** *Islamic Technology. An Illustrated History.* UNESCO/Cambridge University Press, Paris/Cambridge, 1986.

Hill, Donald R. *The Book of Knowledge of Ingenious Mechanical Devices.* Reidel, Dordrecht, Netherlands, 1974.

Hill, Donald R. *Islamic Science and Engineering.* Edinburgh University Press, Edinburgh, 1993.

Al-Jazari. *Al-Jami' bayna al-ilm wa-al-amal al-naf fi sinaat al-hiyal,* or *The Book of Ingenious Devices.* Institute for the History of Arabic Science, Aleppo, Syria, 1979.

Taqi Al-Din. *Sublime Methods of Spiritual Machines,* in A. Y. Al-Hasan, *Taqi al-Din wa-'l-handasa al-mikanikiya al-'arabiya. Ma'a Kitab al-Turuq al-saniya fi-'l-alat al-ruhaniya min al-qarn al-sadis 'ashar.* Institute for the History of Arabic Science, Aleppo, Syria, 1976.

Vision and Cameras

Ibn al-Haytham. *Kitab al-Manazir,* or *Book of Optics,* known in Latin as *De aspectibus,* or *Perspective.* Suleymaniye Library, MS Ayasofya Collection, 2448, Istanbul.

Ibn al-Haytham. *Opticae Thesaurus Alhazeni Arabis libri septem, nunc primum editi. Eiusdem liber de crepusculis et nubium ascensionibus.* David C. Lindberg (editor). Johnson Reprint, New York/London, 1972

Ibn al-Haytham. *Kitab Al Manazir,* A. I. Sabra (editor), Books I-III. The Cultural Council, Kuwait, 1983.

Sabra, A. I. *The Optics of Ibn al-Haytham.* Books I-III, *On Direct Vision.* The Warburg Institute/University of London, London, 1989.

Ibn al-Haytham. *Kitab Al Manazir.* A. I Sabra (editor), Books IV-V. The Cultural Council, Kuwait, 2002.

Kamal al-Din al-Farisi. *Tanqih al-Manazir li-Zawi'l-absar wa'l-Basair.* Suleymaniye Library, Ayasofya Collection, MS 2598, Istanbul.

Kheirandish, Elaheh. *The Arabic Version of Euclid's Optics: Kitab Uqlidis fi ikhtilaf al-manazir.* Springer Verlag, Berlin, 1999.

Kheirandish, Elaheh. "Optics: Highlights from Islamic Lands," in *The Different Aspects of Islamic Culture.* Vol. IV: *Science and Technology in Islam,* Part 1, pp. 337-357. UNESCO, Paris, 2001.

Lindberg, David C. *Theories of Vision from Al-Kindi to Kepler.* University of Chicago Press, Chicago/London, 1976. Reprinted 1996.

Megri, Kheira. *L'Optique de Kamal al-Din al-Farisi.* Presses Universitaires du Septentrion, Villeneuve-d'Ascq, France, 1999.

Rashed, Roshdi. *Géométrie et dioptrique au Xème siècle: Ibn Sahl, al-Quhi et Ibn al-Haytham.* Les Belles Lettres, Paris, 1993.

Schramm, Matthias. "Zur Entwicklung der physiologischen Optik in der arabischen Literatur," in *Sudhoffs Archiv für Geschichte der Medizin und der Naturwissenschaften,* Vol. 43, pp. 289-316. 1959.

Sezgin, Fuat, et al. (editors). *Optics: Texts and Studies.* Natural Sciences in Islam, Vols. 32-34. IGAIW, Frankfurt, 2001.

Smith, A. M. *Alhacen's Theory of Visual Perception: A Critical Edition, with English Translation and Commentary of the First Three Books of Alhacen's De Aspectibus, the Medieval Latin Version of Ibn al-Haytham's Kitab al-Manazir.* 2 volumes. American Philosophical Society, Philadelphia, 2001.

Carpets

Aslanapa, Oktay (editor). *One Thousand Years of Turkish Carpets.* Eren, Istanbul, 1988.

Blair, S., and Bloom J. "Islamic Carpets," in *Islam: Art and Architecture,* M. Hattstein and P. Delius (editors), pp. 530-533, Konemann, Köln, Germany, 2000.

Rabah Saoud. "*The Muslim Carpet and the Origin of Carpeting.*" www.MuslimHeritage.com.

Sardar, Marika. "Carpets from the Islamic World, 1600-1800," in Heilbrunn Timeline of Art History. The Metropolitan Museum of Art, New York, 2000.

SCHOOL

Schools

Al-Ghazzali. *Ihya Ulum al-Din,* or *The Revival of Religious Sciences.* Badwi Tabana (editor). Nusrat Ali Nasri for Kitab Bhavan, New Delhi, 1982.

Gibb, H. A. R. "The University in the Arab-Moslem World," in *The University Outside Europe: Essays on the Development of University Institutions in Fourteen Countries,* pp. 281-298. Edward Bradby (editor). Ayer Publishing, New Hampshire, 1970.

"Al-Madrasa al-Nizāmiyya," in *Encyclopaedia of Islam,* 2nd edition. P. Bearman, Th. Bianquis , C. E. Bosworth, E. van Donzel, and W. P. Heinrichs (editors). E. J. Brill, Leiden, 2010.

Makdisi, George. "Madrasa and University in the Middle Ages," in *Studia Islamica,* No. 32, pp. 255-264. 1970.

Makdisi, George. "Muslim Institutions of Learning in Eleventh-Century Baghdad," in *Bulletin of the School of Oriental and African Studies.* Vol. 24, pp. 1-56. London, 1961.

Makdisi, George. *The Rise of Colleges. Institutions of Learning in Islam and the West.* Edinburgh University Press, Edinburgh, 1981.

Pedersen, J., Rahman, Munibur, and **Hillenbrand, R.** "Madrasa," in *Encyclopaedia of Islam,* 2nd edition. P. Bearman, Th. Bianquis , C. E. Bosworth, E. van Donzel, and W. P. Heinrichs (editors). E. J. Brill, Leiden, 2010.

Libraries and Bookshops

Atiyeh, George N. (editor). *The Book in the Islamic World: The Written Word and Communication in the Middle East.* State University of New York Press, Albany, 1995.

Al-Jahiz. *Al-Bayan wa'l-tabyin,* or *Eloquence and Elucidation.* Hasan Sandubi and Adab al-Jahiz (editors). Yale University Library, New Haven, Conn.

Al-Jahiz. *Al-Bayan wa'l-tabyin,* or *Eloquence and Elucidation,* Hasan al-Sandubi (editor). Ma tba'at al-istiqama, Cairo, 1947; 2nd edition, 1956.

Kohleberg, Etan. *A Medieval Muslim Scholar at Work. Ibn Tawus and His Library.* E. J. Brill, Leiden, 1992.

Al-Muqaddasi. *Ahsan al-Taqasim fi Ma'rifat al-Aqalim,* or *The Best Divisions for Knowledge of the Regions.* G. S. A. Ranking and Rizkallah F. Azoo (translators to English). Bombay, 1897-1910. Reprinted by Fuat Sezgin, Frankfurt, 1989.

Ibn al-Nadim. *Fihrist'al-Ulum,* or *The Catalogue or Index of the Sciences.* Suleymaniye Library, Sehid Ali Pasa 1934, Istanbul.

Ibn al-Nadim. *Kitab al-Fihrist,* mit Anmerkungen hrsg. von Gustav Flügel. 2 volumes. F. C. W. Vogel, Leipzig, Germany, 1871-72.

Mathematics, Trigonometry, and Geometry

Abgrall, Philippe. *Le développement de la géométrie aux IXe-XIe siècles: Abu Sahl al-Quhi.* Blanchard, Paris, 2004.

Ab' Kamil, Shuja Ibn Aslam. *The Algebra of Ab' Kamil: Kitab fi al-jabr wa 'l-muqabala.* Martin Levey (translator); foreword by Marshall Clagett; commentaries by Mordecai Finzi. The University of Wisconsin Press, Madison, Wisconsin, 1966.

Abu al-Wafa'. *Kitab al-Handasa,* or *Book of Geometry.* Cambridge University Library, MS Persian 169, Cambridge, U.K.

Abu al-Wafa'. *Kitab fima yahtaju ilayhi al-sani najara fi amal al-handasiyya,* or *On Those Parts of Geometry Needed by Craftsmen.* Suleymaniye Library, Ayasofya Collection, MS 2753, Istanbul.

Abu Bakr Muhamad ben al-Hasan al-Karaji. *Al-Kafi fi 'l-hisab (Genügendes über Arithmetik) von (4.-5. Jhd/10-11.Jhd.u.).* Ediert und kommentiert von Sami Chalhoub. IHAS, Aleppo, Syria, 1986.

Abu al-Hasan Ahmad b. Ibrahim al-Uqlidisi. *The Arithmetic of al-Uqlidisi. The Story of Hindu-Arabic Arithmetic as Told in 'Kitab al-fusul fi al-hisab al-hindi.* Translated and annotated by Ahmad Salim Saidan. Reidel, Dordrecht, Netherlands, 1978.

Al-Baghdadi. *Kitab al-takmila fi 'ilm al-Hisab,* or *Book of Completion on the Science of Arithmetic.* Suleymaniye Library, Laleli MS 2708/1, Istanbul.

Al-Khwarizmi. *Kitab al Mukhtasar fi'l Hisab al-Jabr wal-Muqabala,* or *Compendious Book of Calculation by Completion and Balancing.* Medina, MS Hikmat jabr 4, 6, Medina, Saudi Arabia.

Banu Musa Brothers. *Tahriru Kitabi Ma'rifat Misahat al-ashkal al-Basitat al-Kuriyya.* Koprulu Library, I. Kisim MS 930/14, Istanbul.

Berggren, Lennart J. "History of Mathematics in the Islamic World: The Present State of the Art," in *Bulletin of the Middle East Studies Association,* Vol. 19, pp. 9-33. 1985.

Berggren, Lennart. *Episodes in the Mathematics of Medieval Islam.* Springer Verlag, Berlin, 1986.

Al-Biruni. *Kitab al-Athar al-Baqiyya 'an al-Qurun al-Khaliyya,* or *Chronology of Ancient Nations,* or *Vestiges of the Past.* Suleymaniye Library, Ayasofya Collection, MS 2947, Istanbul.

Al Biruni. *Kitab al-Athar al-Baqiyya 'an al-Qurun al-Khaliyya,* or *Chronologie orientalischer völker.* Fuat Sezgin (editor); C. Eduard Sachau (publisher). IGAIW, Frankfurt, 1998.

Al-Biruni. *Kitâb Maqalid ilm al-hay'a la triqonometrie spherique chez les arabes de l'est a la fin du X siecle,* or *Kitâb maqalid ilm al-hay'a.* Marie Thérèse Debarnot (editor). Institut Français de Damas, Paris, 1985.

Djebbar, Ahmed. *L'algèbre arabe: Genèse d'un art.* Adapt-Vuibert, Paris, France, 2005.

Al-Farabi. *Maqala fi Ihsa al-Ulum,* or *The Book of the Enumeration of the Sciences.* Koprulu Library, MS 1604/1, Istanbul.

Al-Farabi. *Maqala fi Ihsa al-Ulum,* or *Catálogo de las ciencias.* Imp. de Estanislao Maestre, Madrid, 1932.

Folkerts, M., and **Kunitsch, Paul.** *Die älteste lateinische Schrift über das indische Rechnen nach al-Hwarizmi. Edition, Übersetzung und Kommentar.* Bayerische Akademie der Wissenschaften, Philosophisch-historische Klasse, Abhandlungen, Neue Folge, Heft 113, München, 1997.

Jaouiche, Khalil. *La théorie des parallèles en pays d'Islam. Contribution à la préhistoire des géométries non-euclidiennes.* Vrin, Paris, 1986

Al-Karaji. *Al-Fahri fi'l-Jabr wa'l-Muqabala.* Suleymaniye Library, Husrev Pasa, MS 257/7, Istanbul.

Al-Khayyam, Umar. *Risala fi'l-barahin 'ala masa'il al-jabr wa'l-muqabala,* or *Treatise on Proofs of Problems of Algebra and Balancing.* Riyada, MS 898/3, Cairo.

Ben Miled, Marouane. *Opérer sur le Continu. Traditions arabes du Livre X des Eléments d'Euclide, avec l'édition et la traduction du commentaire d'al-Mahani.* Beit al-Hikma, Carthage, Tunisia, 2005.

"Mathematics and Her Sisters in Medieval Islam: A Selective Review of Work Done from 1985 to 1995," in *HM,* Vol. 24, pp. 407-440. 1997.

Muhammad Ibn Musa al-Khwarizmi. *Le Calcul Indien "Algorismus." Histoire des textes, édition critique, traduction et commentaire des plus anciennes versions latines remaniées du XIIe siècle.* André Allard (editor). Blanchard/Peeters, Paris/Namur, 1992.

Al-Khawarizmi. *Muhammad Ibn Musa al-Khwarizmi (fl.c. 200-815). Texts and Studies.* Collected and reprinted by F. Sezgin et al. 4 volumes. Islamic Mathematics and Astronomy, Vols. 3-6. IGAIW, Frankfurt, 1997.

Lorch, Richard. *Thabit Ibn Qurra: On the Sector-Figure and Related Texts. Edited with translation and commentary.* IGAIW, Frankfurt, 2001.

Rashed, Roshdi. *Entre arithmétique et algèbre. Recherches sur l'histoire des mathématiques arabes.* Les Belles Lettres, Paris, 1984. Traduction anglaise: *The Development of Arabic Mathematics. Between Arithmetic and Algebra.* Kluwer, Dordrecht, Netherlands, 1994.

Rashed, Roshdi, and **Vahabzadeh, Bijan.** *Al-Khayyam mathématicien.* Blanchard, Paris, 1999.

Rosen, Frederick. *The Algebra of Mohammed ben Musa.* Frederick Rosen (editor and translator). Olms Verlag, Hildesheim, Germany, 1986. Reprint of the first edition, Oriental Translation Fund, London, 1831

Al-Siddiqi. *Risala fi'l-handasa,* or *Treatise of Geometry.* Suleymaniye Library, Ayasofya Collection, MS 2736, Istanbul.

Al-Tusi. *Al-Jabr wa'l-Muqabala.* Vatican Library MS 317/2, Vatican City.

Les Mathématiques Infinitésimales du IXe au XIe siécles. Vol. II: *Ibn al-Haytham.* Al-Furqan Islamic Heritage Foundation, London, 1993.

Les Mathématiques infinitésimales du IXe au XIe siécles. Vol. I: *Fondateurs et commentateurs.* Al-Furqan Islamic Heritage Foundation, London, 1996.

Les Mathématiques infinitésimales du IXe au XIe siècles. Vol. III: *Ibn al-Haytham. Théorie des coniques, constructions géométriques et géométrie pratique.* Al-Furqan Islamic Heritage Foundation, London, 2000.

Les Mathématiques infinitésimales du IXe au XIe siècles. Vol. IV: *Méthodes géométriques, transformations ponctuelles et philosophie des mathématiques.* Al-Furqan Islamic Heritage Foundation, London, 2002.

Oeuvre Mathématique d'al-Sijzi: Géométrie des coniques et

théorie des nombres au Xe siècle. Les Cahiers du Mideo, 3. Peeters, Louvain/Paris, 2004.

Chemistry

Jabir ibn Hayyan. *Kitab al-Sab'in,* or *Book of Seventy Treatises on Alchemy.* Istanbul University Library, MS AY 6314, Istanbul.

Jabir ibn Hayyan. *Kitab fi al- Kimiya'.* Vatican Library MS 1485/1, Vatican City.

Jabir Ibn Hayyan. *The Arabic Works of Jabir Ibn Hayyan.* Arabic texts edited by Eric John Holmyard. Paul Geuthner, Paris, 1928.

Jabir ibn Hayyan. *Jabir Ibn Hayyan. Texts and Studies.* Collected and reprinted by Fuat Sezgin et al. 3 volumes. IGAIW, Frankfurt, 2002.

Jabir ibn Hayyan. *Dix traités d'alchimie: les dix premiers traités du livre des soixante-dix.* Sindbad, Paris, 1983.

Jabir ibn Hayyan. *Kitab al-Sab'in,* or *The Book of Seventy.* F. Sezgin (editor). IGAIW, Frankfurt, 1986.

Al-Kindi. *Kitab Kimiya' al-'itr wa't-tas'idat.* See Cleanliness section.

Nomanul Haq, Syed. *Names, Natures and Things: The Alchemist Jabir Ibn Hayyan and His Kitab-al-Ahjar,* or *Book of Stones.* Foreword by David E. Pingree. Kluwer, Dordrecht, Netherlands, 1994.

Al-Razi. *Kitab al Asrar,* or *The Book of the Secret of the Secrets.* Istanbul University Library, Sarkiyat E., MS 77, Istanbul; and the National Library of Medicine, MS A 33 item 9, Bethesda, Md.

Al-Samawi, al-'Iraqi, Abu Al-Qasim Muhammad b. Ahmad. *Kitab Nihayat al-talab fi sharh kitab [al-'ilm] al-muktasab fi zira'at-i al-dhahab.* Paul Geuthner, Paris, France, 1923. Collected and reprinted by F. Sezgin et al. Natural Sciences in *Islam,* Vol. 61. IGAIW, Frankfurt, 2001.

Commercial Chemistry

Al-Kindi. *Kitab Kimiya' al-'itr wa't-tas'idat.* See Cleanliness section.

The Scribe

Abu Jaafar ibn Tophail. *The Improvement of Human Reason, Exhibited in the Life of Hai ebn Yokdhan,* written in Arabic about 500 years ago. Translated from the original Arabic by Simon Ockley. E. Powell, London, 1708.

Abu Bakr Ibn Tufail. *The History of Hayy Ibn Yaqzan,* translated from the Arabic by Simon Ockley, revised, with an introduction by A. S. Fulton. Chapman and Hall, London, 1929.

Conradi, L. I. (editor). *The World of Ibn Tufayl: Interdisciplinary Perspectives on Hayy Ibn Yaczan.* Islamic Philosophy, Theology and Sciences Series, Vol. 24. E. J. Brill, Leiden, 1996.

Goodman, L. "Ibn Tufayl," in *History of Islamic Philosophy.* Chapter 22, pp. 313-29. S. H. Nasr and O. Leaman (editors). Routledge, London, 1996.

Hawi, S. "Ibn Tufayl's Hayy Ibn Yaqzan, Its Structure, Literary Aspects and Methods," *Islamic Culture, Hyderabad Quarterly Review,* Vol. 47: 191-211. 1973.

Hawi, S. *Islamic Naturalism and Mysticism: A Philosophical Study of Ibn Tufayl's Hayy Yaqzan.* E. J. Brill, Leiden, 1974.

Hawi, S. "Beyond Naturalism: A Brief Study of Ibn Tufayl's Hayy Ibn Yaqzan," in *Journal of the Pakistan Historical Society,* Vol. 22, pp. 249-67. 1974.

Hawi, S. "Ibn Tufayl's Appraisal of His Predecessors and Their Influence on His Thought," in *International Journal of Middle East Studies,* Vol. 7, pp. 89-121. 1976.

Hourani, G. "The Principal Subject of Ibn Tufayl's Hayy Ibn Yaqzan," in *Journal of Near Eastern Studies* 15 (1), pp. 40-46. 1956.

Ibn Tufayl. *Ibn Tufayl's Hayy ibn Yaqzān: A Philosophical Tale,* translated with introduction and notes by Lenn Evan Goodman. Twayne, New York, 1972.

Ibn Yaqzan. *The Journey of the Soul: The Story of Hai bin Yaqzan,* as told by Abu Bakr Muhammad bin Tufail, a translation by Riad Kocache. Octagon, London, 1982.

MARKET

Agricultural Revolution

Ibn al-Awwam. *Kitab al-Filaha,* or *Book of Agriculture.* Istanbul University Library, MS 5823, Istanbul, Turkey; and Library of the Topkapi Palace Museum, Hazine MS 429, Istanbul.

Ibn al-Awwam. *Kitab al-Filaha* or *Le livre de l'agriculture.* Fuat Sezgin (editor); J. J. Clement-Mullet (translator from Arabic). Institute for the History of Arabic-Islamic Science at the Johann Wolfgang Goethe University, Frankfurt, 2001.

Ibn Vahshiyya. *El-Filahatü'n-nebatiyye,* or *L'agriculture nabateenne.* Tavfik Fahd (editor). Institut Français de Damas, Damascus, Syria, 1995.

Kusami. *Filahatü'n-Nabatiyye,* or *The Book of Nabatean Agriculture,* Vols. I-VII. Fuat Sezgin (editor). IGAIW, Frankfurt, 1984.

Abu 'l-Jayr al-Ishbili. *Kitab al-filaha, tratado de agricultura.* Instituto de Cooperación con el Mundo Arabe, Madrid, 1991.

Eguaras Ibáñez, Joaquina. *Ibn Luyun: tratado de agricultura.* Patronato de la Alhambra y Generalife, Granada, Spain, 1988.

García Sánchez, Expiración. "El Tratado agrícola del

granadino Al-Tignari," in *Quaderni di studi arabi,* pp. 279-291. Casa editrice Armena, Venezia, Italy, 1988.

García Sánchez, Expiración. "Agriculture in Muslim Spain," in *The Legacy of Muslim Spain,* pp. 987-999. E. J. Brill, Leiden, 1992.

García Sánchez, Expiración, and **Alvarez de Morales, Camilo**, et al. (editors) *Ciencias de la naturaleza en al-Andalus: textos y estudios.* 7 volumes. C.S.I.C./Escuela de Estudios Arabes, Madrid/Granada, 1990-2004.

Guzmán Alvarez, José Ramón. "El compendio de agricultura atribuido a Ibn Wafid al-Nahrawi: nuevas perspectivas sobre su autoría," in *Anaquel de estudios arabes,* No. 16, pp. 83-124, Madrid, 2005.

Abu 'Abd Allah Muhammad b. **Ibrahim Ibn Bassal.** *Libro de agricultura.* Estudio preliminar por Expiración García Sánchez y J. Esteban Hernández Bermejo. Sierra Nevada, Granada, Spain, 1995.

Al-Masudi. *Muruj al-dhahab wa Ma'adin al-Jawhar,* or *The Meadows of Gold and Quarries of Jewels.* Sakarya University, Ilahiyat Faculty Library, MS 193, Sakarya, Turkey.

Al-Masudi. *Muruj al-dhahab wa Ma'adin al-Jawhar,* 4 volumes. Muhammad Muhyiddin Abdulhamid (editor). Al-Maktaba al-Tijariya al-Kubra, Cairo, 1964.

Ibn Wahshiya, *al-Filahah an-Nabatiyyah,* or *The Book of Nabatean Agriculture.* 7 volumes. Introduction in Arabic and English by Fuat Sezgin. IGAIW, Frankfurt, 1993-98.

Varisco, Daniel Martin. *Medieval Agriculture and Islamic Science: The Almanac of a Yemeni Sultan.* The University of Washington Press, Seattle, 1994.

Farming Manuals

Ibn al-Awwam. *Kitab al-Filaha.* See Agricultural Revolution section.

Water Management

Al-Karaji. *Kitab Intibat al-miyah al-khafiyyat,* or *Extraction of Underground Waters.* Oriental Public Library at Bankipore, MS 2468/32, Patna, India.

Al-Karaji. *Kitab inbat al-miyah al-khafiya.* Baghdad Abdul-Mun'im (editor and analysis). Institute of Arabic Manuscripts, Cairo, 1997.

Al-Karaji. *L'Estrazione delle acque nascoste: Trattato tecnico-scientifico di Karaji Matematico-ingegnere persiano vissuto nel Mille* by Giuseppina Ferriello. Kim Williams Books, Turin, Italy, 2007.

Lightfoot, Dale R. "The Origin and Diffusion of Qanats in Arabia: New Evidence from the Northern and Southern Peninsula," in *Geographical Journal,* Vol. 166, No. 3, pp. 215-226, Royal Geographical Society, London, July 2005.

Al-Maqrizi. *Kitab al-Suluk li Ma'rifat Duwal al-Muluk,* or *Book of Entrance to the Knowledge of the Dynasties of the Kings.* Said A. F. Ashour (publisher). Matba'at Dar al-Kutub, Cairo, 1970.

Al-Nuwayri. *Nihayat al-Arab fi Funun al-Adab,* or *The Arab Art of Manners.* Dar al-Kutub al-Misriyah, Cairo, 1923.

Water Supply

Al-Jazari. *Al-Jami' bayn al-'Ilm al-Nafi' wa sina'at al-Hiyal.* See Clocks section.

Taqi al-Din. *Turuq al-Saniyya fi al-Alat al-Ruhaniyya,* or *The Sublime Methods of Spiritual Machines.* Dar al-Kutub, Miqat MS 4, Cairo.

Water-Lifting Devices in the Islamic World. Texts and Studies. Natural Sciences in Islam, Vol. 43. Collected and reprinted by F. Sezgin et al. IGAIW, Frankfurt, 2001.

Krenkow, F. "The Construction of Subterranean Water Supplies During the Abbaside [sic] Caliphate," in *Transactions of the Glasgow University Oriental Society,* Vol. 13, pp. 23-32. 1947-49.

Wiedemann, Eilhard, and **Hauser, Franz.** "Über Vorrichtungen zum Heben von Wasser in der Islamischen Welt," in *Beitrage zür Geschichte der Tecknik und Industrie,* Vol. 8, pp. 121-154. 1921.

Dams

Al-Idrisi. *Nuzhat al-Mushtaq fi 'khtirak al- Afaq* or *Al-Kitab al-Rujari,* or *A Recreation for the Person Who Longs to Traverse the Horizons* or *Book of Roger.* Suleymaniye Library, Husrev Pasa MS 318, Istanbul.

Smith, N. A. F. *A History of Dams.* London, 1971.

Windmills

Al-Masudi. *Muruj al-dhahab wa Ma'adin al-Jawhar.* See Agricultural Revolution section.

Hassan, Ahmad Y., and **Hill, Donald Routledge.** *Islamic Technology: An Illustrated History,* p. 54. Cambridge University Press, Cambridge, U.K., 1986.

Lohrmann, Dietrich. "Von der östlichen zur westlichen Windmühle," in *Archiv für Kulturgeschichte,* Vol. 77, No. 1, pp. 1-30 (8). 1995.

Hill, Donald Routledge. "Mechanical Engineering in the Medieval Near East," in *Scientific American,* pp. 64-69. May 1991.

Lucas, Adam. *Wind, Water, Work: Ancient and Medieval Milling Technology,* E. J. Brill, Leiden, 2006.

Trade

Abulafia, David. "The Role of Trade in Muslim-Christian Contacts During the Middle Ages," in *The Arab Influence in Medieval Europe,* Dionisius A. Agius and Richard Hitchcock (editors). Ithaca Press, Berkshire, U.K., 1994.

Chaudhuri, K. N. *Trade and Civilisation in the Indian Ocean: An Economic History from the Rise of Islam to 1750.* Cambridge University Press, Cambridge, U.K., 1985.

Fischel, Walter. "The Origins of Banking in Medieval Islam: A Contribution to the Economic History of Jews in Baghdad in the Tenth Century," in *Journal of Royal Asiatic Society,* pp. 339-352. 1933.

Fleet, Kate. *European and Islamic Trade in the Early Ottoman State: The Merchants of Genoa and Turkey.* Cambridge University Press, Cambridge, U.K., 1999.

Ghazanfar, S. M. "Scholastic Economics and Arab Scholars: The 'Great Gap' Thesis Reconsidered," in *Diogenese: International Review of Humane Sciences,* No. 154, pp. 117-33, April-June 1991.

Ghazanfar, S. M. "Post-Greek/Pre-Renaissance Economic Thought: Contributions of Arab-Islamic Scholastics During the 'Great Gap' Centuries," in *Research in History of Economic Thought and Methodology,* Vol. 16, pp. 65-90. 1998.

Ghazanfar, S. M. "The Economic Thought of Abu Hamid Al-Ghazali and St. Thomas Aquinas: Some Comparative Parallels and Links," in *History of Political Economy,* Vol. 32, No. 4, pp. 857-888, Fall 2000.

Ghazanfar, S. M., and **A. Azim Islahi.** "Economic Thought of an Arab Scholastic: Abu Hamid Al-Ghazali (AH450-505/AD1058-1111)," in *History of Political Economy,* Vol. 22, No. 2, pp. 381-401, Spring 1990.

Ghazanfar, S. M. (editor). *Medieval Islamic Economic Thought: Filling the "Great Gap" in European Economics,* RoutledgeCurzon Publishers, London, 2003.

Ibn Hawqal. *Kitab al-Masalik wa-al-Mamalik,* or *The Book of the Routes of the Kingdoms,* edited in *Opus geographicum auctore Ibn Haukal (Surat al-ard).* J. H. Kramers (editor). E. J. Brill, Leiden, 1967.

Labib, Subhi. *Handelsgeschichte Agyptens im Spatmittelalter,* or *History of Trade of Egypt in the Late Middle Ages.* Franz Steiner Verlag, Wiesbaden, Germany, 1965.

Labib, Subhi. "Capitalism in Medieval Islam," in *Journal of Economic History,* Vol. 29, pp. 79-96. 1969.

Ritter, Helmut. "Ein Arabisches Handbuch der Handelswissenschaft," or "Handbook of Business Practices, 12th century A.D." in *Der Islam,* Vol. 17, pp.1-97. 1917.

Udovitch, Abraham. "Labor Partnerships in Early Islamic Law," in *Journal of Economic and Social History of the Orient,* Vol. 10, No. 2, pp. 64-80, June 1967.

Udovitch, Abraham. *Partnership and Profit in Medieval Islam,* Princeton University Press, Princeton, N.J., 1970.

Udovitch, Abraham. "Credit as a Means of Investment in Medieval Islam," in *Journal of American Oriental Studies,* Vol. 87. 1967.

Weiss, Walter M., and **Westermann, Kurt-Michael.** *The Bazaar: Markets and Merchants of the Islamic World.* Thames and Hudson, London, 1998.

Paper

Bloom, Jonathan M. *Paper Before Print: The History and Impact of Paper in the Islamic World,* pp. 203-213. Yale University Press, New Haven, Conn., 2001.

Burns, Robert I. "Paper Comes to the West, 800-1400," in *Lindgren, Uta, Europäische Technik im Mittelalter. 800 bis 1400. Tradition und Innovation* (4th edition), pp. 413-42. Gebr. Mann Verlag, Berlin, 1996.

Hunter, Dard. *Papermaking: The History and Technique of an Ancient Craft.* Courier Dover Publications, New York, 1978.

Ibn Hawqal. *Kitab al-Masalik wa-al-Mamalik.* See Trade section.

Levey, Martin. *Mediaeval Arabic Bookmaking and Its Relation to Early Chemistry and Pharmacology.* American Philosophical Society, Vol. 52, Part 4. Philadelphia, 1962.

Loveday, Helen. *Islamic Paper: A Study of the Ancient Craft.* Archetype Publications, Don Baker Memorial Fund, London, 2001.

Quraishi, Silim. "A Survey of the Development of Papermaking in Islamic Countries," in *Bookbinder,* Vol. 3, pp. 29-36. 1989.

Pottery

Al-Maqrizi. *Kitab al-Suluk li Ma'rifat Duwal al-Muluk.* See Water Management section.

Baramki, D. C. "The Pottery from Khirbet El-Mefjer," in *The Quarterly of the Department of Antiquities in Palestine* (QDAP 1942), Vol. 10, pp. 65-103. 1942.

Bernsted, A. K. *Early Islamic Pottery: Materials and Techniques,* Archetype Publications Ltd., London, 2003.

Caiger-Smith, Alan. *Lustre Pottery: Technique, Tradition and Innovation in Islam and the Western World,* Chapters 6 and 7. Faber and Faber, London, 1985.

Cooper, Emmanuel. *Ten Thousand Years of Pottery,* 4th edition. University of Pennsylvania Press, Philadelphia, 2000.

Mason, Robert B. "New Looks at Old Pots: Results of Recent Multidisciplinary Studies of Glazed Ceramics from the Islamic World," in *Muqarnas: Annual on Islamic Art and Architecture,* Vol. 12. E. J. Brill, Leiden, 1995.

Sauer, J. A. "Umayyad Pottery from Sites in East Jordan," in *Jordan,* Vol. 4, pp. 25-32. 1975.

HOSPITAL

Hospital Development

Al-Khujandi. *Al-Talwih li-Asrar al-Tanqih,* or *Tanqih al-Maknun.* Vatican Library, MS 305, Vatican City.

Ibn Abi-Usaybia. *Uyunul-Anba Fi-Tabaqat Al-Atibaa,* or

The Sources of the Knowledge of Classes of Doctors. N. Reda (editor). Dar Maktabat al Hayat, Beirut, 1965.

Ibn Al-Nadim. *Al-Fihrist.* I. Ramadan (editor), 2nd edition. Dar El-Mareefah, Beirut, 1997.

Ibn Jubayr. *Rihlat Ibn Jubayr,* or *The Travels of Ibn Jubayr.* Goodword Books, New Delhi, 2001.

Ibn Juljul. *Tabaqat Al-Atiba' wa-'l-Hukama'.* Sayed Fouad, (editor). French Bureau Publications, Cairo, 1955.

Ibn Sina. *Al-Qanun fi al-Tibb,* or *Canon of Medicine.* Suleymaniye Library, Hekimoglu MS 580, Istanbul.

Ibn Sina. *Al-Qanun fi al-Tibb,* or *Canon of Medicine.* Dar Sadir reprint of Bulaq edition. Cairo, 1877.

Saed Ibn Saed, Al-Andalusi. "Book of the Categories of Nations," in *Science in the Medieval World.* S. I. Salem and A. and Kumar (translators and editors). History of Science Series No. 5. University of Texas Press, Austin, 1996.

Instruments of Perfection

Al-Baghdadi, Muhadhdhab Al-Deen. *Kitab Al-Mukhtarat Fi Al Tibb,* 1st edition. Osmania University, The Bureau, Osmania Oriental Publications, Hyderabad, India, 1942.

Al-Baladi Ahmad ibn Muhammad ibn Yahya. *Kitab tadbeer al habala wa al-atfal wa hifz sehhatihim wa mudawat al amradd al aariddah lahum.* Mahmood Al Hajj Qasim Muhammad (editor), 2nd edition. Dar Al Shueoon Al-Thaqafeyyah Al Aaamah, Baghdad, 1987.

Al-Razi. *Kitab al-Hawi fit-tibb,* or *Liber continens,* 1st edition. Osmania University, The Bureau, Osmania Oriental Publications, Hyderabad, India, 1961.

Ibn al-Jazzar al-Qairawani. *Kitab siasat al-sibiaan wa tadbeeruhum,* or *The Book for Bringing Up and Care for Children,* M. A. Al-Hailah (editor). Al-Dar Al-Tunisiyyah Lil Nashr, Tunis, Tunisia, 1968.

Ibn al-Quff. *Al-Umdah Fi Al-Jiraha.* 2 volumes. Osmania University, The Bureau, Osmania Oriental Publication, 1356 H Hyderabad, India.

Ibn Sina. *Kitab al-Qanun fit-tibb,* or Avicenna's *Canon of Medicine.* Dar Sadir reprint of Bulaq edition. Cairo, 1877.

Ibn Zuhr (Avenzoar). *Kitab al-Taisir fi al-Mudawat wa al-Tadbir,* or *Book of Simplification Concerning Therapeutics and Diet.* M. Al-Khoori (editor). 1st edition. Vols. 1 and 2. Darul Fikr Press for the Arab Educational Scientific and Cultural Organization, Damascus, Syria, 1983.

Spink, M. S., and **Lewis, I. L.** (editors and translators). *Albucassis on Surgery and Instruments.* Wellcome Institute of the History of Medicine, London, 973.

Surgery

Al-Razi. *Kitab al-Hawi,* or *Liber continens.* See Intruments of Perfection section.

Al-Razi. *Ma al-fariq aw al furooq aw kalamun fi al furuuq bain al amradd,* or *What Are the Clues to Differentiate Between Diseases [of Similar Symptoms].* Qattaya Salman (editor). Institute for Arabic Scientific Heritage, Aleppo, Syria, 1978.

Ibn al-Nafis. *Risalat Al-Aadaa,* or *A Treatise in Physiology.* Y. Ziedan (editor). Al Dar Al-Masreyya Al Lubnaneyyah, Cairo, 1991.

Ibn al-Quff. *Kitab al-'Umda fi sina'at al-jiraha,* or *The Foundation.* Suleymaniye Library, Hekimoglu MS 579, Istanbul.

Ibn al-Quff. *Al-Umdah Fi Al-Jiraha.* 2 volumes. Osmania University, The Bureau, Osmania Oriental Publications, 1356 H Hyderabad, India.

Ibn al-Quff. *Al-Shafi fi al-Tibb.* Vatican Library, Appendice 183, Vatican City.

Ibn Rushd. *Al-Kulliyyat Fi Al-Tibb,* or *The Basic Principles of Medicine.* M. A. Al-Jabiry (editor). Arabian Philosophy Heritage Series-Ibn Rushd Works, No. 5. The Institute for Arabic Unity Studies, Beirut, 1999.

Ibn Sina. *Al-Qanun fi al-Tibb.* See Hospital Development section.

Ibn Zuhr. *Kitab al-Taisir fi al-Mudawat wa al-Tadbir,* or *Book of Simplification Concerning Therapeutics and Diet.* M. Al-Khoori (editor). 1st edition, Vols. 1-2. Darul Fikr Press for the Arab Educational Scientific and Cultural Organization, Damascus, Syria, 1983.

Spink M. S., and **Lewis, I. L.** (editors and translators). *Albucassis on Surgery and Instruments.* See Instruments of Perfection section.

Blood Circulation

Abdel-Halim, Rabie E. "Contributions of Ibn Al-Nafis to the Progress of Medicine and Urology: A Study and Translations from His Medical Works," in *Saudi Medical Journal,* pp. 13-22. 2008.

Ibn al-Nafis. *Sharh Tashrih al-Qanun,* or *Commentary on the Anatomy of the Canon of Avicenna.* Suleymaniye Library, Fatih 3626, MS A 21 and MS A 56, Istanbul.

Ibn al-Nafis. *Kitab Sharh Tashreeh Al-Qanun.* Qattaya S. (editor). The Egyyptian Manuscript Editing Bureau, Cairo, 1988.

Iskandar, Albert Z. "Ibn al-Nafis," in *Dictionary of Scientific Biography,* Vol. 9, pp. 602-606, New York, 1974.

Ibn Sina's Bone Fractures

Al-Majusi. *Kamil al-Sina'a al-Tibbiyya.* Library of Topkapi Palace Museum, Ahmed III nr. 2060. Vatican Library MS 314, Vatican City.

Ibn Sina. *Kitab al-Shifa',* or *The Book of Cure, Healing* or *Remedy from Ignorance.* Library of the Topkapi Palace Museum, Ahmed III MS 3261, Istanbul.

Ibn Sina. *Avicenna's De Anima: Being the Psychological Part of Kitab al-Shifa,* or *Kitab al-Shifa: al-fann al-sadis min al-tabiiyyat wa huwa kitab al-nafs.* Fazlurrahman (editor), 3rd edition. University of Durham, Durham, U.K., 1970.

Notebook of the Oculist

Abu al-Farag. *The Abridged Version of The Book of Simple Drugs of Ahmad Ibn Muhammad Al-Ghafiqi.* M. Meyerhof and G. P. Sobhy (translators and editors); Fuat Sezgin (re-editor). IGAIW, Frankfurt, 1996.

Al-Ghafiqi. *Al-Murshid fil al-kuhl,* or *The Right Guide in Ophthalmic Drug.* Laboratoires du Nord de l'Espagne, Barcelona, 1933.

Al-Ghafiqi. *Texts and Studies.* Fuat Sezgin (editor); Mazen Amawi, Carl Ehrig-Eggert, Eckhard Neubauer (publishers). IGAIW, Frankfurt, 1996.

Ibn al-Nafis. *Al-Muhadhab fi tibb al-'Ayn.* Vatican Library MS 307, Vatican City.

Ibn al-Nafis. *Al-Muhadhdhab fi al-Kuhl al-Mujarrab.* M. Z. Wafai andM. R. Qalaji (editors). 2nd edition. Safir Press, Riyadh, Saudi Arabia, 1994.

Ibn Isa. *Tadhkirat al-Kahhalin,* or *Memorandum Book for Ophthalmologists* or *Notebook of the Oculist.* Vatican Library MS 313, Vatican City.

Ali ibn 'Isa. *Memorandum Book of a Tenth-Century Oculist for the Use of Modern Ophthalmologists.* Casey A. Wood (translator), Book I, Chapter 20. Northwestern University Press, Chicago, 1936.

Khalifa ibn Abi Al Mahasin Al Halaby. *Al Kafi Fi Al-Kuhl,* or *The Book of Sufficient Knowledge in Ophthalmology.* Dar al-Fikr, Beirut, 2000.

Herbal Medicine

Al-Dinawari. *Kitab al-Nabat,* or *The Book of Plants.* Bernhard Lewin (editor). A. B. Lundeguistska Bokhandeln, Uppsala-Wiesbaden, Germany, 1953.

Al-Ghafiqi. *Kitab al-adwiya al-Mufrada,* or *The Book of Simple Drugs.* Egyptian University, Cairo, 1932-40.

Al-Ghafiqi. *Kitab jami' al-Mufradat,* or *Materia Medica.* Max Meyerhof and George P. G. Sobhy (editors). Cairo Medical Faculty, Cairo, 1937-38.

Ibn al-Baytar. *Kitab-al-Jami fil Adwiya al-Mufrada,* or *Dictionary of Simples, Remedies and Food.* Suleymaniye Library, Damad İbrahim, MS 929, Istanbul.

Ibn Samajun. *Jami al-adwiya al-Mufrada,* or *Collection of Simples, Medicinal Plants and Resulting Medicines.* Fuat Sezgin (publisher). Institute for the History of Arabic-Islamic Science at the Johann Wolfgang Goethe University, Frankfurt, 1992.

Pharmacy

Al-Biruni. *Kitab al Saydana fi't-tib,* or *Book of Medicines* or *Book of Pharmacology.* Hakim Mohammad Said (publisher and translator into English). Suleymaniye Library, Izmirli I. 4175, Istanbul, 1973.

Al-Baghdadi. *Al Mukhtarat Fi Al-Tibb.* Vols. 1-4, 1362-1364 H. 1941-1944. Osmania University, The Bureau, Osmania Oriental Publications, Hyderabad, India.

Al-Harawi. *Kitab al-Abniya 'an haqa'iq al-Adwiya,* or *The Foundations of the True Properties of Remedies.* Fuat Sezgin (publisher). Institute for the History of Arabic-Islamic Science at the Johann Wolfgang Goethe University, Frankfurt.

Al-Kindi. *Aqrabadhin,* or *Medical Formulary.* Suleymaniye Library, Ayasofya Collection, Turkey.

Al-Razi. *Kitab al-Mansuri,* or *Liber almansoris.* The National Library of Medicine, MS A 28, Bethesda, Md.

Al-Zahrawi. *Al-Tasrif li-man 'ajiza 'an al-taalif.* See Cleanliness section.

Ibn Al-Baitar. *Al-Jamie Limufradat Al-Adwiya Wal-Aghdiya,* or *Materia Medica.* Al-Muthana Bookshop, Baghdad, undated.

Ibn al-Wafid. *Kitab al-Adwiya,* or *The Book of Simple Drugs.* Ahmad Hasan Basaj (publisher). Dar al-Kutub al-'Ilmiyah, Beirut, 2000.

Ibn Sina. *Al-Qanun fi al-Tibb.* See Hospital Development section.

Medical Knowledge

Al-Majusi. *Al-Kihalah (tibb al-'uyun) fi Kitab Kamil al-sina'ah al-tibbiyah al-ma'ruf bi-al-Malaki,* or *The Royal Book,* also known as the *Pantegni.* Muhammad Zafir Wafa'i and Muhammad Rawwas Qal'ah'ji (publishers). Wizarat al-Thaqafah, Damascus, Syria, 1997.

Al-Zahrawi. *Al-Tasrif li-man 'ajiza 'an al-taalif.* See Cleanliness section.

Arabic Science and Medicine: A Collection of Manuscripts and Early Printed Books Illustrating the Spread and Influence of Arabic Learning in the Middle Ages and the Renaissance. Bernard Quaritch catalogue 1186. Introduction by Professor Charles Burnett. Bernard Quaritch, London, 1993.

Ben Miled, Ahmed. *Ibn Al Jazzar. Constantin l'Africain,* Salambô, Tunis, Tunisia, 1987.

Ben Miled, Ahmed. *Histoire de la médecine arabe en Tunisie.* Dar al-Gharb al-Islami, Beirut, 1999.

Ben Miled, Ahmed. *Ibn Al Jazzar. Médecin à Kairouan.* Al Maktaba Al Tounisia, Tunis, Tunisia, 1936.

Ibn al-Dhahabi. *Kitab al-Ma'a,* or *The Book of Water.* Dr. Hadi Hamoudi (editor). Ministry of National Heritage and Culture, Oman, 1996.

Ibn al-Jazzar. *Zad al-Musafir,* or *The Guide for the Traveler Going to Distant Countries,* or *Traveler's Provision,* known in Latin as the *Viaticum*. Gerrit Bos (editor and translator). Kegan Paul International, London and New York, 2000.

Ibn al-Nafis. *Al-Shamil fi al-Tibb*, or *Comprehensive Book on the Art of Medicine*. Koprulu Library, I. kisim, nr. 987/1, Istanbul; and Vatican Library MS 306, Vatican City.

Ibn al-Nafis. *Al-Mujaz Fi Al-Tibb*. Al-Ezbawy A. (editor). 4th edition. Islamic Heritage Revival Committee, Supreme Council for Islamic Affairs, Ministry of Endowments, Cairo, 2004.

Ibn al-Nafis. *Al-Shamil Fi Al-Sinaa Al-Tibbiyyah*. Y. Ziedan (editor). Al-Mujammaa Al-Thaaqfi, Abu Dhabi, 2000.

TOWN

Public Baths

Dunn, Ross E. *The Adventures of Ibn Battuta–A Muslim Traveler of the 14th Century*. University of California, Berkeley, 2004. Ibn Battuta. *Al-Rihla*, or *The Journey*. Public Library of Cambridge, Library No. 890.8 O7p no., Cambridge, Mass.

Fountains

Banu Musa Brothers. *Kitab al-hiyal al-Handasiyah*. See Trick Devices section.

WORLD

Planet Earth

Al-Battani. *Kitab al-Zij al-Sabi'*, or *De scientia stellarum -De numeris stellarum et motibus*, or *The Sabian Tables*. Zaytuna, MS 2843, Tunis, Tunisia.

Al-Biruni. *Kitab fi ifrad al-Maqal fi amr al-azlal*, or *Shadows* or *Gnomonics*. Oriental Public Library at Bankipore, 2468/36, Patna, India.

Al-Biruni. *Al-Qanun al-Mas'udi fi'l-hay'a wa'l-nujum*, or *Mas'udic Canon on Astronomy and Astrology*. Suleymaniye Library, Carullah, MS 1498, Istanbul.

Al-Khujandi. *Risala fi tashih al-mayl wa 'ard al-balad*, or *Treatise on Determining the Declination and Latitude of Cities with More Accuracy*. Greek Orthodox School Library, 364/1, Beirut.

Ibn Hazm. *Al-Fasl fil-Millal wa al-ahwa wa'n-nihal*, or *Conclusion on the Nations*. Cambridge University Library, Library/Call No. Moh.121.b.50, Cambridge, U.K.

Ibn Yunus. *Al-Zij al-Hakimi*, or *The Hakemite Tables*. In C. Caussin. "Le livre de la grande table hakemite," in *Notices et extraits des manuscrits de la Bibliothèque Nationale* Vol. 7, 1804, pp. 16-240.

Nallino, C. A. *Al-Battani sive Albatenii, Opus Astronomicum*. 3 volumes. Osservatorio astronomico di Breva, Milan, Italy, 1899, 1903, 1907.

Earth Science

Abu Ridah, M. A. H. (editor). *The Letters (Rasa'il) of al-Kindi al-falsafiyya*. Matbaatu Hassan, Cairo, 1978.

Al-Biruni. *Kitab Al-Jamahir fi Ma'rifat al-Jawahir*, or *Treatises on How to Recognize Gems*. Library of the Topkapi Palace Museum, Ahmed III 2047. Fuat Sezgin (editor). Institute for the History of Arabic-Islamic Science at the Johann Wolfgang Goethe University, Frankfurt, 2001.

Al-Biruni. *Al-Qanun al-Mas'udi fi'l-hay'a wa'l-nujum*. See Planet Earth section.

Al-Kindi. *Risala fi anwa al-jawahir al-thaminah wa ghayriha*, or *Treatise on Various Types of Precious Stones and Other Kinds of Stones*.

Ibn Sina. *Kitab al-Shifa'*. See Ibn Sina's Bone Fractures section.

Ikhwan al-Safa'. *Rasa'il*, or *Epistles*. Vatican Library 1608/1, Vatican City, Italy; and Princeton University Library, Library No. 1129 (Garrett Collection), Princeton, N.J.

Masawayh. *Kitab al-Jawahir wa-Sifatiha wa-fi ayyi Baladin Hiya, wa-Sifat al-Ghawwasin wa-al-Tujjar*, or *Gems and Their Properties*. The Wellcome Trust Library, Library Number Wellcome MS Arabic 468 (Haddad Collection), London.

Natural Phenomena

Al-Biruni. *Al-Qanun al-Mas'udi fi'l-hay'a wa'l-nujum*. See Planet Earth section.

Al-Kindi. *Risala fi'l-illa al-fa'ila li'l-madd wa'l-jazr*, or *Treatise on the Efficient Cause of the Tidal Flow and Ebb*. Bodleian Library, I 877/12, Oxford, U.K.

Al-Kindi. *Risala fi 'illat al-lawn al-azraq alladhhi yura fi'l-jaww fi jihat al-sama*, or *Treatise on the Azure Colour Which Is Seen in the Air in the Direction of the Heavens and Is Thought to Be the Colour of the Heavens*. Suleymaniye Library, Ayasofya 4832/2, Istanbul.

Ibn al-Haytham. *Kitab al-Manazir*, or *Book of Optics*. See Vision and Cameras section.

Ibn Hazm. *Al-Fasl fil-Millal*. See Planet Earth section.

Geography

Al-Bakri. *Kitab al-Masalik wa'l-Mamalik*, or *Book of Highways and of Kingdoms*. Cambridge University Library, Library/Call No. 590:01.b.17.1, Cambridge, U.K.

Al-Biruni. *Alberuni's India: An Account of the Religion, Philosophy, Literature, Geography, Chronology, Astronomy, Customs, Laws, and Astrology of India About A.D. 1030, Vols. I-II*. Edward C. Sachau (translator), Fuat Sezgin (editor). IGAIW, Frankfurt, 1993.

Al-Biruni. *The Determination of the Coordinates of Positions for the Correction of Distances Between Cities,* *or Kitab Tahdid nihayat al-amakin li-tashih masafat al-masakin.* Fuat Sezgin (editor); in collaboration with Mazen Amawi, Carl Ehrig-Eggert, and Eckhard Neubauer. IGAIW, Frankfurt, 1992.

Al-Idrisi. *Nuzhat al-Mushtaq fi 'khtirak al-Afaq.* See Dams section.

Al-Jahiz. *Kitab al-Buldan.* Matbaat al-Hukumah, Baghdad, 1970.

Al-Khwarizmi. *Kitab Surat al-Ardh min al-mudun wa'l-jibal wa'l-bihar wa'l-jaza'ir wa'l-anha,* or *Book of Geography: A Picture Book of the Earth, Cities, Mountains, Seas, Islands, and Rivers,* or *The Form of the Earth.* German translation titled *Das Kitab Surat al-ard, des Abu Ga'far Muhammad ibn Musa al-Khuwarizmi, herausg.* Unikum des Bibliotheque de l'Universite et Regionale in Strasbourg, Austria, 1926.

Hill, Donald R. *Islamic Science and Engineering.* Edinburgh University Press, Edinburgh, 1993.

Ibn al-Nadim. *Fihrist'al-Ulum.* See Libraries and Bookshops section.

Ibn Jubayr. *Rihlat Ibn Jubayr.* See Hospital Development section.

Mahmud Kashghari. *Divanu Lugat-it-Turk,* or *Compendium of Turkish Dialects.* Istanbul, 1915-17.

Maslama al-Majriti. *Rutbat Al-Hakim,* or *The Rank of the Wise.* Ali Emiri-Arabi, 2836/2, Istanbul.

Al-Muqaddasi. *Ahsan al-Taqasim fi Ma'rifat al-Aqalim.* See Libraries and Bookshops section.

Al-Ya'qubi. *Kitab al-Buldan,* or *Book of Countries.* Istanbul University, Islam Arastirmalari Library, 1262, Istanbul, Turkey; and Yale University Library, Library/Call No. Geography. Folio B4737, New Haven, Conn.

Yaqut. *Mu'jam al-Buldan,* or *Dictionary of Countries.* Cambridge University Library, Library/Class No. Moh. 280.b.1, Cambridge, U.K.

Maps

Al-Idrisi. *Nuzhat al-Mushtaq fi 'khtirak al- Afaq.* See Dams section.

King, David A. *World-Maps for Finding the Direction and Distance to Mecca: Innovation and Tradition in Islamic Science.* Al-Furqan Islamic Heritage Foundation, E. J. Brill, Leiden, 1999.

King, D. A. "Two Iranian World Maps for Finding the Direction and Distance to Mecca," in *Imago Mundi. The International Journal for the History of Cartography,* Vol. 49, pp. 62-82. 1997.

King, D. A., and **Lorch, R.** "Qibla Charts, Qibla Maps, and Related Instruments," in *The History of Cartography,* Vol. 2, Book 1: *Cartography in the Traditional Islamic and South Asian Societies,* pp. 189-205. J. B. Harley and D. Woodward (editors). University of Chicago Press, Chicago, 1992.

Piri Reis. *Kitab i-Bahriye,* or *The Book of Sea Lore,* or *The Book of the Mariner,* or *The Naval Handbook.* Ertugrul Zekai Okte (editor); Vahit Cabuk (transcription); Vahit Cabuk and Tulay Duran (Turkish text); Robert Bragner (English text). Culture and Tourism Ministry, Ankara, Turkey, 1988.

Travelers and Explorers

Abu al-Fida'. *Taqwim al-Buldan,* or *Survey of Countries.* Library of the Topkapi Palace Museum, Ahmed III 2855, Istanbul.

Al-Biruni. *Tahdidu nihayat al-amakin li't-tashihi masafat al-masakin.* Fuat Sezgin (editor); Mazen Amawi, Carl Ehrig-Eggert, Eckhard Neubauer (publishers). IGAIW, Frankfurt, 1992.

Flowers, Stephen E. *Ibn Fadlan's Travel-Report: As It Concerns the Scandinavian Rûs.* Rûna-Raven, Smithville, Texas, 1998.

Frähn, Christian Martin, *Die ältesten arabischen Nachrichten über die Wolga-Bulgaren aus Ibn-Foszlan's Reiseberichte.* Mémoires de L'Académie Impériale des Sciences de St. Petersbourg, VIème série, 1823.

Frye, Richard N. (editor). *Ibn Fadlan's Journey to Russia: A Tenth Century Traveler from Baghdad to the Volga River.* Markus Wiener Publishers, Princeton, N.J., 2005.

Ibn Battuta. *Al-Rihla.* See Public Baths section.

Ibn Fadlan. *Voyage chez les Bulgares de la Volga.* Marius Canard (French translation), Sinbad, Paris, 1988.

Ibn Fadlan. *Collection of Geographical Works by Ibn al-Faqih, Ibn Fadlan, Abu Dulaf Al-Khazraji.* Fuat Sezgin (editor), IGAIW, Frankfurt, Germany, 1987.

Ibn Fadhlan, Ahmad b. al-'Abbas b. Rashid b. Hammad. *Reisebericht. Rihlat Ibn Fadlan.* Ahmed Zeki Validi Togan (editor and translator into German). Deutsche Morgenländische Gesellschaft, Abhandlungen für die Kunde des Morgendlandes. XXIV, 3. F. A. Brockhaus, Leipzig, Germany, 1939. Reprinted Institute for the History of Arabic-Islamic Science, Frankfurt, 1994.

Ibn Jubayr. *Rihlat Ibn Jubayr.* See Hospital Development section.

Ibn Khurradadhbih. *Al-Masalik wal Mamalik,* or *Book of Roads and Provinces,* or *Le livre des routes et des provinces.* Casimir Barbier de Meynard (editor). Cambridge University Library, Library/Class No. Moh.280.c.28, Cambridge, U.K.

Magmu' fi 'l-gughrafiya: tubi'a bi-'t-taswir 'an makhtut al-Maktaba ar-Radawiya fi Mahhad 5229. Mimma allafahu

Ibn-al-Faqih wa-Ibn-Fadhlan wa Abu Dulaf al-Khazraji. F. Sezgin et al. (editors). Institute for the History of Arabic-Islamic Science, Frankfurt, 1987.

Al-Muqaddasi. *Ahsan al-Taqasim fi Ma'rifat al-Aqalim.* See Libraries and Bookshops section.

Al-Ya'qubi. *Kitab al-Buldan.* See Geography section.

Yaqut. *Mu'jam al-Buldan.* See Geography section.

Navigation

Al-Masudi. *Muruj al-dhahab wa Ma'adin al-Jawhar.* See Agricultural Revolution section.

Al-Qibjaqi. *Kitab Kanz al-Tujjār fi ma'rifat al-Ahjar,* or *The Book of Treasure for Merchants Who Seek Knowledge of Stones.* B. A. Rosenfeld and E. Ihsanoglu, No. 649, IRCICA, Istanbul, 2003.

Homsi, H. "Navigation and Ship-building," in *The Different Aspects of Islamic Culture,* Vol. IV: *Science and Technology in Islam,* Parts I-II. Ahmad Y. Al-Hassan, Yusuf Iskandar, Albert Zaki, and Ahmad Maqbul (editors). UNESCO, Paris, 2001.

Ibn Majid, Shihab al-Dein. *Arab Navigation in the Indian Ocean Before the Coming of the Portuguese,* or *Kitab al-Fawa'id fi usul al-bahr wa'l-qawa'id.* G. R. Tibbets (translator). The Royal Asiatic Society of Great Britain and Ireland, London, 1981.

Nadwi, Allama Syed Sulaiman. *The Arab Navigation.* Syed Sabahuddin Abdurahman (translator). Sh. Muhammad Ashraf, Lahore, Pakistan, 1966.

Piri Reis. *Kitab-i-bahriyye.* See Maps section.

Global Comunication

Al-Nuwayri. *Nihayat al-Arab fi Funun al-Adab.* See Water Management section.

War and Weaponry

Ibn Aranbugha al-Zardkash. *Armoury Manual.* Fuat Sezgin (editor). Institute for History of Arabic-Islamic Science at the Johann Wolfgang Goethe University, Frankfurt, 2004.

Al-Rammah. *Kitab Al-Furusiyya wa Al-Manasib Al-Harbiyya,* or *The Book of Horsemanship and Ingenious War Devices.* Suleymaniye Library, Ayasofya 3799, and Nurosmaniye Library 2294, Istanbul.

Omeri. *A Muslim Manual of War,* or *Tafrij al-kurub fī tadbir al-hurub.* George T. Scanlon (editor and translator). The American University at Cairo, Cairo, 1961.

Social Science and Economics

Ibn Khaldun. *Muqaddimah,* or *The Introduction to History.* Istanbul University Library, Arabic, 2743, 835, Istanbul, Turkey; and The Library of Congress, Library/Call No. D.16.7.123.1879, Washington, D.C.

Ibn Khaldun. *The Muqaddimah: An Introduction to History.* N. J. Dawood (editor); Franz Rosenthal (translator). Routledge and Kegan Paul, London, 1978.

UNIVERSE

Astronomy

Al-Battani. *Al-Zij al-Sabi.* See Planet Earth section.

Al-Biruni. *Kitab al-Tafhim li-awa'il sina'at al-tanjim,* or *The Book of Instruction in the Elements of the Art of Astrology.* R. Ramsay Wright (translator); reprinted by Fuat Sezgin. IGAIW, Frankfurt, 1998.

Carmody, Francis J. *Alfragani differentie in quibusdam collectis scientie astrorum.* Berkeley, California, 1943.

Al-Farghani. *Kitab fi Harakat al-Samawiyah wa Jawami Ilm al Nujum,* or *Compendium of Astronomy.* Suleymaniye Library, Ayasofya 2843/2, Istanbul.

Al-Farghani and **Al-Battani.** *Texts and Studies.* Collected and reprinted by Fuat Sezgin in collaboration with Mazen Amawi, Carl Ehrig-Eggert, and Eckhard Neubauer. IGAIW, Frankfurt, 1998.

Goldstein, Bernard R. *Al-Bitrūjī: On the Principles of Astronomy.* 2 volumes. Yale University Press, New Haven, Conn., 1971.

Ibn al-Shatir al-Muwaqqit. *Kitab Nihayat al-sul fi Tashih al-Usul,* or *Limit of Desire in Correcting Principles.* Teymur riyada 154, Cairo.

Ibn Rushd. *Tahafut al-Tahafut,* or *The Incoherence of the Incoherence.* Translated from the Arabic with introduction and notes by Simon van den Bergh, 2 volumes, pp. 311–316. Luzac & Co., London, 1954.

Ibn Yunus. *Al-Zij al-Kabir al-Hakimi, or the Hakemite Tables.* See Planet Earth section.

King, David A. "The Astronomical Works of Ibn Yūnus." Ph.D. dissertation, Yale University, New Haven, Conn., 1972.

King, David A. "Ibn Yūnus' Very Useful Tables for Reckoning Time by the Sun," in *Archive for History of Exact Sciences* 10, pp. 342–394. 1973.

Puig, Roser. *Al-Šakkāziyya: Ibn al-Naqqāš al-Zarqālluh. Edición, traducción y estudio.* Instituto Millas Vallicrosa de Historia de la Ciencia Araba, Barcelona, 1986.

Al-Zarqali. *Kitab al-a'mal bi'l-safiha al-Zijiyya,* or *Book of Operations by Means of Tympanum of Zijes.* Suleymaniye Library, Esad Efendi 2671/1, Istanbul.

Observatories

Abu Mansur. *Al-Zij al Mumtahan,* or *The Verified Tables.* Library of the St. Laurentius Monastery, II, 927, Escorial, Spain.

Dizer, M. (editor). *Proceedings of the International Symposium on the Observatories in Islam* (September 19-23, 1977). Millî Egitim Basımevi, Istanbul, 1980.

Sayili, Aydin. *The Observatory in Islam.* Türk Tarih Kurumu Basimevi, Publications of the Turkish Historical Society, Ankara. Réimpression Arno Press, New York, 1981.

Astronomical Instruments

Al-Battani. *Al-Zij al-Sabi.* See Planet Earth section.

Al-Halabi. *Bughyat al-Tulab fil'amal bi'l rub al-astrulab,* or *Aims of Pupils on Operations with the Quadrant of Astrolabe.* University Library 1001/8, Leiden.

Al-Hamawi. *Ad Durr al-Gharib fil amal bi dairat al-tayyib,* or *Rare Pearls on Operations with the Circle for Finding Sines.* University Library 187b/4, Leiden.

'Izz al-Din al-Wafa'i. *Al-Nujum al-Zahirat fi amal bi'l rub al-Muqantarat,* or *Brilliant Stars on Operations with the Almucantar Quadrant.* Suleymaniye Library, Fatih 3448, Istanbul.

Jabir ibn Aflah. *Kitab al-Hai'a,* or *Book of Cosmology.* Berlin MS 5653, No. 5479, catalogue Die Handschriften-Verzeichnisse der Königlichen Bibliothek zu Berlin, 23 volumes, 1853-1914.

Jabir ibn Aflah. *Islah al Majisti,* or *Correction of the Almagest of Ptolemy.* Berlin, Staatsbibliothek-State Library, 5653, Berlin.

Al-Khujandi. *Al-Talwih li-Asrar al-Tanqih.* See Hospital Development section.

Shihab al-Din al-Hamawi. *Masail Handasya,* or *Geometrical Problems.* Riyada 694, Cairo.

Al-Sufi. *Suwar al-Kawakib al-Thabit,* or *Book of Fixed Stars.* Suleymaniye Library, Fatih 3422, Istanbul.

Taqi al-Din. *Turuq al-Saniyya fi al-Alat al-Ruhaniyya.* See Water Supply section.

Ragep, F. J. (editor and translator). *Nasir al-Din al-Tusi's Memoir on Astronomy 'al-Tadhkira fi 'ilm al-hay'a'.* 2 volumes. Springer Verlag, Berlin, 1993.

Astrolabe

Al-Biruni. *Al-Isti'ab fi San'at al-Usturlabe.* Diyarbakir Public Library, 403/3, Diyarbakir, Turkey.

Al-Bitruji. *Kitab-al-Hay'ah,* or *Kitab al-murta'ish fi'l-hay'a,* or *Book of Cosmology.* Library of the Topkapi Palace Museum, 3302/1, Istanbul.

Al-Farghani. *Kitab fi san'at al-astrolabe.* Kastamonu Public Library, 794-5, Kastamonu, Turkey.

Al-Farghani. *Kitab fi Harakat al-Samawiyah wa Jawami Ilm al Nujum.* See Astronomy section.

Ibn Isa. *Risala fi al-Usturlab.* Vatican Library, Codici Borgiani Arabi 217/3, Vatican City.

Jamal al-Din al-Tariqi. *Risala fi ma'rifat al-Taqwim wa ma'rifat al-usturlab wa mawaqit wa 'ilm ahkam al-Nujum.* Vatican Library 1398/3, Vatican City.

Masha'Allah. *Al-Kitab al-ma'ruf bi'l-sabi' wa-'l ishrin,* or *The Book Known as Twenty-seventh,* or *De scientia motus orbis, Massahalae de scientia motus orbis.* Nuremberg, 1504.

Masha'Allah. *Kitab san'at al-asturlabat wa'l-'amal biha,* or *Book on the Construction of Astrolabes and Their Operations,* or *De compositione et utilitate astrolabii.*

Al-Zarqali. *Kitab al-a'mal bi'l-safiha al-Zijiyya.* See Astronomy section.

Armillary Sphere

Dawud ibn Sulayman. *Kitab dhat al-halaq,* or *Book on the Armillary Sphere.* Miqat 969/1a, Cairo.

Jabir ibn Aflah. *Islah al-Majisti.* See Astronomical Instruments section.

Lunar Formations

Abu al-Fida'. *Mukhtasar Tarikh Al-Bashar,* or *Concise History of Humans.* Corum Hasan Pasa Public Library, 1178, Corum, Turkey.

Abu al-Fida'. *Taqwim al-Buldan.* See Travelers and Explorers section.

Masha'Allah. *Al-Kitab al-ma'ruf bi'l-sabi' wa-'l ishrin.* See Astrolabe section.

Masha'Allah. *Kitab san'at al-asturlabat wa'l-'amal biha.* See Astrolabe section.

Al-Sufi. *Suwar al-Kawakib al-Thabit.* See Astronomical Instruments section.

Al-Tusi. *Tarcama-i Kitab-i Suwar al-kawakib.* Suleymaniye Library, Ayasofya Collection, 2595, Istanbul.

Al-Tusi. *Al-Tadhkira fi al-Hay'a.* Vatican Library 319/1, Vatican City.

Ulugh Beg. *Al-Zij,* or *Astronomical Tables.* Suleymaniye Library, Ayasofya Collection, MS 2692, Istanbul.

Constellations

Al-Sufi. *Suwar al-Kawakib al-Thabit.* See Astronomical Instruments section.

Flight

Al-Firdawsi. *Shahnameh,* or *Book of Kings.* Ankara National Library, B 530, Ankara, Turkey.

Ibn Jubayr. *Rihlat Ibn Jubayr.* See Hospital Development section.

05 FURTHER READING

GENERAL

Abattouy, Mohammed. *L'Histoire des sciences arabes classiques: une bibliographie sélective commentée*. Foundation of King Abdulaziz, Casablanca, 2007.

Abattouy, Mohammed (editor). *La science dans les sociétés islamiques: approches historiques et perspectives d'avenir*. Foundation of King Abdulaziz, Casablanca, 2007.

Abattouy, Mohammed, Renn, Jürgen, and Weinig, Paul (editors). *Science in Context*, Vol. 14. Special double issue: *Intercultural Transmission of Scientific Knowledge in the Middle Ages: Graeco-Arabic-Latin*. Cambridge University Press, Cambridge, U.K., 2001.

Al-Khalili, Jim. *Pathfinders: The Golden Age of Arabic Science*. Allen Lane, London, 2010.

Al-Qurashi, Diya Al-Din Muhammad Ibn Muhammad Al-Shafii, known as Ibn Al-Ukhuwwa. *The Ma'alim Al-Qurba Fi Ahkam Al-Hisba*. Reuben Levy (editor). Abstract of contents, glossary, and indices. Cambridge University Press/Luzac and Co., Cambridge/London, 1938.

Arabick Roots. A catalogue of exhibition manuscripts and letters of early founders of the Royal Society revealing their connections with Arabic. The Royal Society, London, August-November 2011. The Royal Society, London, 2011.

Avicenne, Al-Husayn Ibn Abdullah Ibn Sina. *Poème de la médicine, Urguza fi al-tibb (Cantica Avicennae)*, Texte Arabe, Traduction Française, Traduction Latine du XIIIe siècle, avec Introduction, notes, et index, établi et présenté par Henri Jahier et Abdul-Kader Noureddine, Les Belles Lettres, Paris, 1956.

Berggren J. Lennart. "Historical Reflections on Scientific Knowledge: The Case of Medieval Islam," in *Knowledge Across Cultures: Universities East and West*, pp. 137-153. Ruth Hayhoe (editor). Hubei Educational Press/OISE Press, Toronto, 1993.

Brockelman, Carl. *Geschichte der arabischen Litteratur*. 3 volumes plus 2 supplements. 3rd edition, E. J. Brill, Leiden, 1943-49.

Carra de Vaux, Bernard. *Les penseurs de l'Islam*. 5 volumes. Geunther, Paris, 1921-26.

Casulleras, Josep, and Samsó, Julio (editors). *From Baghdad to Barcelona. Studies in the Islamic Exact Sciences in Honour of Professor Juan Vernet*. 2 volumes. Instituto Millás Vallicrosa de Historia de la Ciencia Arabe—Anuari de Filología, Universitat de Barcelona, XIX b-2, 1996.

Catalogue of Arabic Science and Medicine: A Collection of Manuscripts and Early Printed Books Illustrating the Spread and Influence of Arabic Learning in the Middle Ages and the Renaissance. Vol. 1186 de Bernard Quaritch catalogue. Introduction by Professor Charles Burnett. Bernard Quaritch Ltd., London, 1993.

Dallal, Ahmad. "Science, Medicine, and Technology: The Making of a Scientific Culture," in *The Oxford History of Islam*, pp. 155-213. Edited by John L. Esposito. Oxford University Press, Oxford, U.K., 1999.

Djebbar, Ahmed. *L'âge d'or des sciences arabes*. Editions Le Pommier/La Cité des sciences et de l'industrie, Paris, 2005.

——. *Une histoire de la science arabe*. Entretiens avec Jean Rosmorduc. Editions du Seuil, Paris, 2001.

Endress, Gerhard. "*Die wissenschaftliche Literatur*," in *Grundriß der Arabischen Philologie*, pp. 399-506. Edité par Helmut Gätje. Band II: *Literatur wissenschaft*. Dr. Ludwig Reichert Verlag, Wiesbaden, 1987.

Gillispie, Charles (editor). *Dictionary of Scientific Biography*. 18 volumes. Charles Scribner's Sons, New York, 1970-90.

Hamilton, Michael M. "Lost History: The Enduring Legacy of Muslim Scientists, Thinkers, and Artists," in *National Geographic*, June 19, 2007.

Hartner, Willy. "La science dans le monde de l'islam après la chute du Califat." *Studia Islamica*, Vol. 31, pp. 135-151. 1970.

Hassan, al-, Ahmad Y., Iskandar, Yusuf, Zaki, Albert, and Maqbul, Ahmad (editors). *The Different Aspects of Islamic Culture*. Vol. IV: *Science and Technology in Islam*, Parts I-II. UNESCO, Paris, 2001.

Hayes John (editor). *The Genius of Arab Civilization: Source of Renaissance*. The MIT Press, Cambridge, Mass., 1983.

Hogendijk, J. P., and Sabra, A. I. [Abdelhamid Ibrahim] (editors). *The Enterprise of Science in Islam. New Perspectives*. The MIT Press, Cambridge, Mass., 2003.

[Ibn al-Haytham]. *Al-Hasan ibn al-Hasan ibn al-Haytham (d. 430-1039). Texts and Studies*. Collected and reprinted by F. Sezgin et al. 2 volumes. Institut für Geschicte der Arabisch-Islamischen Wissenschaften, Frankfurt, 1998.

Ibn al-Nadim. *The Fihrist of al-Nadim. A Tenth-Century Survey of Muslim Culture*. English translation by Bayard Dodge. 2 volumes. Columbia University Press, London/New York, 1970.

——. *Kitab al-Fihrist*. Mit Anmerkungen hrsg. von Gustav Flügel. 2 volumes. F. C. W. Vogel, Leipzig, 1871-72.

Ibn Khaldun. *The Muqaddimah. An Introduction to History*. English translation by F. Rosenthal. 3 volumes. Princeton University Press, Princeton, N.J., 1967.

Ihsanoglu, Ekmeleddin (editor). *Catalogue of Islamic Medical Manuscripts (in Arabic, Turkish, Persian) in the Libraries of Turkey.* Prepared by Ramazan Sesen, Cemil Akpinar, and Cevat Izgi. IRCICA, Istanbul, 1984.

——. *Osmanli Astronomi Literatürü Tarihi. History of Astronomy Literature During the Ottoman Period.* Prepared by Ekmeleddin Ihsanoglu, Ramazan Sesen, Cevat Izgi, Cemil Akpinar, and Ihsan Fazlioglu. 2 volumes. IRCICA, Istanbul, 1997.

——. *Osmanli Matematik Literatürü Tarihi. History of Mathematical Literature During the Ottoman Period.* Prepared by Ekmeleddin Ihsanoglu, Ramazan Sesen, and Cevat Izgi. 2 volumes. IRCICA, Istanbul, 1999.

Ihsanoglu, Ekmeleddin, and Günergun, Feza (editors). *Science in Islamic Civilization.* Proceedings of the Science Institutions in Islamic Civilization and International Symposia Science and Technology in the Turkish and Islamic World. IRCICA, Istanbul, 2000.

Kahn, A. S. *A Bibliography of the Works of Abu 'l-Rayhan al-Biruni.* New Delhi, 1982.

Kennedy, Edward Stewart. "The Arabic Heritage in the Exact Sciences," in *Al-Abhath,* Vol. 23, pp. 327-344. The American University of Beirut, Beirut, 1970.

Kennedy, E. S., Colleagues, and Former Students. *Studies in the Islamic Exact Sciences.* D. A. King and M. H. Kennedy (editors). The American University of Beirut, Beirut, 1983.

Makdisi, George. "Muslim Institutions of Learning in Islam and in the West," in *Bulletin of the School of Oriental and African Studies,* Vol. 24, pp. 1-56. Travail pionnier sur le système d'éducation développé dans la civilisation islamique. University of London, London, 1961

——. *The Rise of Colleges: Institutions of Learning in Islam and in the West.* Edinburgh University Press, Edinburgh, 1981.

Matvievskaia, Galina P., and Rozenfeld, Boris A. *Matematiki i astronomi musulmanskogo srednevekovya i ikh trudi* (VII-XVII vv). 3 volumes. Nauka, Moscou, 1983.

Mieli, Aldo. *La Science arabe et son rôle dans l'évolution scientifique mondiale.* E. J. Brill, Leiden, 1st édition, 1938; 2nd édition, 1966.

Morelon, Régis, and Hasnawi, Ahmed (editors). *De Zénon d'Élée à Poincaré. Recueil d'études en hommage à Roshdi Rashed.* Peeters, Louvain/Paris, 2004.

Nasr, Seyyed Hosein. *Science and Civilization in Islam.* Harvard University Press, Cambridge, Mass., 1968. Reprints: New American Library, New York, 1968, 1970; The Islamic Texts Society, Cambridge, U.K., 1987.

Rashed, Roshdi (editor). *Encyclopedia of the History of Arabic Science.* Edited with the collaboration of Régis Morelon. 3 volumes. Vol. 1: Astronomy—Theoretical and Applied; Vol. 2: Mathematics and the Physical Sciences; Vol. 3: Technology, Alchemy, and Life Sciences. Routledge, London/New York, 1996.

Rosenfeld, Boris A., and Ihsanoglu, Ekmeleddin. *Mathematicians, Astronomers, and Other Scholars of Islamic Civilization and Their Works (7th-19th Centuries).* IRCICA, Istanbul, 2003.

Sabra, A. I. "Situating Arabic Science: Locality Versus Essence," in *Isis,* Vol. 87, pp. 654-670. Chicago University Press, Chicago, 1996.

Said, Hakim Mohammed. *Al-Biruni: Commemorative Volume.* Proceedings of the international congress held in Pakistan on the occasion of millenary of Ab´ RaiúŒn Muúamad Ibn Aúmad al-B¥r´n¥ (November 26, 1973-December 12, 1973). Times Press, Karachi, 1979.

Said, Hakim Mohammed (editor). *Ibn Al-Haitham.* Proceedings of the celebrations of 1,000th anniversary held under the auspices of Hamdard National Foundation, Pakistan. Times Press, Karachi, 1969.

Saliba, George. *Islamic Science and the Making of the European Renaissance.* The MIT Press, Cambridge, Mass., 2007.

Saliba, George. "Arabic Planetary Theories After the Eleventh Century A.D.," in *Encyclopedia of the History of Arabic Science,* pp. 58-127. Routledge, London, 1996.

——. *A History of Arabic Astronomy: Planetary Theories During the Golden Age of Islam,* New York University Press, 1994.

——. *Rethinking the Roots of Modern Science: Arabic Scientific Manuscripts in European Libraries,* Occasional Paper, Center for Contemporary Arabic Studies, Georgetown University, Washington, D.C., 1999.

——. "A Sixteeenth-Century Arabic Critique of Ptolemaic Astronomy: The Work of Shams al-Din al-Khafri," in *Journal for the History of Astronomy,* Vol. 25, pp. 15-38. 1994.

Samsó, Julio. *Las Ciencias de los antiguos en Al-Andalus.* Mapfre, Madrid, 1992.

Sarton, George. *Introduction to the History of Science.* 3 volumes: Vol. 1: *From Homer to Omar Khayyám*; Vol. 2: *From Rabbi Ben Ezra to Roger Bacon*; Vol. 3: *Science and Learning in the Fourteenth Century.* The Williams and Wilkins Company for the Carnegie Institution, Baltimore, 1927-48.

Savage-Smith, Emilie. "Gleanings from an Arabist's Workshop: Current Trends in the Study of Medieval Islamic Science and Medicine," in *Isis,* Vol. 79, pp. 246-72. 1988.

Schacht, J., and Bosworth, C. E. *The Legacy of Islam.* Oxford University Press, 1974; 2nd edition, 1979.

Selin, Helaine (editor). *Astronomy Across Cultures: The History of Non-Western Astronomy.* Kluwer, Dordrecht, 2000.

——. *Encyclopaedia of the History of Science, Technology, and Medicine in Non-Western Cultures.* Kluwer, Dordrecht, 1997.

——. *Mathematics Across Cultures: The History of Non-Western Mathematics.* Kluwer, Dordrecht, 2000.

Sezgin, Fuat. *Geschichte des Arabischen Schriftums*. 12 volumes. E. J. Brill, Leiden, 1967-2000.

Spink, M. S., and Lewis, I. L.((editors and translators). *Albucassis on Surgery and Instruments: A Definitive Edition of the Arabic Text with English Translation and Commentary*. Wellcome Institute of the History of Medicine, London, 1973.

Süter, Heinrich. *Beiträge zur Geschichte der Mathematik und Astronomie im Islam*. Nachdruck seiner Schriften aus den Jahren 1892-1922. 2 volumes. Fuat Sezgin (editor). Institut für Geschicte der Arabisch-Islamischen Wissenschaften, Frankfurt, 1986.

[Thabit ibn Qurra]. *Thabit ibn Qurra. Texts and Studies*. Collected and reprinted by F. Sezgin et al. Frankfurt: Institut für Geschicte der Arabisch-Islamischen Wissenschaften, 1997.

Vernet, J., and Samsó, J., et al. *El Legado científico andalusi*. Ministerio de Cultura, Madrid, 1992.

Wiedemann, Eilhard. *Aufsätze zur Arabischen Wissenschaftsgeschichte*. 2 volumes. Hildesheim, New York: G. Olms, 1970.

——. *Gesammelte Schriften zur arabisch-islamischen Wissenschaftsgeschichte*. Gesammelt und bearb. Von Dorothea Girke. 3 volumes. IGAIW, Frankfurt, 1984.

Woepcke, Franz. *Etudes sur les mathématiques araboislamiques*. Nachdruck von Schriften aus den Jahren 1842-1874. 2 volumes. Institut für Geschicte der Arabisch-Islamischen Wissenschaften, Frankfurt, 1986.

Young, M. J. L., Latham, J. D., and Serejant, R. B. *Religion, Learning and Science in the Abbasid Period*. Cambridge University Press, Cambridge, 1990.

HOME

BBC 2. *What the Ancients Did for Us: The Islamic World*. February 16, 2004.

Channel 4 TV. *An Islamic History of Europe*. August 5-19, 2005.

Ellis, John. *An Historical Account of Coffee with an Engraving, and Botanical Description of the Tree: To Which Are Added Sundry Papers Relative to Its Culture and Use, as an Article of Diet and of Commerce*. Edward Dilly and Charles Dilly, London, 1774.

Friedman, D., and Cook, E. *A Miscellany*. www.daviddfriedman.com/Medieval/miscellany_pdf/Miscellany.htm.

Hart-Davies, Adam. *What the Past Did for Us: A Brief History of Ancient Inventions*. BBC Books, London, 2004.

Lindberg, D. C. *Studies in the History of Medieval Optics*. Varorium, London, 1983.

——. "The Western Reception of Arabic Optics," in *Encyclopaedia of History of Arabic Science*. R. Rashed (editor). Routledge, London, 1996.

Omar, S. B. *Ibn al-Haytham's Optics*. Bibliotheca Islamica, Chicago, 1977.

Ree, Hans. *The Human Comedy of Chess*. Russell Enterprises, Milford, Conn., 1999.

Sopieva, Natasha. *Ibn al-Haytham, the Muslim Physicist*. www.MuslimHeritage.com, 2001.

SCHOOL

Al-Ghazali. *Dear Beloved Son*. Translated from Arabic by K. El-Helbawy, Awakening U.K., Swansea, 2000.

Burnett Charles. *Leonard of Pisa (Fibonacci) and Arabic Arithmetic*. www.MuslimHeritage.com, 2005.

Dodge, B. *Muslim Education in Medieval Times*. The Middle East Institute, Washington, D.C., 1962.

Haskins, C. H. *Studies in the History of Mediaeval Science*. Frederick Ungar Publishing Co., New York, 1967.

Ihsanoglu, Ekmeleddin. *Primary Schools Under the Ottomans*. www.MuslimHeritage.com, 2005.

Mackensen, R. "Moslem Libraries and Sectarian Propaganda," in *The American Journal of Semitic Languages*, 1934-35.

Makdisi, George. "On the Origin and Development of the College in Islam and the West," in *Islam and the Medieval West*. Khalil I. Semaan (editor). State University of New York Press, Albany, 1980.

Nakosteen, M. *History of Islamic Origins of Western Education AD 800–1350*. University of Colorado Press, Boulder, 1964.

Pedersen, J. *The Arabic Book*, Geoffrey French (translator). Princeton University Press, Princeton, N.J., 1984.

Pinto, O. "The Libraries of the Arabs During the Time of the Abbasids," in *Islamic Culture* 3. 1929.

Ribera, J. *Disertaciones y Opúsculos*. 2 volumes. Imprenta de Estanislao Maestre, Madrid, 1928.

Sardar, Z., and Davies, M. W. *Distorted Imagination*. Grey Seal Books, London, 1990.

Sarton, G. *Introduction to the History of Science. 3 volumes. The* Carnegie Institution, Washington, D.C., 1927.

Tibawi, A. *Islamic Education*. Luzac and Company Ltd., London, 1972.

Watt, W. M. *The Influence of Islam on Medieval Europe*. Edinburgh University Press, Edinburgh, 1972.

Wilds, E. H. *The Foundation of Modern Education*. Rinehart and Co., New York, 1959.

Zaimeche, Salah. *Education in Islam: The Role of the Mosque*. www.MuslimHeritage.com, 2002.

MARKET

Artz, F. B. *The Mind of the Middle Ages*. Revised third edition. University of Chicago Press, Chicago, 1980.

Bolens, L. "Agriculture," in *Encyclopedia of the History of Science, Technology, and Medicine in Non-Western*

Cultures. Helaine Selin (editor). Kluwer Academic Publishers, Dordrecht/Boston/London. 1997.

Channel 4 TV. *An Islamic History of Europe*. August 5-19, 2005.

De Vaux, Baron Carra. *Les Penseurs de l'Islam*. Vol. 2, Geuthner, Paris, 1921.

Hill, D. R. *Islamic Science and Engineering*. Edinburgh University Press, Edinburgh, 1993.

Idrisi, Zohor. *The Muslim Agricultural Revolution and Its Influence on Europe*. www.MuslimHeritage.com, 2005.

Le Bon, G. *La Civilisation des Arabes*. IMAG, Syracuse, Italy, 1884.

Scott, S. P. *History of the Moorish Empire in Europe*. 3 volumes. J. B. Lippincott Company, London. 1904.

Watson, A. M. *Agricultural Innovation in the Early Islamic World*. Cambridge University Press, Cambridge, U.K., 1983.

Zaimeche, Salah. *A Review on Muslim Contribution to Agriculture*. www.MuslimHeritage.com, 2002.

HOSPITAL

Abdel-Halim, R. E. "Contributions of Ibn Al-Nafis (1210-1288 A.D.) to the Progress of Medicine and urology: A Study and Translations from His Medical Works," in *Saudi Medical Journal*, Vol. 29, pp. 13-22. 2008.

——. *Experimental Medicine 1,000 Years Ago*. Vol. 3, pp. 55-61. Urol Ann, Paris, 2011.

——. *Lithotripsy: A Historical Review*. Matouschek E., editor. Endo-urology—Proceedings of the Third Congress of the International Society of Urologic Endoscopy, Karlsruhe; August 26-30, 1984. Bau-Verlag Werner Steinbruck, Baden, Germany, 1985. (Also available at http://www.hektoeninternational.org/Lithotripsy.html.)

——. "Obesity: 1,000 Years Ago." *Lancet*, 366:204. 2005. (Also available at http://www.rabieabdelhalim.com/Obesity1000YsAgo.htm.)

Abdel-Halim, R. E., Altwaijiri, A. S., Elfaqih, S. R., and Mitwalli, A. H. "Extraction of Urinary Bladder Stone as Described by Abul-Qasim Khalaf Ibn Abbas," in *Saudi Medical Journal*, Vol. 24, pp. 1283-91. 2003. A translation of original text and a commentary.

Al-Mazroa, A. A., and Abdel-Halim, R. E. "Anaesthesia 1,000 Years Ago—I," in *The History of Anaesthesia*, pp. 46-8. R.S. Atkinson and T. B. Boulton (editors). Royal Society of Medicine Services and the Parthenon Publishing Group, London and New York, 1989. (Also available at http://rabieabdelhalim.com/anesthesia1.html.)

——. "Anesthesia 1,000 Years Ago—II," in *Middle East Journal Anesthesiology*, Vol. 15, pp. 383-92. 1991, 2000. (Also available at http://www.rabieabdelhalim.com/anaesthesia2.html.)

Burnett, Charles. *Arabic Medicine in the Mediterranean*. www.MuslimHeritage.com, 2004.

Campbell, D. *Arabian Medicine, and Its Influence on the Middle Ages*. Philo Press, Amsterdam, 1974.

Channel 4 TV. *An Islamic History of Europe*. August 5-19, 2005.

Cumston, C. G. "Islamic Medicine," in *An Introduction to the History of Medicine from the Time of the Pharaohs to the End of the XVIII Century*. Kegan Paul, Trench, Trubner, and Co. Ltd., London; and Alfred A. Knopf, New York, 1926.

FSTC. *Pharmacology in the Making*. www.MuslimHeritage.com, 2001.

Ghalioungui, Paul. "Ibn Nafis," in *Studies in the Arabic Heritage*. The Ministry of Information of Kuwait, Kuwait, 1970.

Hirschberg, J., Lippert, J., and Mittwoch, E. *Die arabischen Lehrbucher der Augenheilkunde*, Abhdl, Der Preussischen Akademie, Berlin, 1905.

Iskandar, A. Z. *A Catalogue of Arabic Manuscripts on Medicine and Science in the Wellcome Historical Medical Library*. The Wellcome Historical Medical Library, London, 1967.

Keys, T. E., and Wakim, K. G. *Contributions of the Arabs to Medicine*, Vol. 28. Proceedings of the Staff Meeting, Mayo Clinic, Rochester, Minn., 1953.

Kirkup, J. R. *The History and Evolution of Surgical Instruments*. I. Introduction. Annals of the Royal College of Surgeons of England, 1981.

Leclerc, L. *Histoire de la médecine arabe*. Ernest Ledaux, Paris, 1876.

Levey, M. *Early Arabic Pharmacology*. E. J. Brill, Leiden, 1973.

Lindberg, D. C. "The Western Reception of Arabic Optics," in *Encyclopedia of History of Arabic Science*. R. Rashed (editor). Routledge, London, 1996.

Meyerhof, M. "Ibn Nafis and His Theory of the Lesser Circulation," in *Isis*, Vol. 23. 1935.

Sarton, G. *Introduction to the History of Science*. Carnegie Institution, Washington, D.C., and Williams and Wilkins Company, Baltimore, 1927-31. Reprinted Robert E. Krieger Publishing Co. Inc., New York, 1975.

Shaikh, Ibrahim. *Who Discovered Pulmonary Circulation, Ibn al-Naphis or Harvey?* www.MuslimHeritage.com, 2001.

Ullmann, M. *Islamic Medicine*. Islamic surveys No. 11. Edinburgh University Press, Edinburgh, 1978.

TOWN

Channel 4 TV. *An Islamic History of Europe*. August 5-19, 2005.

Forbes, R. J. *Studies in Ancient Technology*, Vol. 2. E. J. Brill, Leiden, 1965.

Frothingham, A. W. *Lustreware of Spain*. The Hispanic Society of America, New York, 1951.

Glick, T. *Islamic and Christian Spain in the Early Middle Ages*. Princeton University Press, Princeton, N.J., 1979.

Harvey, J. *The Master Builders*. Thames and Hudson, London, 1973.

Haskins, C. H. *Studies in the History of Mediaeval Science*. Frederick Ungar Publishing Co., New York, 1967.

Hobson, R. L. *A Guide to the Islamic Pottery of the Near East*. British Museum, London, 1932.

Lambert, E. *Art Musulman et Art Chrétien dans la Peninsul Iberique*. Editions Privat, Paris, 1958.

Lane, A. *Early Islamic Pottery*. Faber and Faber, London, 1947.

Male, E. *Art et Artistes du Moyen Age*. Librairie Armand Colin, Paris, 1928.

Saoud, R. *Introduction to the Islamic City*. www.MuslimHeritage.com, 2001.

Wren, Christopher. *Parentalia, or Memoirs of the Family of the Wrens*. Mathew Bishop, T. Osborn, and R. Dodsley, London, 1750.

WORLD

Alhabshi, Syed Othman. *Mapping the World*. www.MuslimHeritage.com, 2001.

Briffault, R. *The Making of Humanity*. George Allen, London, 1928.

Channel 4 TV. *An Islamic History of Europe*. August 5-19, 2005.

Feber, S. (editor). *Islam and the Medieval West*. A loan exhibition at the University Art Gallery, State University of New York (April 6-May 4), 1975.

Fuat Sezgin in Zusammenarbeit mit Eckhard Neubauer. *Wissenschaft und Technik im Islam: Einführung in die geschihcte der Arabisch-Islamischen Wissenschaften*. Vols. I-V. Institut für Geschichte der Arabisch-Islamischen Wissenschaften, Frankfurt am Main, 2003.

Glick, T. *Islamic and Christian Spain in the Early Middle Ages*. Princeton University Press, Princeton, N.J., 1979.

Harley, J. B., and Woodward, D. (editors). *The History of Cartography*. Vol. 2, Book 1: *Cartography in the Traditional Islamic and South Asian Societies*. University of Chicago Press, Chicago, 1972.

Holt, P. M., Lambton, A. K. S., and Lewis, B. (editors). *The Cambridge History of Islam*. Vol. 2, Cambridge University Press, Cambridge, U.K., 1970.

Kimble, G. H. T. *Geography in the Middle Ages*. Methuen and Co. Ltd., London, 1983.

Roshdi, Rashed (editor). *Encyclopaedia of the History of Arabic Science*. Routledge, London, 1996.

Scott, S. P. *History of the Moorish Empire in Europe*, 3 volumes. J. B. Lippincott Company, London, 1904.

Watt, M. *The Influence of Islam on Medieval Europe*. Edinburgh University Press, Edinburgh, 1972.

Zaimeche, Salah. *A Review of Muslim Geography*. www.MuslimHeritage.com, 2002.

UNIVERSE

Arnold, Sir Thomas, and Guillaume, Alfred. *The Legacy of Islam*. First edition. Oxford University Press, Oxford, U.K., 1931.

Artz, F. B. *The Mind of the Middle Ages*. Third revised edition. University of Chicago Press, Chicago, 1980.

BBC 4. *An Islamic History of Europe*. August 5-19, 2005.

Bedini, Silvio A. *The Pulse of Time*. Leo S. Olschki, publisher (found in the Biblioteca Di Nuncius), 1991.

Briffault, R. *The Making of Humanity*. George Allen, London, 1928.

De Vaux, Baron Carra. *Les Penseurs de l'Islam*. Vol. 2. Geuthner, Paris, 1921.

Glubb, John. *Short History of the Arab Peoples*. Hodder and Stoughton, London, 1969.

Hitti, P. K. *History of the Arabs*. Tenth edition, Macmillan and St. Martin's Press, London, 1970.

Hogendijk, Jan P., and Abdelhamid, I. Sabra. *The Enterprise of Science in Islam: New Perspectives*. The MIT Press, Cambridge, Mass., 2003.

Ronan, C. "The Arabian Science," in *The Cambridge Illustrated History of the World's Science*. Cambridge University Press, Cambridge, U.K., 1983.

Saliba G. *Islamic Science and the Making of the European Renaissance*. The MIT Press, Cambridge, Mass., 2007.

Savage-Smith, Emilie. "Celestial Mapping," in *The History of Cartography 2*, Book 1. J. B. Harvey and David Woodward (editors). University of Chicago Press, Chicago, 1992.

Savory, R. M. *Introduction to Islamic Civilization*. Cambridge University Press, Cambridge, U.K., 1976.

Sayili, Aydin. *Observatories in Islam*. Republished from Dizer, M. (editor), International Symposium on the Observatories in Islam (September 19-23, 1977), Istanbul. FSTC. www.MuslimHeritage.com, 2005.

Sedillot, L. A. *Memoire sur les instruments astronomiques des Arabes*. Imprimerie royale, Paris, 1841

Selin, Helaine. *Encyclopaedia of the History of Science, Technology and Medicine in Non-Western Cultures*. Kluwer Academic Publishers, London, 1997.

Smith. D. E. *History of Mathematics*, Vol. 2. Dover Publications, New York, 1953.

Zaimeche, Salah. *A Review on Missing Contribution to Astronomy*. www.MuslimHeritage.com, 2002.

06 GLOSSARY

Abbasid dynasty A dynasty that ruled the Muslim caliphate from Iraq between 750 and 1258. The Abbasids are renowned for fostering learning and science. Their most distinguished caliphs are Harun al-Rashid (ruled 786-809) and his son Al-Ma'mun (ruled 813-833), who made Baghdad the center of science and learning. They founded the House of Wisdom, a famous library and scholarship center. Harun al-Rashid is renowned for gifting Charlemagne a water clock and an organ in 797.

Aghlabids Muslim dynasty that ruled from 800 to 909, and was semi-independent of Baghdad. The Aghlabids' capital, Al-Qayrawan, was a vibrant city during that time. Among their famous legacies is the water reservoir of Al-Qayrawan. From Al-Qayrawan, Tunisia, they ruled Sicily and Malta.

Allah "Allah" is the Arabic word for God, the supreme and only God, the Creator, who according to the Quran is the same God as of the Bible.

Allahu Akbar *Allahu Akbar* is Arabic for "God is the greatest." The phrase is said during each stage of both obligatory and voluntary prayers. The Muslim call to prayer, or *athan*, or azan, and call to commence the prayer, or *iqama*, also contain the phrase. The actual title of this phrase is *takbir*. In the Islamic world, instead of applause, Muslims often shout "*takbir*" and the crowd responds "*Allahu Akbar*" in chorus to show agreement and satisfaction.

Almohad One of the greatest medieval dynasties, which ruled North Africa (and much of Spain) from circa 1147 until the rise to power of the Merenids around 1269. The Almohad dynasty (from the Arabic *Al-Muwahhidun*, "the monotheists" or "the Unitarians," the name being corrupted through Spanish), were a Berber Muslim religious power who founded the fifth Moorish dynasty in the 12th century, uniting North Africa as far as Egypt, together with Muslim Spain.

Al-Andalus The Arabic name given to the Iberian Peninsula when it was ruled by Muslims from 711 to 1492. Al-Andalus once encompassed the area extending from the Mediterranean to northern Spain, bordering the kingdom of Aragon in the north. Today, Andalusia is used to denote the southern region of Spain. Different meanings have been suggested for Al-Andalus, the most famous ones being the "gardens" (in Arabic) and the "land of the Vandals," rulers who inherited the Roman Empire and ruled Spain before the Muslims.

Arab and/or Muslim The term "Arab" is applied to those people who are of Arab origin, regardless of whether they are/were Muslims. "Muslim" is used to refer to the people who adhere to the Muslim religion, which includes Arabs and non-Arabs, such as those from Iran, Pakistan, or Indonesia, for example.

Asabiyah This is an Arabic word, which can mean "solidarity" or "group consciousness" but is usually translated as "group feeling." At the most basic level, *asabiyah* is something that a person feels for his family, a kind of "brotherhood." According to Ibn Khaldun, the successful ruler is he who manages to spread and maintain the *asabiyah* to all members of the society, so that all think of one another as they would think of their own brothers.

Ayyubids A dynasty founded by the Muslim Kurdish general Salah al-Din al-Ayyubi (d. 1193), known to Christians in Europe as Saladin. Salah al-Din established the Ayyubid dynasty in 1169. The Ayyubids united Egypt and Syria and other parts of the Muslim East, which enabled them to defeat the Crusaders at Hattin and recover Jerusalem.

Al-Azhar A university connected to the mosque in Cairo named in honor of Fatima al-Zahraa, the daughter of Prophet Muhammad, from whom the Fatimid Dynasty claimed descent. The mosque was built in two years from 971 to 973, and the school of theology connected with it was founded in 988. It remains to this day. It is one of the oldest operating universities in the world.

Baidaq Pawn, in chess.

Al-Barrani Al-Barrani consists of a large dome-covered hall in a bathhouse, incorporating a drum (below the dome) with stained glass windows. The Damascenes spent much of their talent on lavishly staining the walls of Al-Barrani with elegant tiles of dazzling colors, reflecting mirrors, and calligraphy plates welcoming clients and citing Arabic proverbs. It is here that they got ready to proceed to other sections of the *hammam* and where they retired after bathing.

Al-Baydah A village near Qaim in Iraq.

Caliph Literally means "one who replaces someone who left or died." In Islamic context, this means a successor to the Prophet Muhammad as a political, military, and administrative leader of the Muslims, but does not include a prophetic role.

Caliphate The Islamic state or government, whose head is the caliph.

C.E. The Common Era, formerly known as the Christian Era. It is the agreed international dating terminology.

Chatrang *Chatrang* is Persian for chess, and the oldest form of the game.

Dinars Basic currency unit, consisting of 1,000 fils.

Eid A Muslim celebratory festival, of which there are two, one after fasting in the month of Ramadan (called *Eid al-Fitr*), and the other in celebration after *Hajj* (called *Eid al-Adhha*).

Faqih An expert in Islamic law.

Faras Arabic term for mare or horse, and the knight in chess.

Fatimid A dynasty, named after Fatima al-Zahraa, the daughter of Prophet Muhammad, which rose to political domination in North Africa in 909. The Fatimids are the founders of Islamic Cairo, the capital city of Egypt, in 969.

Al-Fihrist Literally this means "a table of contents" or "an index." *Al-Fihrist* is an index of all books written in Arabic by both Arabs and non-Arabs. It was written by Abu al-Faraj Muhammad ibn Ishaq ibn Muhammad ibn Ishaq, also called ibn al-Nadim. He began to make this catalog of authors and the names of their writings for use in his father's bookstore. As he grew older, he became interested in the many subjects he read about in books, or those he learned about from friends and chance acquaintances. So, instead of being merely the catalog for a bookshop, *Al-Fihrist* became an encyclopedia of medieval Islamic culture.

Fiqh Literally meaning "knowledge and understanding," it is the understanding and applications of Sharia (divine law) from its sources.

Al-Fustat Al-Fustat is the first capital of Islamic Egypt established in 642 by 'Amru ibn Al- 'As, and was probably named after the Roman military term *fossatum*, or encampment.

Hadith Narrations of the sayings of the Prophet Muhammad, which form one of the major sources of Islamic law. Each hadith is composed of a basic text the authenticity of which was guaranteed by a chain of witnesses and narrators.

Hajj Pilgrimage to Mecca in Saudi Arabia.

Hammam Arabic public bath.

Haram Sacred, holy, and/or prohibited.

Ifriqiya In medieval history, Ifriqiya or Ifriqiyah was the area comprising the coastal regions of what are today western Libya, Tunisia, and eastern Algeria. In modern Arabic, the term means "Africa."

Imam One who leads the prayers.

Jabal al-'arus A mountain in Córdoba, Spain.

Ka'bah Literally, "a high place of respect and regard." It is the sacred building in the center of the Al-Haram al-Shareef Mosque at Mecca, Saudi Arabia. It is the center toward which Muslims around the world pray. It houses the divine black stone.

Kiswa Literally "a cover." The holy Ka'bah is covered with a new *kiswa* (textile cover) every year on the tenth *Dhul Hijjah*, which coincides with *Hajj*. Every year, the old *kiswa* is removed, cut into small pieces, and gifted to certain individuals, visiting foreign Muslim dignitaries, and organizations.

Koshk Turkish for kiosk.

Kutubiyun The word *kutubiyun* is a Moroccan Arabic name for bookbinders.

Madrasa The word madrasa means a school, and evolved originally from the lectures organized in mosques before schools became independent entities. These days madrasa has a different meaning, and thousands of madrasas around the world are said to be educational institutions, usually teaching Islamic sciences or law.

Maghreb The Arabic world was traditionally divided into two parts, the Mashriq or eastern part and the Maghreb or western part (literally, "the west" or "where the sun sets"). Geographically it is defined as the region of the continent of Africa north of the Sahara and west of the Nile—specifically, the modern countries of Morocco, Western Sahara (annexed and occupied by Morocco), Algeria, Tunisia, Libya, and to a much lesser extent Mauritania.

Mamluk Originally Turkish slaves who formed part of the Abbasid army. The Mamluks were a member of the Turkish-speaking cavalry that went on to rule Egypt and Syria under the 13th-century Mamluk dynasty.

Manarah Arabic for minarets of the mosque. Literally means "lighthouse."

Mihrab A niche in the wall of the mosque that indicates the direction in which one should pray, toward Mecca.

Minaret A tower from which the muezzin, or crier, calls people to prayer.

Minbar A pulpit for the imam, or prayer leader.

Miswak A cleaning stick, actually a twig from certain trees, essentially the arak tree botanically known as *Salvadore persica,* used for cleaning the teeth. Investigations by Swiss pharmaceutical company Pharba Basel, Ltd. found that it contains antibacterial substances that destroy harmful germs in the mouth.

Mithqals Weights.

Mosque A public place for worship and prayer for the Muslims.

Mu'allim Islamic teacher.

Muhandis Engineer or architect.

Al-Muhtasib *Al-Muhtasib* is literally "a judge" *(qadi)* who takes decisions on the spot, in any place and at any time, as long as he protects the interests of the public. His responsibilities are almost endless in order to implement the foregoing principle: commanding the good and forbidding the evil of wrongdoing. *Al-Muhtasib* and/or his deputies, like a full judge, must have high qualifications and be wise, mature, pious, well poised, sane, free, just, empathic, and a learned scholar, or *faqih.* He has the ability to ascertain right from wrong, and the capability to distinguish the permissible, *halal,* from the nonpermissible, *haram.* So, *Al-Muhtasib* is entrusted to secure the common welfare and to eliminate injuries to society as a whole, even if such honorable tasks require him to take a stance against the ruling governance. In short, he must be an appointee (fully authorized), pious, and just.

Muwaqqit Timekeeper, a wise man given the task to observe and decide on the times of prayers.

Pbuh Peace (and blessings of Allah) be upon him (Prophet Muhammad); a vow of devotion and belief that Muhammad was the Prophet of God (Allah). This phrase is repeated by Muslims every time they pronounce or hear the name of Prophet Muhammad.

Qadi A Muslim judge.

Al-Qahwa Arabic term for coffee.

Al-Qayrawan It is a town in northeast Tunisia and a revered city of Islam. Founded in 670 by Uqbah bin Nafi, an Arab leader, it was the seat of Arab governors in North Africa until 800. Under the Aghlabid dynasty (800-909), it became the chief center of commerce and learning, and remained so during the Fatimid rule (909-921). The city was ruined (1057) by Bedouin invaders, the Banu Hilal tribe, and subsequently was supplanted by Tunis.

Al-Qali This word was derived from *qalai* (to dry or roast in a pan). *Al-qali* is "the substance that has been roasted" or "ashes of the plant saltwork." In most languages of Europe, both substances were named *natron.*

Qamara A dark room, also a ship's cabin.

Qanat It is a type of underground irrigation canal between an aquifer on a piedmont zone to a garden on an arid plain. The word is Arabic, but the system is best known from ancient Iran.

Qibla An Arabic word referring to the direction of Mecca, Saudi Arabia, that Muslims should face toward when they pray.

Rajab The seventh month in the Islamic lunar calendar.

Ramadan It is the ninth month in the Islamic calendar, best known as the holy month of fasting for Muslims.

Rawdiya The inhabitants of the early Islamic world were enchanted by greenery. This love of plants is clearly shown in a genre of poetry, the *rawdiya,* or garden poem, probably of Persian origin, which came to be one of the main poetic forms in the Abbasid orient from the eighth to the tenth centuries.

Rihla Literally means "journey, travel, and travelogue." It is a piece of writing about travel.

Safavid dynasties The Safavids, an Iranian dynasty that ruled from 1501 to 1736. They had their origins in a long-established Sufi order, which had flourished in Azerbaijan since the early 14th century. Its founder was Sheikh Safi al-Din (1252-1334), after whom it is named.

Al-Saratan Arabic term for cancer.

Seljuks A Turkish dynasty that ruled across Persia, Anatolia, and Turkey between 1038 and 1327. They are best known for their great promotion of learning, arts, and trade. The Seljuks gave the madrasa (school) its final shape and definition, as it became a completely separate building from the mosque. They were also behind the rise of the caravansaries, hostel complexes providing free accommodation, food, and services for trading caravans. In the arts, they are best remembered for the introduction of the *iwan* plan and *muqarnas* vaulting.

Shadoof Machine for lifting water, consisting typically of a long, pivoted wooden pole acting as a lever, with a weight at one end. The other end is positioned over a well. The *shadoof* was in use in ancient Egypt, and is still used in Arab countries today.

Sheikh A social title of respect given to an elderly, wise, or a religious person in the community.

Sharia Sharia is the law system inspired by the Quran and the sayings of Prophet Muhammad. Sharia is often referred to as Islamic law.

Souk The marketplace.

Sufism Mystical belief and practice in which the truth of divine love and knowledge of God is sought.

Al-Tasrif Literally means "conducting" or "handling a certain issue." Here it is a medical encyclopedia written by Abul Qasim Khalaf ibn al-Abbas al-Zahrawi, also known as Abulcasis. The complete title is *Al-Tasrif li-man 'ajiza 'an al-taalif*, or *The Method of Medicine*, translated as *The Arrangement of Medicine*. It had 1,500 pages, showing that Abulcasis was not only a medical scholar, but also a great practicing physician and surgeon. It influenced the progress of medicine in Europe. See the section Medical Knowledge in the Hospital chapter to learn more about it.

Tawaf The circumambulation or walking counterclockwise around the Ka'bah in Mecca.

Thikr The action of remembering God (Allah), consisting of the repetition of words in praise of God.

Al-Ud The *'ud* (also spelled *oud*) is a musical instrument common to the Arab culture. It is a stringed instrument slightly smaller than a guitar, with eleven strings in six courses. Some *'uds* may have more or fewer strings; common are versions with thirteen strings in seven courses, or ten strings in five courses.

'Ulama Scholars of the Islamic sciences.

Vizier/wazir Chief minister of the Abbasid caliphs and also government official in Islamic states.

Waqf Religious charitable institutions that manage various gifted and donated financial assets. The *waqfs* finance mosques, madrasas, fountains, and other public services. Their role has been greatly undermined by modern state intervention.

Waraq Paper.

Warraq Paper manufacturer/bookbinder/scribe.

Wudhu Performance of the ritual of ablution. Before offering the prayer, one must be in good shape and pure condition. It is necessary to wash the parts of the body that are generally exposed to dirt or dust or smog, like the hands, mouth, nose, face, arms, hair, ears, and feet. This is called ablution, and the person who has performed it is ready to start his prayer.

07 Illustrations Credits

Front Matter: 4, Courtesy of Topkapi Palace Library, Istanbul; 9, Bridgeman Art Library (British Library, London, U.K./British Library Board. All Rights Reserved.) (Or2784 fol.96). 13, 1001 Inventions Ltd.

Chapter 1: 14-23, MuslimHeritage.com.

Chapter 2: 34, The Trustees of the Chester Beatty Library, Dublin; 37, Guildhall Library, London; 38, V&A Images/Victoria and Albert Museum; 39, Courtesy of Topkapi Palace Library, Istanbul; 42(br), MuslimHeritage.com (Aidan Roberts); 43(bl), MuslimHeritage.com (Sayed Al Hashmi); 45, MuslimHeritage.com (Sayed Al Hashmi); 46(b), Courtesy of the Royal Asiatic Society, London; 47(l), Courtesy of Suleymaniye Library, Istanbul; 47(r), Ricky Jay; 48(t), Courtesy of Topkapi Palace Library, Istanbul; 48(b), Tips Images; 49(b), Mukhtar and Soraya Sanders, Inspiral Design, London; 50, MuslimHeritage.com; 51(t), by permission of the British Library (T.12646); 51(bl), MuslimHeritage.com; 52, MuslimHeritage.com; 53, MuslimHeritage.com; 54, University of Chicago Press; 55, History of Science Collections, University of Oklahoma Libraries; 57, MuslimHeritage.com (Ali Hasan Amro); 59, Courtesy of Topkapi Palace Library, Istanbul.

Chapter 3: 62, Mukhtar and Soraya Sanders, Inspiral Design, London; 64, Saudi Aramco World/PADIA (Nik Wheeler); 65, Library of Congress, Prints and Photographs Division; 66, Corbis (Chris Hellier); 67, Courtesy of Topkapi Palace Library, Istanbul; 68(r) MuslimHeritage.com (Ahmed Salem); 69, Corbis (Bettmann); 70, Bibliothèque nationale de France, Paris; 71, Bridgeman Art Library (Musée Atger, Montpellier); 72, Library of Congress, Prints and Photographs Division; 74, MuslimHeritage.com; 75, MuslimHeritage.com; 76, www.worldreligions.co.uk; 79, Bibliothèque nationale de France, Paris; 80, Department of Printing and Graphic Arts, Houghton Library, Harvard College Library (Typ 620.47.452 F); 83, Castilla-La Mancha University (Spain); 89, MuslimHeritage.com (Ali Hasan Amro); 90, Bibliothèque nationale de France, Paris; 91, Mary Evans Picture Library; 93, by permission of the British Library (Add.25724 f.36); 97, Mukhtar and Soraya Sanders, Inspiral Design, London; 99, Library of Congress, Prints and Photographs Division; 100, Mukhtar and Soraya Sanders, Inspiral Design, London; 101(l), National Museums Liverpool (Liverpool Museum); 101(tr), MuslimHeritage.com (Wai Yin Chang); 102(l), Mukhtar and Soraya Sanders, Inspiral Design, London; 102(r), Courtesy of Topkapi Palace Library, Istanbul; 103, Den Islamske Informasjonforeningen, Oslo, Norway; 105, Mamure Oz of Topkapi Palace Museum Studio (Gilding) and Huseyin Oksuz (Calligraphy);

Chapter 4: 108, Bibliothèque nationale de France, Paris; 114, Edinburgh University Library; 116, Corbis (Arthur Thévenart); 117, MuslimHeritage.com (Ali Hasan Amro); 118, Chris Barton; 119, Corbis (Roger Wood); 120, www.worldreligions.co.uk; 121, Courtesy of Topkapi Palace Library, Istanbul; 122, Corbis; 118, MuslimHeritage.com (Sayed Al Hashmi); 119, MuslimHeritage.com (Sayed Al Hashmi); 124, Courtesy of Topkapi Palace Library, Istanbul; 125, MuslimHeritage.com (Sayed Al Hashmi); 126, Corbis (Paul Almasy); 127, Corbis (M. ou Me. Desjeux); 130, Saudi Aramco World/PADIA (Robert Azzi); 131, Bibliothèque nationale de France, Paris; 132, Courtesy of Walter B. Denny; 133, Saudi Aramco World/PADIA (Norman MacDonald); 135, by permission of the British Library (3754-05); 136, Corbis (Kazuyoshi Nomachi); 137, Bibliothèque nationale de France, Paris; 138(t), Science Museum/Science & Society Picture Library; 138(b), by permission of the British Library (Add.Or.1699); 139, Hikmut Barutcugil of Ebristan, Istanbul, Turkey; 140(t), Mukhtar and Soraya Sanders, Inspiral Design, London; 140(b), V&A Images/Victoria and Albert Museum; 143, V&A Images/Victoria and Albert Museum; 144(b), V&A Images/Victoria and Albert Museum; 145(r), Eric Tischer; 146(t), The Trustees of The British Museum; 146(b), Hussein Gouda www.egypthome.net; 147, Forschungsbibliothek Gotha; 148, Saudi Aramco World/PADIA (Norman MacDonald); 149 (bl), The Trustees of The British Museum; 149, Princess Wijdan Fawaz Al-Hashemi; 151, The Trustees of The British Museum.

Chapter 5: 152, Millet Library, Istanbul; 156, Aga Khan Visual Archive, M.I.T. (Kara Hill, 1989); 157(l), Courtesy of Topkapi Palace Library, Istanbul; 157(r), Aga Khan Visual Archive, M.I.T. (Jamal Abed, 1987); 159, University of St Andrews Library; 160, Bildarchiv Preussischer Kulturbesitz/Art Resource, NY; 161, MuslimHeritage.com (Ali Hasan Amro); 163, Millet Library, Istanbul; 166, Bridgeman Art Library (Bibliothèque de la Faculté de Médecine, Paris); 167, MuslimHeritage.com (Sayed Al Hashmi); 169(t), History of Science Collections, University of Oklahoma Libraries; 169(b), Bridgeman Art Library (Biblioteca Universitaria, Bologna); 171, National Library of Medicine; 173, Science Museum/Science & Society Picture Library; 174, Jonathan C. Horton MD PhD; 176, Library of Congress, Prints and Photographs Division; 177, Turkish Postal Authority; 178, Courtesy of Topkapi Palace Library, Istanbul; 179(br), Courtesy of Suleymaniye Library, Istanbul; 179, Courtesy of Topkapi Palace Library, Istanbul; 180, Courtesy of Suleymaniye Library, Istanbul; 182(b), Werner Forman Archive/Metropolitan Museum, New York; 182(t), Science Museum/Science & Society Picture Library; 183, The Royal Library, Copenhagen; 184, Dr. James T. Goodrich; 185, Crown Publishers Inc., a division of Random House Inc., 1977.

Chapter 6: 186, Courtesy of Topkapi Palace Library, Istanbul; 188, José Vicente Resino; 189, Mary Evans Picture Library; 190, Erich Lessing; 191, University Library, Istanbul; 192, Photo Scala, Florence, 1990; 193(t), Richard Seaman; 193(b), Renata Holod; 195(b), Aga Khan Visual Archive, M.I.T. (M. al-Asad, 1986); 195, MuslimHeritage.com (Sayed Al Hashmi); 196, David Alcock www.thecraven-image.co.uk; 197(m), Eddie Gerald; 198, JP Lescourret; 199, José A. Entrenas, Infocordoba.com; 201, Mukhtar and Soraya Sanders, Inspiral Design, London; 202(t), Dean & Chapter; 206, Mashreq Maghreb; 207(t), Art and Architecture; 207(b), Izzet Keribar/Images & Stories; 211(t), Courtesy of Suleymaniye Library, Istanbul; 212, Izzet Keribar/Images & Stories; 214, Peter Sanders; 215, Courtesy of Topkapi Palace Library, Istanbul; 216, Courtesy of Topkapi Palace Library, Istanbul; 217, Courtesy of Topkapi Palace Library, Istanbul; 219, Courtesy of Topkapi Palace Library, Istanbul; 220, Bridgeman Art Library (Topkapi Palace Museum, Istanbul, Turkey); 222, V&A Images/Victoria and Albert Museum; 225, MuslimHeritage.com (Sayed Al Hashmi).

Chapter 7: 229, MuslimHeritage.com (Sayed Al Hashmi); 233(t), MuslimHeritage.com (Sayed Al Hashmi); 234, Saudi Aramco World/PADIA (Michael Winn); 236, MuslimHeritage.com (Ali Hasan Amro); 239, Bodleian Library; 241, Saudi Aramco World/PADIA (S M Amin); 242(l), Library of Congress, Prints and Photographs Division; 243, Courtesy of Topkapi Palace Library, Istanbul; 245, Courtesy of Topkapi Palace Library, Istanbul; 247, Bibliothèque nationale de France, Paris; 248, Bibliothèque nationale de France, Paris; 250, Saudi Aramco World/PADIA (Norman MacDonald); 251, Saudi Aramco World/PADIA (Norman MacDonald); 254, National Library Board, Singapore; 255, National Library Board, Singapore; 256, Jan Adkins; 259(l), Enigma Museum, http://w1tp.com/enigma; 260(l), Bibliothèque nationale de France, Paris; 260(m), Courtesy of Topkapi Palace Library, Istanbul; 260(r), Courtesy of Topkapi Palace Library, Istanbul; 262, Princeton University Press; 263, MuslimHeritage.com (Ali Hasan Amro).

Chapter 8: 264, University Library, Istanbul; 268, Anna Pietrzak, Nicholaus Copernicus Museum; 269, University Library, Istanbul; 272, University Library, Istanbul; 273(l), Aga Khan Visual Archive, M.I.T. (Hatice Yazar, 1990); 276, Erich Lessing; 277, The Trustees of The British Museum; 279, University Library, Istanbul; 281, Saudi Aramco World/PADIA (Robert Azzi); 283, Saudi Aramco World/PADIA (Robert Azzi); 284, National Maritime Museum, London; 285, National Maritime Museum, London; 286(b), Courtesy of Suleymaniye Library, Istanbul; 287, University Library, Istanbul; 288, Mukhtar and Soraya Sanders, Inspiral Design, London; 289, Fe-noon Dr Ahmed Moustafa, Research Centre for Arab Art and Design; 291(t), by permission of the British Library (16325); 292, Gothard Astrophysical Observatory; 293, Ralph Aeschliman; 294(br), Beinecke Rare Book and Manuscript Library, Yale University; 294(t), Courtesy of Suleymaniye Library, Istanbul; 294(bl), Courtesy of Suleymaniye Library, Istanbul; 298, MuslimHeritage.com (Ali Hasan Amro); 300, Library of Congress, Prints and Photographs Division; 301, Library of Congress, Prints and Photographs Division.

Reference: 302, Bridgeman Art Library (British Library, London, UK/British Library Board. All Rights Reserved.) (Or.2838 f.20v); 304-309, MuslimHeritage.com (Ali Hasan Amro).

08 INDEX

Boldface indicates illustrations.

A

Aadani, Al- 46
Abano, Pietro d' 183
Abawayh (potter) 140–141
Abbas I, Shah 61
Abbas II, Shah 126
Abbas ibn Firnas *see* Ibn Firnas
Abbasids 25, 140–141, 199
Abd al-Malik ibn Marwan (caliph) 149–150
'Abd al-Rahman I (Umayyad ruler) 49
'Abd al-Rahman II (Umayyad ruler) 49, 145
'Abd al-Rahman III (Umayyad ruler) 201
Abdul Aziz, Sultan (Ottoman Empire) 261
'Abdullah (Sufi mystic) 59
Abgali, Mohammed Ben Ali 33
Ablutions 50
Abraham 136
Abu al-Faraj 249
Abu al-Fida' 247, 293
Abu al-Jud 97
Abu al-Majid al-Bahili 157
Abu al-Wafa' al-Buzjani 86, 87, 89, 98, 291
Abu 'Ali al-Kattani 70
Abu Bakr (first caliph) 24
Abu 'Inan, Sultan (Fez) 251
Abu Mansur 25
Abu Nasr al-Farabi *see* Farabi, Al-
Abu Ruh Muhammad ibn Mansur ibn Abdullah (Al-Jurjani) 174–175
Abu Yaqub Yusef (caliph) 103
Abu Zayd Hasan 246, 249
Abubacer *see* Ibn Tufayl
Abul-Hasan 'Ali ibn Nafi' *see* Ziryab
Abulcasis *see* Zahrawi, Al-
Acre harbor 211
Adelard of Bath 81
Aeronautics *see* Flight
Aga, Cassem 33, 177
Aghlabids 126
Agriculture 19
agricultural revolution 110–113, **117**
chicken eggs 33
farming manuals 28, 114–117, 119
fertilization 109
irrigation 112, 118, 119, **122,** 124, 128, 130
windmills 130
Ahmad ibn Tulun, Emir (Egypt) 26, 211
Ahmad ibn Tulun Hospital, Cairo 154, 190
Albategnius *see* Battani, Muhammad al-
Alchemy *see* Chemistry
Alcohol 94, 95
Alembic stills 15, 92, 94–95
Aleppo, Syria 77
Alexandria, Egypt 30, 132
Alfonso VI, King (Spain) 83
Alfonso VIII, Prince (Spain) 151
Alfonso X (el Sabio), King (Spain) 29, 46, 47, 48, 49, 271, **271**
Algebra 15, 75, 84–85, **85,** 107
Alhambra Jar 142
Alhambra Palace, Granada, Spain **97,** 101, 187, 201, **201,** 209, **221,** 224
Alhazen *see* Ibn al-Haytham
Ali, Gelibolulu Mustafa 39
Ali ibn Isa (oculist) 174
Ali Macar Reis 242
Aljizar *see* Ibn al-Jazzar al-Qayrawani
Almadén, Spain 146
Almagest (Ptolemy) 55, 88, 267, 286
Almeria, Spain 134, 145
Almohad dynasty 78
Alpago, Andrea 167
Amalfitan merchants 196
Amicable numbers 85
Andalusia
home 38, 40–41, 47, 49, 58–59
hospital 162, 180
market 110, 111, 112, 134, 139, **140,** 142, 143, 145, 150
town **188,** 197
Andromeda galaxy 294
Animal husbandry 113
Apollonios of Perga 96
Arabesque **100,** 100–101
Arabic numerals 29, 86–87, **87**
Arches 14, **14,** 194–197, **195**
Archimedes 96–97, 210
Architecture 21, 192–193
arches 14, **14,** 194–197, **195**
castles 14, 210–211
domes 193, 202–205
influential ideas 208–209
kiosk to conservatory 218–219
public baths 212–215
spires 206–207
vaults 14, **14,** 198–201
Aristotle **9,** 54, 72, 73, 74, 75, 82–83
Arithmetic *see* Mathematics
Arlandes, Marquis d' 300
Armillary spheres **286,** 286–287, **287**
Arnold, Thomas 104
Art **97,** 98–99, 100–101, **140**
Arzachel *see* Zarqali, Al-
Astrolabes **25, 269, 280,** 280–285, **284, 285**
Astronomical phenomena 288–289
Astronomy **272**
and agriculture 112
instruments 22, **266,** 276–279
lunar eclipses 27, **27**
measurements 26
motivation for studying 266
Quran on 288, **288**
and trigonometry 88
see also Astrolabes; Observatories; Universe
Astrulabi, Maryam al-Ijliya al- *see* Ijliya al-Astrulabi, Maryam al-
Avenzoar *see* Ibn Zuhr, Abu Marwan
Averbak, Yuri 46
Averroes *see* Ibn Rushd
Aviation *see* Flight
Avicenna *see* Ibn Sina
Awfi, Muhammad al- 252
Azarquiel *see* Zarqali, Al-
Azdi, Al- 185
Al-Azhar Mosque, Cairo **190**
Al-Azhar University, Cairo 27, **27,** 68, 69, 78

B

Bab Mardum Mosque, Toledo 27, **27, 195,** 199–200, 201
Babur, Emperor (India) **222**
Bacon, Roger 29, 54, 57, 91, 298
Baghdad **72**
as Abbasid capital 25
baths 214
hospitals 26, 154, 156
as intellectual city 72
libraries and bookshops 77, **79**
Mongol rule 29
observatory 270, 274
paper mill 25
reconstruction of medieval plan 234–235
schools 28, **65**
textiles 136, 137, **137**
see also House of Wisdom
Baghdad Koshk, Topkapi Palace, Istanbul 218
Bakri, Al- 126, 235–236
Balloon flight 299–300, **301**
Bands, musical 49, **49**
Banking system 150
Banu Musa brothers 15, 26, **52,** 52–53, **53,** 73, 83, 85, 224–225, **225**
Barbican 106, 210–211
Baths 51, **51, 212,** 212–215, **214, 215,** 294
Battani, Muhammad al-
astronomy 228, 266, 267, 268, 274, 276, 292
birth 26
translations 33
trigonometry 88-89
Battlements 211
Baylak al-Qibjaqi 252
Beatrice, Queen (Portugal) 134

Bedouin 26, 60, 246
Berggren, J. L. 97–98
Berggren, Len 229
Biruni, Al-
 accomplishments 27
 astronomy 267, 291
 The Book of Pharmacology 183
 Chronicles of India 231
 Chronology of Ancient Nations **114**
 Earth science 227, 229, 230, 231
 natural phenomena studies 232, 233
 on spread of Islam 246
 trigonometry 89
Bitruji, Nur al-Din ibn Ishaq al- 293
Black Death 30
"The Blackbird" *see* Ziryab
Blériot, Louis 300
Blindness prevention 174
Blood circulation 14, 166–167, **167**
Blue Mosque, Istanbul **202**
Boat travel 246–247, **248**
Bologna, Italy 30, 139
Bonding columns 211
Bone fractures 168–169
Bonoeil, John 137
Book of Roger 14
Book production 139
Bookshops *see* Libraries and bookshops
Botany **29, 178, 179, 180**
 see also Herbal medicine
Bowden, Lord B. V. 8
Boyle, Robert 23, 32
Brahe, Tycho 271, 278, 291
Brighton, England 51, **51**
Brothers of Purity *see Ikhwan al-Safa'*
Bulbous dome 203, **204,** 205
Burnet, John 214
Burrows, William E. 299
Bursa, Turkey 137, 142
Bursevî, Mehmed 216
Burton, Decimus 217
Busbecq, Count Ogier de 222
Buyid family 96
Byzantine coins 149–150

C

Caetani, Francisco 222
Cagaloglu Hamami, Istanbul **212**
Cairo, Egypt
 Black Death 30
 hospitals 26, 29, 30, **154,** 154–155, **156,** 166
 libraries 78
Calendars 115, 266, 290
Calligraphy **102,** 102–105, **105**
Camel caravans 133, **133**
Camera obscura 14, 56, **57**
Camshafts 121, 123, **123**
Cancer 165
Cannons 261
Canterbury, England 143, **143**
Canute *see* Knut the Great
Caravans 133, **133, 247**
Caravansaries **132,** 132–133
Cardano, Geronimo 54
Cardwell, Donald 8, 10
Carnations 223
Carpets **29, 60,** 60–61, **61, 101,** 134
Castell, Edmund 32
Castles 14, **15,** 189, **210,** 210–211, **211**
Cataract treatments 172, 175
Catgut 162, **162,** 182
Cauterization 158, **159**
Celebi, Hasan **102**
Celestial globes 276, **276**
Ceramics *see* Pottery
Charlemagne 25
Charles, Prince of Wales 10
Charles I, King (Great Britain and Ireland) 23, 33
Chaucer, Geoffrey 282, **282**
Chaucer, Lewis 282
Checks 150
Chemistry 15, **90,** 90–95, 180
Chess 18, **46,** 46–47, **47**
China
 currency 148
 map 249
 papermaking 138
 trade 133, 246–247, 254–256
 weaponry 260
Christina, Queen (Denmark) 38
Church of Cristo de la Luz, Toledo *see* Bab Mardum Mosque, Toledo
Cinnabar mines 146
Circulatory system *see* Blood circulation
Cities *see* Towns
Cleanliness 50–51
Clepsydras 42, **42,** 44, 118
Clocks and clockmaking 25, **42,** 42–43, **43, 68, 194**
 Elephant Clock **16,** 17, 43, 44, **44, 45**
 water-powered 30, **30,** 42, **42,** 43, **43, 68**
Cloth **58–59**
 see also Textiles
Clothing 58–59, 137
Cluny, France 196
Codes *see* Cryptography
Coffee 15, 33, **33, 34,** 36–37, **36–37,** 39
Coins 25, **25,** 148, **149, 151**
 see also Currency
Colombo, Realdus 167
Columbus, Christopher 31, 241, **242,** 244
 boats **242, 256**
Communication, global 258–259
Compasses 252, **252**
Conic sections 96–97
Conservatories **218,** 218–219
Constantine the African 27, **27,** 71, 184, **184,** 196
Constellations **294,** 294–295, **295**
Contracts 113
Cookbooks 40–41
Copernicus, Nicolaus 31, **31,** 266, 268, **268,** 274
Corals and coral reefs 147
Córdoba, Spain **24**
 home 49, 58
 market 115, 127, 134
 school 65, 77, 78
 town 14, 25, 189, **189,** 190, **194**
Corrosive sublimate 95
Cosmetics 50
Cotter, John 142, 143
Cotton **110,** 112, 113, 136, 137
Cotton paper 138
Counting systems 86
Courtyard houses 130
Cowpox inoculations 33, **33,** 176, **176,** 177
Cowrie shells 148
Cranks 122
Crateuas (physician) 178
Crichton, Michael 250
Crops **110,** 111–112, **112**
Crusades 28, 192–193, 210, 211, 213, 261
Cryptography 15, 258, 259, **259**
Crystal **38,** 38–39, 145
Currency 148–151
 see also Coins

D

Da Vinci, Leonardo *see* Leonardo da Vinci
Dakhwar, Al- 157
Damascus, Syria
 hospital 28, 154, 155, 156, **157**
 market 138, 144
 school 65
 universe 25, 270, 274, 276
Damiani, Pier (Cardinal of Ostia) 47
Dams 126–127
Daniel of Morley 28, 81–82
Daniell, Thomas 217
Dar al-Islam (Muslim world)
 map 14–15
Dark Ages 8, 10
De Grandville, Richard 208
Decimal fractions 87
Defoe, Daniel 103
Dhaifa Khatoon, Queen 29
Dimashqi, Al- 131, 249–250
Dinawari, Abu Hanifa al- 180
Diocles of Carystus 178
Dioscorides (physician) 178, 179, 181, 182
Distillation 15, **25, 91,** 92, **93,** 94, 95
Diyabakir, Turkey **191**
Doctor's code 170–171
Dome of the Rock Mosque, Jerusalem 24, **24,** 29, 203, **205**
Domes 193, **201, 202,** 202–205, **203, 204, 205**
"Drinking Bull" robot **52,** 52–53
Drip irrigation 119
Durham Cathedral, England 192, 193
Dyeing 60, **135**

E

Earth (planet) **228,** 228–229, **236,** 236–237, **237,** 238
Earth science 230–231
East India Company 137
Economics 8, 14, 262–263
 see also Currency; Market

Education *see* School
Edward I, King (England) 29, 137, 143, 205, 208
Edward Lloyd's Coffee House, London **33,** 36, 37, **37**
Egypt
chicken eggs 33
Fatimid rule 26
glassmaking 144, **144,** 145
Mamluk dynasty 29
papermaking 138
water management **118,** 119, 120, **122**
Eilmer of Malmesbury 298
Eleanor of Castile, Queen (England) 29, 137, 143, 205, 208
Elephant Clock **16,** 17, 43, 44, **44, 45**
Elgood, Cyril 174
Elyot, Sir Thomas 181
Emerson, Ralph Waldo 78
Emery 146
Encryption *see* Cryptography
Energy 127
see also Windmills
Engineering *see* Dams; Water management; Water-raising machines
Engines 44, 123
England
coffee 33, **33**
coins 151
Norman Conquest 28
pottery 142, 143
silk 137
English words with Muslim roots 106–107
Enigma (code machine) 258, **259**
Erdogan, Recep Tayyip 12–13
Ermessind, Countess of Barcelona 47
Escher, M. C. 101
Euclid 54, 55, 96
Euler, Leonhard 85
Evliya Celebi 299
Experimentation 80, 181
Exploration *see* Naval exploration; Travelers and explorers
Eyck, Hubert Van 61

F

Fakhr al-Dawla 277
Farabi, Al- 27, 48, 98
Farghani, Al- 268, 292
Farisi, Kamal al-Din al- 54, 56, 85, 233
Farman, Henri 300
Farming *see* Agriculture
Fashion 58–59, **58–59**
Fatih, Muhammad al- 218
Fatih Kulliye, Istanbul 66
Fatimids 26, 27, 73, **149,** 150
Fazari, Al- 25, 286
Feldman, Anthony 8, 10
Ferrand, Gabriel 249
Fertilization 109, 116
Fez, Morocco 136
Fibonacci, Leonardo 29, 87
Fihri, Fatima al- 18, 26, 69
Fine dining **38,** 38–39, **39**
Finger-reckoning 86
Fiorina, Carly 11
Firdawsi, Al- 296, **296**
Fistula treatment 158–159
"Flask with Two Spouts" (game) 53, **53**
Flight 296–301
birds **297**
human 26, **26,** 296–299, **298, 299**
rockets 33, **33**
Flying buttresses 196
Food and drink *see* Agriculture; Coffee; Fine dining; Three-course menu
Ford, Peter 8, 10
Foster, John 196
Fountain pens 104, **104**
Fountains 191, **224,** 224–225, **225**
Frederic II, King (Denmark) 278
Frequency analysis 258, 259

G

Galen (physician) 54, 73, 166–167, 168
Galilei, Galileo 32, **80**
Gama, Vasco da 31
Games 53, **53**
see also Chess
Gan Fredus 103
Gardens 191, 220–223, **222**
see also Botany; Herbal medicine
Gems 231, **231**
see also Jewels
Gentile da Fabriano 104
Geoffrey Langley 208
Geography 234–237
see also Mapmaking; Maps; Travelers and explorers
Geology *see* Earth science
Geometry 96–99, **97,** 100, **140,** 291
George IV, King (England) 51, 217
Gerard of Cremona, translations by 75, 83, 91, 93, 159, 161, 169, 170, 185
Ghafiqi, Muhammad ibn Qassum ibn Aslam al- 175, 181
Ghazali, Al- 28
Gillray, James 33, 176
Giotto 30
Glass industry **108, 144,** 144–145, **145**
see also Crystal
Global communication 258–259
Goats 36, **36**
Gold jewelry **146**
Gold Mancus coin 25, **25**
Golden ratio 86, 98, **98**
Gothic arch *see* Ogee arch
Gothic architecture 193, 196, 197, 199–201, 203, 205
Gothic rib vaulting 14, **14,** 199–201, **200**
Great Mosque, Córdoba 25, **194,** 195, 196, 197, **197,** 198, **199**
Great Mosque, Damascus *see* Umayyad Great Mosque, Damascus
Greeves, John 33
Gregory VII, Pope 47
Gregory IX, Pope 70
Grosseteste, Robert 29
Guangzhou, China 133, 246–247
Gunpowder 260–261, **261**
Gynecology 165

H

Hadi, Al- (caliph) 72
Hafsah (daughter of Umar I) 76
Hafsid dynasty 77
Hagia Sophia, Istanbul 271, **273**
Hajj (pilgrimage) 234, **241,** 246
Hajji Khalifa *see* Katib Celebi
Hakam, Al-, II (Umayyad caliph) 27, 77, 78
Hakim, Al- (Fatimid caliph) 73
Hakluyt, Richard 61
Halabi, Ahmad al- 277
Halley, Edmund 23, 33
Hama, Syria **120,** 121
Hamdani, Al- 231
Hammams see Baths
Hapgood, Charles 241
Harun al-Rashid (caliph) 25, 72, 74, 150, 199
Harvey, Sir William 166, 167
Hasan al-Rammah 260, 261
Hazarfen Ahmed Celebi 299, **299**
Hegira calendar 290, 291
Hemp paper 138
Henna 51, 213
Henry I, King (England) 208
Henry II, King (England) 81
Henry VII, King (England) 208–209
Henry VIII, King (England) 60, 100, **101**
Heptagon 96–97
Herbal medicine 170, **178,** 178–181, **179, 180**
Herman the German 83
Hevelius, Johannes 23, 32, 33, 80
Hippocrates 73, 74
Hirschberg, Julius 172, 174
History, methodology 263
Home 18, 34–61
carpets **29, 60,** 60–61, **61, 101,** 134
chess 18, **46,** 46–47, **47,** 107
cleanliness 50–51
clocks 42–43
coffee trail 36–37
fashion and style 58–59
fine dining **38,** 38–39, **39**
music **48,** 48–49, **49**
three-course menu 40–41
trick devices 15, 52–53
vision and cameras **54,** 54–57, **55, 56, 57**
Horseshoe arch 14, **14, 195,** 195–196
Hospitals 14, 19–20, 26, 153, 154–157, **164,** 190
see also Medicine
Hot air balloons 299–300, **301**
Hou Hsien 255
House of Wisdom, Baghdad 15, 18, 25, 52, 72–75, **74, 75,** 292

House of Wisdom, Cairo 73
Houses 130
Hugh of Cluny, Saint 196
Hugh the Illuminator 209
Hulagu Khan (Mongol ruler) 270, 293
Hulwan library, Baghdad **79**
Hunayn ibn Ishaq al-'Ibadi 20, 73, 74, 173
Huo Lung Ching 260
Hydropower 127

I

Ibn 'Abd al-Dhahir 259
Ibn Abdallah, Mustafa 294
Ibn 'Abdun 59
Ibn abi al-Mahasin al-Halabi 29
Ibn Abi Mansour *see* Yahya ibn Abi Mansur
Ibn Abi Usaybi'ah 157
Ibn al-Awwam 115, 116, 119
Ibn al-Baytar 29, 179, 180–181, 183
Ibn al-Bitriq al-Turjuman, Yuhanna 72–73
Ibn al-Faqih 249
Ibn al-Hajj 66
Ibn al-Haytham **80**
  at Al-Azhar University, Cairo 68
  birth 27
  experimentation 80
  legacy 32, 33, 292–293
  natural phenomena studies 232, 233
  optics 20, 33, 54–55, **55,** 56, 83
  trigonometry 89
  Wilson's theorem 85–86
Ibn al-Jazzar al-Qayrawani 26, 184–185
Ibn al-Nadim 46, 78
Ibn al-Nafis 14, 29, 157, 166–167, 168, 185
Ibn al-Quff 29, 165
Ibn al-Saffar 235
Ibn al-Sarraj al-Hamawi, Abu Bakr 276–277
Ibn al-Shatir 31, 268
Ibn al-Thahabi *see* Azdi, Al-
Ibn al-Wafid 183
Ibn Ali, Musa 200
Ibn Amirshah, Mehmed 67
Ibn 'Arabi 59
Ibn Badis 60, 95, 139
Ibn Bakhtishu 9
Ibn Bassal 28, 112, 114, 115
Ibn Battuta **30**
  on baths 214
  on China 247
  on currency 148
  dictating *Rihla* **250**
  on education 66
  on glassmaking 144
  legacy 250
  on pearl-diving 147
  supplications **251**
  on travel to Syria 246
  travels 14, **14,** 21, 30, 132, **250,** 250–251
Ibn Fadlan 250, 251
Ibn Firnas, 'Abbas 26, **26,** 38–39, 145, 296–298, **298,** 299, 300
Ibn Haddu, Muhammed 33
Ibn Hawqal, Muhammad 64, 133
Ibn Hazm 229, 232
Ibn Hilal, Muhammad 276
Ibn Isa *see* Ali ibn Isa
Ibn Jubayr 156, 235, 236
Ibn Juljul 181
Ibn Juzayy 251
Ibn Khaldun 14, 30, **30,** 46, 68, **262,** 262–263, **263**
Ibn Khurradadhbih 247
Ibn Majid 31, 252
Ibn Masawayh Yahya 231
Ibn Muqla, Abu-'Ali 102
Ibn Nafis *see* Ibn al-Nafis
Ibn Rushd (Averroes) 28, 82, 268
Ibn Rustah 249
Ibn Sahl 55
Ibn Sa'id al-Maghribi 249
Ibn Salih, Mu'awiya 112
Ibn Samajun 180
Ibn Shawka al-Baghdadi, Mahmud 284
Ibn Sina (Avicenna)
  bone fractures 168–169, 170
  The Book of Cure 230
  Canon of Medicine 27, **27,** 32, **32,** 165, 168–170, **169, 171,** 180, 185
  commentaries on 166
  Hayy ibn Yaqzan 103
  herbal medicine 180
  introduction 20
  legacy 32, 185
  pharmacology 182
Ibn Sinan, Ibrahim 97–98
Ibn Tufayl 28, 29, 32, 103
Ibn Tulun Mosque, Cairo 14, 103–104, **154,** 196
Ibn Wahhab 247
Ibn Yunus 27, 89, 228, 267
Ibn Zamrak 224
Ibn Zuhr, Abu Marwan 28, 159, 160
Idrisi, Al-
  birth 28
  on carpets 60–61
  on coral gathering 147
  on Córdoba dam 127
  mapmaking 14, 22, 28, **28, 236,** 236–237, 238, 239, 242
Ijliya al-Astrulabi, Maryam al- 22, 282
Ikhwan al-Safa' (Brothers of Purity) 99, 231
India
  Ganges River Basin **230**
  jewels **146**
  map 230
  trade 137
Indian chintz 137
Indian numerals 86
Inks 104, 193
Inoculation 33, **33,** 176–177
Intersecting arches **195,** 196
Iran
  pigeon keeps 116
  water management 118
Iraq, water management 118–119
Irises 223
Iron Muslim "robot" 47, **47**
Irrigation 112, 118, 119, **122,** 124, 128, 130
Isfahan, Iran 22, 116, **116, 119,** 274
Ishaq (son of Hunayn ibn Ishaq al-'Ibadi) 73
An Islamic History of Europe (BBC program)
  on Al-Idrisi 237
  on architecture 208
  on astrolabes 282
  on astronomy and agriculture 112
  on bathhouses 213
  on Ibn Rushd 83
  on medicine 175
  on town planning 189
  on Ziryab 58
Ismail 136
Istanbul
  observatory 32, **32, 264,** 273, 274, **275, 279**
  schools 66, **67**
Iznik, Turkey 142–143
'Izz al-Din al-Wafa'i 277

J

Jabir ibn Aflah 28, 268, 276, 286, 293
Jabir ibn Hayyan 15, 25, 90, 91, 92, 94–95
Jaburi, Al- 77
Jahangir, Emperor (India) 149
Jahiz, Al- 77
Jamal al-Din 274
James I, King (Great Britain) 134, 137
Janissary band 49, **49**
Jazari, Al- **19**
  The Book of Knowledge of Ingenious Mechanical Devices **7,** 43, 50
  clockmaking 42–43
  Elephant Clock **16,** 17, 44, **44, 45**
  engineering contributions 121
  mechanical marvel 44–45, **44–45**
  reciprocating pump 122–123, **124, 125**
  water-raising machines 15, **121,** 121–123, 124, **124, 125**
  wudhu (washing) machine 50, **50**
Jenner, Edward 33, **33, 176,** 177
Jerome, Saint 70
Jewels **146,** 146–147, **147**
  see also Gems
Al-Jeyushi Mosque, Cairo 207
Johannes de Sancto Amando 183
Josephine, Empress (France) 217
Juhari, Al- 298, 299
Jurjani, Al- *see* Abu Ruh Muhammad ibn Mansur ibn Abdullah (Al-Jurjani)

K

Ka'bah, Mecca 136, **136**
Kalandar Pasha 59
Kamani Khidir Aga 48

Karaji, Muhammad al- 75, 84, 85, 87, 118
Kashghari, Mahmud al- 235
Kashi, Jamshid al- 87
Katib Celebi 286
Kebar dam, Iran 126
Keeps (castles) 210–211
Kemal Reis 242, 244
Kempelen, Wolfgang de 47
Kepler, Johannes 32, 57, 278
Khaju Bridge, Isfahan 126, **126**
Khalid the goat herder 15, 36
Khayyám, Omar *see* Umar al-Khayyam
Khirbat al-Mafjar, Jordan 192, **192**
Khujandi, Al- 228–229, 277–278
Khwarizmi, Al- **28, 84**
algebra 15, 25, 84
astrolabe treatise 280
geography 234
House of Wisdom 73
translations 28, 87
trigonometry 89
Kindi, Al- **25, 72**
accomplishments 25
chemistry 91, 92, 93
cryptography 15, 227, 258, 259
Earth science 231
on knowledge 17
music 48
natural phenomena studies 232–233
optics 54
perfume distillation 51, 92, 95
pharmacology 182
spherical geometry 291
translations by 74
translations of 83, 91
Kingsley, Sir Ben 13, **19**
Kiosks 218–219
Kircher, Athanasius 139
Kiswa (cloth) 136, **136**
Knights Templar Order 209
Knut the Great, King (Denmark) 47
Konya, Turkey **132,** 142
Koran *see* Quran
Kufic script 102, **102,** 103, 104, 150
Kuhi, Abu Sahl al- 96–97

L

La Boullaye-le-Gouz, François de 216
La Touche, David 217
Labna 27
Lagari Hasan Celebi 33, **33,** 299, **299**
Lagrange, Joseph-Louis 86
Lalys (artist) 208
Lambert, Elie 200
Landownership 113
Language 106–107
Law schools 70
Le Gouz de La Boullaye, François 216
Le Strange, Guy 113, 235
Leather 136
Leo the African 146
Leonardo da Vinci 31, **31,** 57, 99, 100, 298–299
Leonardo of Pisa *see* Fibonacci, Leonardo
Leprosy 155
Levey, Martin 183
Lewis, Geoffrey 159
Libraries and bookshops 76–79
Library of Sicily 73
Lilienthal, Otto 300
Linen paper 138
Lion Fountain, Alhambra 30, **30,** 224, **224**
Liotard, Jean-Etienne 137
Lippert, J. 174
Longmarket Excavation, Canterbury, England 142, 143
Loopholes (arrow slits) 210
Lotfollah Mosque, Isfahan **100, 201**
Louis, IX, King (France) 210, 261
Louis XIV, King (France) 216
Lunar calendar 266, 290
Lunar eclipses 291
Lunar formations 292–293
Luster (pottery process) 141, 143

M

Machicolations 211
Madinat al-Zahra', Spain 141–142
Magins (architect) 195
Mahamli, Sutaita al- 27
Mahdi, Mohammad al- (caliph) 72
Mahmud of Ghazni, Sultan 126–127
Mahomed, Sake Dean 51
Mail 258–259
Maitland, Charles 177
Majolica ware 39, 143
Majriti, Maslama Ibn Ahmad al- *see* Maslama
Majusi, Ali ibn Abbas al- 184
Málaga, Spain 39, 132, 134, 142, 143, 145
Maldives, currency 148
Malikshah, Sultan 22, 270, 274
Mamluk dynasty 29, 119, 136, 140
Ma'mun, Al- (caliph)
astronomy 270, 274, 292
Earth science 229, 238
House of Wisdom 25, 72, 73, 74, 266, 292
legacy 75
translations 76
Mancus (coin) 151
Mansur, Abu Amir al- 162
Mansur, Al- (caliph) 25, 150, 151
Al-Mansuri Hospital, Cairo 30, 155, **156**
Mapmaking 14–15, 31, 73, **236,** 238, 240–245
Maps
America 31, 245
China 249
Cyprus 243
Al-Idrisi's map 239
moon 293
world 28, 226
Zheng He's navigation chart 255
Maqdisi, Al- *see* Muqaddasi, Al-
Maqrizi, Al- 119, 140
Maragha Observatory, Iran 22, **269,** 270–271, 274, 276, 293
Marbled paper 139, **139**
Marco d'Aviano, Padre 37
Marcus Vitruvius Pollio *see* Vitruvius
Maria Theresa, Empress (Hungary) 47
Marj al-Lil, Tunisia **127**
Market 18–19, **108,** 108–151, **111**
agricultural revolution 110–113, **117**
currency 148–151
dams 126–127
farming manuals 28, 114–117, 119
glass industry **108, 144,** 144–145, **145**
jewels **146,** 146–147, **147**
paper 25, 30, 78, 103, **138,** 138–139
pottery 39, **108, 140,** 140–143, **141, 142, 143,** 182
textiles **134,** 134–137, **135, 136, 137**
trade 19, 132–133, 246–247
water management 118–119
water supply 120–125
windmills 24, **24,** 128–131, **130, 131**
Marsad Falaki (astronomy center) 73
Masha'Allah 280, 292
Maslama 235
Massio, Niccolo di *see* Gentile da Fabriano
Masudi, Al- 19, 27, 111, 128, 130
Mathematics 29, 84–87, 112
Mausoleum of Mustapha Pasha, Cairo 209
Mawsili, al- 172, 174, 175
Mecca 136, **136,** 234, **247**
Medicine 152–185
blood circulation 14, 166–167, **167**
doctor's code 170–171
first licensing regulation 26
herbal medicine 178–181
hospital development 154–157
Ibn Sina's bone fractures 168–169
inoculation 33, **33,** 176–177
medical knowledge 184–185
medical schools 65, 70, **71,** 156–157
mental diseases 26
mobile hospital services 26
oculists 172–175
pediatrics 26
pharmacy 182–183
pulmonary circulation 29
surgery 26, 29, 162–165
surgical instruments 14, **14,** 158–159
surgical precision 160–161
textbooks 32, **32**

transfer to Europe 27
see also Hospitals; Surgery; Vision and cameras
Medina 65
Mehmed II, Sultan (Ottoman Empire) 261
Menniger, Karl 87
Mental illness 156
Menzies, Gavin 254, 255
Meton (astronomer) 290
Meyerhof, Max 181
Military musical bands 49, **49**
Minaret 203, 206–207
Mirza Ali ibn Hacemkulî III 219
Mogul Empire 149
Mohamed Zakariya 281, 283
Mohammed V, Sultan (Spain) 30, 224
Money *see* Currency
Mongol conquest 29
Montagu, Lady Mary Wortley 33, **33,** 137, **176,** 176–177, 218
Montgolfier brothers 299–300, **301**
Montpellier University, France 71, **71**
Montreuil, Eudes de 210
Moon 232, **232,** 233, **233,** 266, **290,** 290–293, **291**
Moorish arch *see* Horseshoe arch
Morley, Daniel of *see* Daniel of Morley
Mosaic glass 144, **145**
Mosques
architecture 14, **14,** 21, 27, **27**
libraries 77
orientation 234
as schools 64
in town planning 188, 189, 190
universities 68
Moustafa, Ahmed 289
Mozarabs 195, 201
Muhammad (prophet)
on cleanliness 213
cosmetics 51
death 24
on legacies 80
mosques as schools 64
prohibition of humans and animals in art 100
Quran 76
on seeking knowledge 249, 251
Muhammad I, Sultan (Nasrid dynasty) 145, 149
Muhammed XII (Nasrid dynasty) 149
Muhammed Baqir Yazdi 85
Mu'izz, Al- (caliph) 104
Muktafi, Al- (caliph) 72
Müller, Johann *see* Regiomontanus
Multi-foil arch 197
Mumtaz Mahal 209
Muqaddasi, Al- 78, 235, 247
Muqarnas (honeycomb dome) 201, **201,** 202
Muqtader, Al- (caliph) 26
Murad III, Sultan (Ottoman Empire) **215, 219,** 273
Murad IV, Sultan (Ottoman Empire) 218, 299
Murad Efandi 77
Musa ibn Ali 200
Music **48,** 48–49, **49**
Muslim world, map 14–15
Mustafa Ali, Gelibolulu 39
Mustafa Pasha, Lala **39**
Mu'tadhid, Al- (caliph) 72
Al-Mutawakkil Mosque, Samarra 197
Muwaffaq , Abu al-Mansur 182–183

N

Napoleon I, Emperor (France) 49
Nash, John 205, 217
Nasir al-Din, Artuq King of Diyarbakir 43
Nasir al-Din al-Tusi *see* Tusi, Nasir al-Din al-
Al-Nasiri Hospital, Cairo 29
Naskh script 102
Nasrid coins **149,** 150
Natron 147
Natural phenomena 232–233, 288–289
Naval exploration 254–257
Navigation 31, 252–253
Nayla Khatun 77
Nelmes, Sarah 177
Newton, Isaac 33
Nilometer, Egypt **118**
Nizam al-Mulk 28, 65
Nizamiyah school, Baghdad 28, 65
Norias (waterwheels) 112, **120,** 120–121, 127
Norman Conquest 28, 205, 208
Nu'man ibn Muhammad, Qadi abu Hanifah al- 104
Number theory 85
Nur al-Din Bimaristan, Damascus **157**
Nur al-Din Zangi (Nureddin) 28, 157
Nuri Hospital, Damascus 14, 28, 154, 155, 156, 157
Nuwayri, al- 119, 259

O

Observatories 22, **31,** 32, **264, 269, 270,** 270–273
Oculists 172–175
Offa, King of Mercians 25, 151
Ogee arch **195,** 197, **197**
Oil distillation 94
Oil paint 101
Omaar, Rageh
on Al-Idrisi 237
on the Alhambra 101
on architecture 208
on astrolabes 282
on astronomy and agriculture 112
on bathhouses 213
on Ibn Rushd 83
on medicine 175
on Toledo 82
on town planning 189
on Ziryab 58
Omar (second caliph) *see* Umar I (second caliph)
Omar Khayyám *see* Umar al-Khayyam
1001 Inventions (exhibition) **12,** 12–13, **20, 21, 23**
1001 Inventions and the Library of Secrets (film) 13, **19**
Onion-shaped dome *see* Bulbous dome
Ophthalmology 29, 172–175, **173,** 174, **175**
Optics 32, 33, **54,** 54–57
Ottoman Turk (chess robot) *see* Iron Muslim "robot"
Oz, Mamure 105

P

Palatine Chapel, Palermo **208**
Paper 25, 30, 78, 103, **138,** 138–139
Papyrus 102–103
Parchment 102–103
Paris, France 71, 81, 83
Parnell, John 217
Pearls 147
Pedersen, Johannes 64, 139
Pens 104, **104**
Perfumes 51, **91,** 95
Perkins, George 169
Perry, Charles 40
Pharmacology 93, **182,** 182–183, **183**
Philosophy 28
Phipps, James 177
Piazza Ducale, Italy 207
Pigeons 109, 116, **116,** 258–259
Pilâtre de Rozier, Jean-François 300, **301**
Piri Reis
mapmaking 22, 31, **31,** 240–242, 244
maps 243, 245
navigation 252–253
Planetariums 39, 273
Planets **266,** 267
Plato 209
Pleiades star group **295**
Pococke, Edward 32
Podzamcze, Poland **211**
Pointed arch 14, **14, 195,** 196–197
Polo, Marco 30
Pottery 39, **108, 140,** 140–143, **141, 142, 143,** 182
Ptolemy
Almagest 55, 88, 267, 286
astronomy 228, 266–267, 268, 270
on moon's movements 290, 291
Public baths *see* Baths
Pulmonary circulation 29, 166–167
Pumps 44, 122–123, **123,** 124, **124, 125**

Q

Qadi Aqib ibn Mahmud ibn Umar, al- 69
Qal'at of Benu Hammad, Algeria 206, **206**
Qalawun, al-Mansur 155
Qanats (underground canals) 112, 118, **119**
Qanun (table zither) 48, 107
Al-Qarawiyin University, Fez 18, 26, **26, 68,** 68–69

Al-Qayrawan, Tunisia 70, 126, **127,** 133, 155, **155,** 184
Qazwini, Zakariya' ibn Muhammad al- 249, 291
Qitara (guitar) **48**
Quadrants 277, **277,** 278, **278**
Quran
- in art **100**
- on astronomical phenomena 288, **288**
- blessings to those who read and write it 102
- calligraphy **102, 105**
- on currency 148
- economic principles 8
- on gardens 220
- hajj 246
- history 76
- names of Allah 146
- in schools 65
- on society 189
- translations 28

Qusaybah dam, near Medina 126
Qutb al-Din al-Shirazi 271

R

Rababah 27, **27,** 48
Rainbows 233, **233**
Ramadan 290
Ramla, Palestine 199
Rammah, Al- *see* Hasan al-Rammah
Rashid, Al- (caliph) *see* Harun al-Rashid
Razi, Al- **26, 91**
- accomplishments 26
- *Al-Hawi* 32
- blindness prevention 174
- chemistry 15, 90–91, 92, 93, 95
- chess 46
- *Comprehensive Book* 185
- herbal medicine 180
- hospital work 156
- pharmacology 182
- surgery 162, 164–165
- translations 75, 83, 91

Reciprocating pump 122–123, **124, 125**
Regiomontanus 266
Repton, Humphrey 218
Rhazes *see* Razi, Al-
Rib vaulting **198,** 198–201
Ribat, Susa, Tunisia 199
Riccioli, Joannes Baptista 292
Rice 112, 115–116
Richard I (the Lionheart), King (England) 185
Robert of Chester 28, 87
Robertus Sculptor 208
Robots **7,** 47, **47, 52,** 52–53
Rockets 33, **33,** 260, 261, 299, **299**
Roger II, Norman King of Sicily 28, 208, 236, **236,** 238, 239, 242
Rose windows 192–193
Rosee, Pasqua 36
Rubies **146,** 147

S

Sabuncuoglu, Serefeddin 30, 153, 163
Sabur ibn Sahl 182
Sa'da (architect) 200
Safar (Jordanian soldier) 112
Safavid dynasty 61
Sa'id ibn Harun al-Katib 73
St. Miguel de Esacalda, Spain 195
St. Paul's Cathedral, London **202,** 203, 205, 207
St. Philibert, Tournus, France 199
Sakk (check) 150
Saladin *see* Salah al-Din al-Ayyubi
Saladin of Ascolo 183
Salah al-Din al-Ayyubi 29, 185, 210
Salerno, Italy 27, 71, 156, **169,** 181, **184,** 185
Salt 146
Samarkand Observatory 271
Samarra, Iraq 140, 144–145, **145,** 197
Samawal, al- 84, 85
Samso, Julio 282
Sanad ibn Ali al-Yahudi 73
Sankore Mosque, Timbuktu 69, **69**
Santiago de Compostela 201
Sapphires, artificial 147
Saqati, Al- 59
Saracenic Theory 203
Sarton, George 54
Sassa ibn Dahir 46
Sayf al-Dawla, Prince (Syria) 77
School 18, 63–107
- art and the arabesque 100–101
- chemistry 15, **90,** 90–95, 180
- geometry 96–99, **97,** 100, **140,** 291
- House of Wisdom 15, 18, 25, 52, 72–75, **74, 75,** 292
- libraries and bookshops 76–79
- mathematics 84–87
- schools **64,** 64–67, **65**
- scribes 77, 102–105
- translating knowledge 80–83
- trigonometry 88–89, **89**
- universities **66,** 68–71, **70**
- word power 106–107

Scot, Michael 75, 82–83
Scott, S. P. 110
Scribes 77, 102–105
Seasons 228, **229**
Selim I, Sultan (Ottoman Empire) **216**
Selimiye Mosque, Edirne 31, 193, **193,** 204
Seljuk architecture 201
Selkirk, Alexander 103
Semicircular dome 202–203
Servetus, Michael 167
Seville, Spain 29, 190
Sexagesimal system 86, **87**
Sextants **264,** 271, **273,** 277–278, **278**
Shadoof (water contraption) 120, **122**
Shah Jahan, Emperor (India) 149, 209
Shakkaz, Ali ibn Khalaf al- 282
Al-Shammasiyah Observatory, Baghdad 25, 270, 273
Sharaf al-Din al-Tusi 85
Shifa, Al- 24
Shiraz, Iran 78, 136
Shoes 59
Sicily, Italy 26, 73
Sidi Omar 87
Sijzi, Abd ul-Jalil al- 97
Silk 134, **134,** 136, 137
Simon Simeon 209
Sinan (architect) 21, 193, 204
Sinan ibn Thabit ibn Qurra 26
Sistan province, Persia 128, 130
Smallpox 33, 176, 177
Snell's law 55
Social science 262–263
Sociology 14
Solar apogee 267
Solberg, Karima 103
Solmization 48
Souk 189
Spherical geometry 291
Spink, Martin 159
Spires 203, 206–207
Squinches 202
Stanislas of Lorraine, King (Poland) 218
Ste.-Marie-Madeleine, Vézelay **198,** 199
Steam engines 123
Sufi, 'Abd al-Rahman al- 267, 276, 284, 292, 294
Sugarcane 111, 112, 113, **113**
Suger of Saint-Denis 196
Suhrab (geographer) 234–235
Suleyman I (the Magnificent) **217, 220,** 222
Suleymaniye Mosque, Istanbul 21, 31, 193, **193**
Suli, Abu Bakr al- 46, 47
Sultan Ahmed Mosque, Istanbul **202**
Sultan Qalawun, Cairo **156**
Summerhouses 218–219
Sundials **42,** 97, 98
Surgery 29, 161, **161, 162,** 162–165
- eye surgery 172
- instruments 14, **14,** 158–159, 160, **160, 162**
- procedures 158–159, 162, **163,** 164–165
- textbooks 30

Surveying 235, **235**
Susa, Tunisia 199
Sutaita al-Mahamli 27
Sylvester I, Pope 86–87
Sylvester II, Pope 137
Synesios (translator) 184
Synthetic chemistry 95
Syria
- glassmaking 144
- Mongol rule 29
- travel restrictions 246

T

Taj Mahal, India 209, **209**
Tallas, Battle of (751) 138
Tamerlane 22, 263
Taqi al-Din **275**
- automated machinery 121–122
- instruments 22, 278, **279**

observatory 32, **32, 264,** 273, 274
six-cylinder pump 19, 123, **123,** 124
Tattawi, Muhyi al-Deen al- 166
Templars 29
Temple Church, London, England 29, 209
Tents 216–217, **217**
Terrasse, Henri 49, 58
Textiles **108, 134,** 134–137, **135, 136, 137**
Thabit ibn Qurra 85, 96, 292
Theon of Alexandria 280, 284
Thierry of Chartres 71
Three-course menu 38, 40–41
Tides 232, 233, **233**
Time line 24–33
Timoni, Emmanuel 177
Tirmidhi, al- 81
Toledo, Spain 14, 27, **27,** 28, 75, **81, 82,** 82–83, **83**
Topkapi Palace, Istanbul **101,** 105, **186,** 218
Topkapi scroll 100, **101**
Toscanelli, Ludovico dal Pozzo 183
Towers **206,** 206–207, **211**
Towns 21, 186–225
arches 14, **14,** 194–197, **195**
architecture 192–193
castles 14, **15,** 189, **210,** 210–211, **211**
domes 193, **201, 202,** 202–205, **203, 204, 205**
fountains 191, **224,** 224–225, **225**
gardens 191, 220–223, **222**
influential ideas 208–209
from kiosk to conservatory 218–219
public baths 51, **51, 212,** 212–215, **214, 215,** 294
spires 203, 206–207
tents 216–217, **217**
town planning 188–191
vaults **198,** 198–201
Toys, executive 15
Trade 19, 132–133, 246–247
Translations 25, 72–73, 74–75, 80–83, 106–107
Travelers and explorers **234,** 240, 246–251
see also Hajj (pilgrimage); Ibn Battuta
Trebuchet **260**
Treffy, Sir John 214
Trick devices 15, 52–53
Trigonometry 88–89, **89**
Tsarskoe Selo, Russia **203**
Tudor architecture 208–209
Tulips 220, 222–223, **223**
Turia River, Spain 127
Turkey
pottery 142
smallpox inoculation 33, 176–177, **177**
textile industry 137
Tusi, Nasir al-Din al- 31, 32, 88, 99, 269, 271, 274, 293

U

Ukhaydar Palace, Iraq 199
Ulugh Beg 31, 33, 270, 271, 274, 293
Ulugh Beg Observatory, Uzbekistan **31, 273,** 274
Umar al-Khayyam 84–85, 87, 99
Umar I (second caliph) 24, 76, 128, 290
Umayyad dynasty 24, 25, 77, 149–150, 192
Umayyad Great Mosque, Damascus 14, 195, 206, 207, **207**
Umayyad Mosque, Aleppo 77
Universe 22, 264–301
armillary spheres **286,** 286–287, **287**
astrolabes **25, 269, 280,** 280–285, **284, 285**
astronomical instruments 22, **266,** 276–279
astronomical phenomena 288–289
constellations **294,** 294–295, **295**
flight 296–301
lunar formations 292–293
moon 232, **232,** 233, **233,** 266, **290,** 290–293, **291**
observatories 22, **31,** 32, **264, 269, 270,** 270–275
Quran on 288
see also Astronomy
Universities **66,** 68–71, **70**
Urethral stones 159, 164–165
'Uthman ibn 'Affan (third caliph) 76

V

Vaccination *see* Inoculation
Valencia, Spain 39, 127
Vaults **198,** 198–201
Victoria, Queen (United Kingdom) 261
Vision and cameras **54,** 54–57, **55, 56, 57**
see also Ophthalmology; Optics
Vitruvius (Roman architect and engineer) 99, 120, 210, 211
Vogel, Sebastian 54
Voices from the Dark (BBC program) 83

W

Wallis, John 23, 32
Wang Ching-Hung 255
Waqf (donations) 66, 154, 155
War and weaponry **260,** 260–261
Warraq (profession) 78
Washing machines **50**
Water clocks 30, **30,** 42, **42, 68,** 118
see also Elephant Clock
Water management 118–119
Water-raising machines 15, 112, **121,** 121–123, **123,** 124, **124, 125,** 180
Water supply 120–125
Waterwheels *see Norias*
Watson, Andrew 110
Weaponry *see* War and weaponry
Webster, Jason 58
What the Ancients Did for Us (BBC documentary) 51, 141, 142, 260, 261
Wheler, Sir George 223
Whitaker, Brian 73
Whittock, Nathaniel 217
William IV, King (Great Britain and Ireland) 51
Williams, Harold 280
Wilson, John 85–86
Wilson's theorem 85–86
Windmills 24, **24,** 128–131, **130, 131**
Witelo (physicist) 54
Wood, Casey 174
Word power 106–107
World 21–22, 226–263
cultural crossroads 238–239
Earth (planet) **228,** 228–229, **236,** 236–237, **237,** 238
Earth science 230–231
geography 234–237
global communication 258–259
map 226
mapmaking 240–245
natural phenomena 232–233
naval exploration 254–257
navigation 31, 252–253
social science and economics 8, 14, 262–263
travelers and explorers 246–251
war and weaponry **260,** 260–261
Wren, Sir Christopher 202, 203, 205, 206
Wright brothers 300, **300**
Wudhu (ablutions) 50, **50**

Y

Yahya ibn Abi Mansur 73, 270
Ya'qubi, Al- 235, 247, 249
Yaqut al-Hamawi 237, 247
Yong Le, Emperor (China) 255

Z

Zahrawi, Al- **158**
birth 26, **26**
cauterization 158, **159**
cosmetics 50
performing surgery **161**
pharmacology 182
surgical instruments 14, 20, 158–159, 160, **160,** 161
surgical procedures 158–159, 162, 164–165
translations 31, 75, 83, 185
Zain, Amani 141, 142, 260, 261
Zanatiyeh, Maryam al- 30
Zardkash, Ibn Aranbugha al- 260
Zarqali, Al- 266, 268, 282, 284, 293
Zaytuna Mosque, Tunisia 70, 76, **76,** 77
Zero (mathematics) 86
Zheng He 21, 30, **30,** 241, **254,** 254–257
Ziryab (Abul-Hasan 'Ali ibn Nafi') 38, 39, 47, 48–49, 58

09 ACKNOWLEDGMENTS BY PROFESSOR SALIM T. S. AL-HASSANI

This book is published as part of the award-winning 1001 Inventions educational initiative created by the U.K.-based Foundation for Science, Technology and Civilisation (FSTC) and as the official companion to the touring 1001 Inventions exhibitions. A full list of references used in this book and in the development of the 1001 Inventions exhibition can be found online at http://www.1001inventions.com/references.

This book is based on the two previously published editions,which would not have come to fruition without the dedication and perseverance of the staff of 1001 Inventions and FSTC and their key associates.

A special dedication is made to our dear friend and colleague Peter Raymond (1938-2011), co-founder and trustee of FSTC, whose leadership and commitment helped create 1001 Inventions. May he rest in peace.

Special thanks are due to the Abdul Latif Jameel Community Initiatives; the Home Office: Cohesion & Faiths Unit (U.K.); the Science Museum (U.K.); the University of Manchester (U.K.); the Wellcome Trust (U.K.); the Northwest Regional Development Agency (U.K.); the Qualifications and Curriculum Development Agency; the Office of Science and Technology DTI (U.K.); the Bin Hamoodah Group (U.A.E.); and the British Science Association (U.K.).

Much of the material for this book is based on peer-reviewed papers, articles, and presentations published on our academic portal, www.Muslim-Heritage.com. Chief among these are written by the following scholars, arranged in alphabetical order: Professor Mohammed Abattouy (Science and Philosophy); Professor Rabie Abdel Halim (Medicine); Professor Abdulkader M. Abed (Materials Science); HRH Princess Wijdan Ali (Art and Islamic coins); Dr. Salim Ayduz (Ottoman Science); Dr. Subhi al-Azzawi (Architecture); Professor Charles Burnett (Islamic Influences on Europe); Dr. Mahbub Gani (Mathematics and Numbers); Professor S. M. Ghazanfar (Economics); Professor Salim T. S. al-Hassani (Engineering); Dr. Zohor Idrisi (Agriculture and Codes); Professor Ekmeleddin Ihsanoglu (History of Science); Dr. Abdul Nasser Kaadan (Medicine); Professor Mustafa Mawaldi (Mathematics); Professor Jim al-Khalili (Physics); Dr. Munim al-Rawi (Geology); Professor George Saliba (Science and Astronomy); Dr. Rabah Saoud (Architecture and Town Planning); Professor Nil Sari (Ottoman Medicine); Professor Aydin Sayili (Muslim Observatories); Dr. Ibrahim Shaikh (Surgery); Professor Sevim Tekeli (Engineering and Mapping); Dr. Rim Turkmani (Astronomy).

Extended appreciation for the assistance, support, and contributions of the renowned historian Professor Ekmeleddin Ihsanoglu, secretary general of the Organisation of the Islamic Conference; the members of the Muslim Heritage Awareness Group; Professor Charles Burnett, Warburg Institute; Professor Emilie Savage-Smith, Oriental Institute Oxford; Dr. Anne-Maria Brennan, London South Bank University; Professor Mohamed El-Gomati, York University; Lord William Waldegrave, Professor Chris Rapley, Heather Mayfield, and Dr. Sue Mossman, Science Museum (U.K.); Dr. Ian Griffin, Oxford Trust; Paul Keeler, CEO, Golden Web Foundation, Cambridge; Mohammed Qujja, Syrian Archaeological Society; Dr. Rim Turkmani, Imperial College London; Yaqub Yousuf, London; Bettany Hughes, London; Marianne Cutler, the Association for Science Education, Hatfield; Peter Fell, Professor Stephen Parker, and Professor John Pickstone University of Manchester; Muhammad Hafiz; Ian Fenn; Zeki Poyraz; Samar El Sayed, Director of El Sayed Foundation; Diana El-Daly; Hannah Becker; Margaret Morris; and Kaouthar Chatioui. Last but not least, to my family, whose sacrifice and devotion to this project words fail to describe.

$14—

1001 Inventions: The Enduring Legacy of Muslim Civilization

Professor Salim T. S. al-Hassani, Chief Editor and Chairman of 1001 Inventions and the Foundation for Science, Technology and Civilisation (FSTC), United Kingdom

Published by the National Geographic Society
John M. Fahey, *Chairman of the Board and Chief Executive Officer*
Timothy T. Kelly, *President*
Declan Moore, *Executive Vice President; President, Publishing*
Melina Bellows, *Executive Vice President; Chief Creative Officer, Books, Kids, and Family*

Prepared by the Book Division
Hector Sierra, *Senior Vice President and General Manager*
Barbara Brownell Grogan, *Vice President and Editor in Chief*
Jonathan Halling, *Design Director, Books and Children's Publishing*
Marianne R. Koszorus, *Design Director, Books*
Lisa Thomas, *Senior Editor*
Carl Mehler, *Director of Maps*
R. Gary Colbert, *Production Director*
Jennifer Thornton, *Managing Editor*
Meredith C. Wilcox, *Administrative Director, Illustrations*

National Geographic Staff for This Book
Barbara Payne, *Project Editor*
Sanaa Akkach, *Art Director*
Judith Klein, *Production Editor*
Mike Horenstein, *Production Manager*
Robert Waymouth, *Illustrations Specialist*
Linda Makarov, Jodie Morris, and Noelle Weber, *Design Assistants*

FSTC and 1001 Inventions Staff for This Book
Rebecca Mileham, *Editor*
Elizabeth Woodcock, *Editor*
Dr. Rabah Saoud, *Editor*
Diana Eldaly-Rookledge, *Editor*
Professor Mohammed Abattouy, *Historical Verification*
Professor Rabie Abdel Halim, *Medical Sources Verification*
Dr. Salim Ayduz, *Historical Verification*
Yassir Salem, *Editorial Logistics*
Hannah Becker, *Research Assistant*
Nosheen Ladha, *Illustrations Researcher*

For content contributors, see Acknowledgments

Manufacturing and Quality Management
Christopher A. Liedel, *Chief Financial Officer*
Phillip L. Schlosser, *Senior Vice President*
Chris Brown, *Technical Director*
Nicole Elliott, *Manager*
Rachel Faulise, *Manager*
Robert L. Barr, *Manager*

The National Geographic Society is one of the world's largest nonprofit scientific and educational organizations. Founded in 1888 to "increase and diffuse geographic knowledge," the Society works to inspire people to care about the planet. National Geographic reflects the world through its magazines, television programs, films, music and radio, books, DVDs, maps, exhibitions, live events, school publishing programs, interactive media and merchandise. *National Geographic* magazine, the Society's official journal, published in English and 33 local-language editions, is read by more than 40 million people each month. The National Geographic Channel reaches 370 million households in 34 languages in 168 countries. National Geographic Digital Media receives more than 15 million visitors a month. National Geographic has funded more than 9,600 scientific research, conservation and exploration projects and supports an education program promoting geography literacy. For more information, visit www.nationalgeographic.com.

For more information, please call 1-800-NGS LINE (647-5463) or write to the following address:

National Geographic Society
1145 17th Street N.W.
Washington, D.C. 20036-4688 U.S.A.

For information about special discounts for bulk purchases, please contact National Geographic Books Special Sales: ngspecsales@ngs.org

For rights or permissions inquiries, please contact National Geographic Books Subsidiary Rights: ngbookrights@ngs.org

ISBN: 978-1-4262-0934-5 (paperback)
ISBN: 978-1-4262-0947-5 (hardcover)

Printed in Hong Kong

13/THK/2